On the Trail of the Ancient Opium Poppy

FIGURE 1. The opium poppy, *Papaver somniferum* L. *(from Rhind 1868: plate XXXI)*.

On the Trail of the Ancient Opium Poppy

Mark David Merlin

Rutherford • Madison • Teaneck
Fairleigh Dickinson University Press

London and Toronto: Associated University Presses

Associated University Presses
440 Forsgate Drive
Cranbury, NJ 08512

Associated University Presses
25 Sicilian Avenue
London WC1A 2QH, England

Associated University Presses
2133 Royal Windsor Drive
Unit 1
Mississauga, Ontario
Canada L5J 1K5

Library of Congress Cataloging in Publication Data

Merlin, Mark David.
On the trail of the ancient opium poppy.

Bibliography: p.
Includes index.
1. Opium poppy—History. 2. Opium poppy—Origin.
3. Neolithic period. 4. Bronze age. 5. Plant remains
(Archaeology) I. Title.
SB295.065M47 1983 633.7'5'09 80-70994
ISBN 0-8386-3097-9

This book is dedicated to Ollie and Casey,
who faithfully guarded my family during my prolonged absence.

Printed in the United States of America

Contents

Tables

Figures

Acknowledgments

Over the six-year period during which I was involved in research pertinent to this study, numerous individuals inspired me, provided useful information, and offered helpful criticism. Without their assistance, my work would have suffered greatly.

First I would like to thank those members of my doctoral dissertation committee who systematically directed my research and helped solve the innumerable problems which I encountered along the way toward completion. They include Dr. John Street, Dr. Lyndon Wester, Dr. Donald Fryer, and Dr. Forrest Pitts, all faculty members of the Geography Department at the University of Hawaii.

Many other colleagues and friends helped bring this lengthy effort to fruition. I will not name them all, but their help was deeply appreciated. One individual who must be recognized is Mr. Joseph Kennedy, my long-time friend, who has been with me in spirit from the very beginning of this research. His criticisms and encouragement were very important to me throughout the years. A special debt of gratitude must also go to Denise Peltier, my dear friend, who spent so many hours helping edit and type my manuscript. Her support was invaluable.

I would also like to thank the numerous scholars with whom I corresponded during the past five years of research. They include Paul Åström (Klassiska Institutionen, Göteborgs Universitet, Göteborg, Sweden), R. D. Barnett (archaeologist, London, England), K. Lundstrom Baudais (archaeologist, Besançon, France), Kathrin Bühler (Bernisches Historisches Museum, Bern, Switzerland), Annie Caubet (Conservateur au Dept. des Antiquités Orientales, Musée du Louvre, Paris, France), John Chadwick (Downing College, Cambridge University, Cambridge, England), C. F. K. Cook (Direktor der Botanischer Garten und Institut für Systematische Botanik der Universität Zurich), Ara Der Marderosian (Philadelphia College of Pharmacy and Science), James A. Duke (Chief, Plant Taxonomy Laboratory, United States Department of Agriculture, Agricultural Research Center), Herbert L. Edlin (Publications

Officer of Forestry Commission, London, England), William Emboden (California State University, Northridge), Bent Fredskild (Universitets Botaniske Museum, Copenhagen, Denmark), Marija Gimbutas (The Institute of Archaeology, University of California at Los Angeles), Harry Godwin (The Botany School, Cambridge University, Cambridge, England), Peter Goldblatt (B. A. Krukoss Curator of African Botany, Missouri Botanical Garden, St. Louis, Missouri), James Greig (Department of Botany, University of Birmingham, Birmingham, England), Hans Helbaek (archaeologist and botanist, Helsinge, Denmark), Vassos Karageorghis (Director of the Department of Antiquities, Nicosia, Cyprus), Kempton Corey Keeble (Assistant Curator, Royal Ontario Museum, Toronto, Canada), John C. Kramer (California College of Medicine, Medical Pharmacology and Therapeutics, University of California, Irvine, California), A. D. Krikorian (Department of Cellular and Comparative Biology, State University of New York at Stony Brook, New York), Seton Lloyd (archaeologist, Swindon, England), R. S. Merrillees (Foreign Affairs Office, Australia House, London, England), P. R. S. Moorey (Ashmolean Museum, Oxford, England), Richard Evans Schultes (Director, Botanical Museum, Harvard University, Cambridge, Massachusetts), Jürgen Schultze-Motel (Akademie der Wissenschaften der DDR, Gatersleben, East Germany), F. H. Schweingruber (Eidgenössische Anstalt für das Forstliche Versuchswesen, Birmensdorf, Switzerland), A. F. Shore (School of Archaeology and Oriental Studies, University of Liverpool, England), Michael Vickers (Department of Antiquities, Ashmolean Museum, Oxford, England), J. M. J. de Wet (Crop Evolution Laboratory, University of Illinois at Urbana-Champaign), Tamara S. Wheeler (Department of Metallurgy and Materials Science, University of Pennsylvania, Philadelphia, Pennsylvania), C. K. Williams II (Director, Corinth Excavations, American School of Classical Studies at Athens, Greece), W. van Zeist (Rijksuniversiteit, Groningen, Netherlands), and Bernhard Zepernick (Botanischer Garten and Botanisches Museum, Berlin, West Germany).

In addition to sharing their own time and personal expertise, a number of these individuals graciously provided me with useful illustrations. Figures 3, 4, 7, 10, 11, 27, 28, 29, 31, 51, 53, 65, 81, 83, 84, and 100 were photographed by the author.

1 • Introduction

The well-known opium poppy (*Papaver somniferum* L.) is an annual herb that is widely cultivated in several temperate and subtropical regions of the world (cf. Addens 1939:93–101; Wishart 1974:14–25; and Krikorian and Ledbetter 1975:30–103). It has been commonly used as an ornamental plant and tasty seed crop, but is probably best known as a source of powerful, narcotic drugs. In fact, this species has been referred to as one of the most important medicinal plants known to man (Duke 1973:390); and its valuable alkaloidal sap, opium, has been immoderately described as the "World's Greatest Pain-Killer" (Taylor 1965:206), "God's own medicine" (Brecher 1972:8), and "the most hypnotic, the most potent" of all the drugs in the pharmacopeia (Wright 1961:22).

Some suggest that humans have "always been engaged in a search for greater comfort," and that "high priority has been granted to the development of drugs to relieve pain" (Cohen 1969:76; compare Aaronson and Osmond 1971:4; Malcolm 1971:65; and Leake 1947:7). Oddly enough, it has also been suggested that if the experience of pain did not exist, we would be deprived of the comfort felt when the suffering is relieved: "For ages that comfort has been provided by opium and its byproducts, but not without the handicaps of physical dependence and habituation" (Cohen 1969:76). In spite of the fact that opiate drugs have been used in beneficial ways for their analgesic, hypnotic, and antidiarrhoea effects since early times, indiscriminate use of these same drugs has often resulted in addiction, especially during this century.

While many doctors still legally prescribe opiate drugs for pain relief, worldwide legislative efforts designed to suppress the illicit use of these drugs continue to be largely unsuccessful. Today, the legal and illegal use of opiate drugs affects a large number of physicians, patients, politicians, criminals, and law-enforcement personnel in various parts of the world. Although I hope that my study will provide a useful perspective, it does not deal with the contemporary problems of opiate use. Instead, it concentrates attention on the enigmatic natural origins of the opium poppy and the early relations between man and this famous drug plant.

Past Research and Present Objectives

Early research in the field of culturally oriented biogeography was strongly influenced by Alphonse de Candolle. His classic work, *L'origine des plantes cultivées* (1883), is still a basic reference book. The studies of de Candolle and others such as Victor Hehn (see *The Wanderings of Plants and Animals from Their First Home*, 1885) encouraged many scholars to determine the natural origins, domestication, and early cultural histories of the major economic plants. My study follows in this tradition.

According to La Barre (1970:74), "Only four or five narcotic cultigens are commercially important; and they are unknown in the wild state, indicating long association with man." One of these exceptional drug plants is the opium poppy, which Krikorian (1975:92) has referred to as "the most important of those relatively few plants that yield drugs which have been used since antiquity and have never been matched by a synthetic product." I will attempt to show that in areas where the opium poppy was available during the prehistoric period, man used this plant for a number of purposes, the most important of which was probably pain relief. However, before I discuss man's early associations with this species in the latter part of this study, I first survey and synthesize the diverse literature relevant to the natural origins of the opium poppy. In chapter 2, I address myself to questions concerning the taxonomy of the species, *Papaver somniferum* L. This review provides background information pertinent to the origin of the species, which is discussed in chapter 3.

Here, I describe various hypotheses and data that are relevant to the botanical ancestry of the opium poppy. Past and present research involving fossil evidence of the opium poppy, crossbreeding experiments, chromosomal investigations, and other relevant biological aspects of the species are considered in detail.

In chapter 4, I present what I believe is the first speculative discussion of man's initial relationships with the opium poppy, emphasizing the many uses that this plant could have had for Mesolithic and early Neolithic peoples such as human food, animal fodder, lubricating oil, medicines, and psychoactive drugs for ritualistic purposes. In addition, I offer a hypothesis to explain how these very early relationships may have developed.

In chapter 5, I survey the early archaeological record for the opium poppy and evaluate the fossil evidence. A striking aspect of this record is that practically all of the prehistoric evidence for the opium poppy comes from Neolithic and Early Bronze Age settlements along the lakeshores and bogsides of the Swiss Foreland in west central Europe. The implications of this distribution are discussed in a later section of this chapter which deals with the differential preservation of macrofossil plant remains. After surveying the Neolithic and Early Bronze Age sites which have yielded fossil opium poppy evidence, I address myself, in the latter part of chapter 5, to questions concerning the status of the plant as a cultivated or weedy species and its probable use by the prehistoric peoples of the region.

In chapter 6, I evaluate the various possible places of origin for the use of the opium poppy, including those in southeastern Europe, the Near East, northwest Europe, southwest Europe, and Switzerland. In the process of presenting the alternative possibilities, I discuss the limited fossil evidence for the opium poppy which has been found in archaeological sites outside of the Swiss Foreland (a plateau region located between the Jura Mountains and the Alps). I also discuss the etymology of some words for the opium poppy in hopes that this will shed additional light on the cultural origins of the species.

Several short historical studies of human utilization of opium drugs have been published (e.g., Macht 1915:477–81; Collom 1957:614–26; Wright 1968:62–70; and Cohen 1969:76–85). However, these studies have employed a broad perspective and consequently have not surveyed the ancient periods in any detail. Some research has focused on the possible use of the opium poppy in specific regions during the late prehistoric and early historic periods. For example, Gabra (1954–55:39–56) presented evidence indicating that the opium poppy was known and used in ancient Egypt; Merrillees (1962:287–92) studied the prehistoric trade between Cyprus, the Levantine Coast, and Egypt and suggested that opium was used and traded between these areas during the middle of the second millennium B.C. (i.e., the Late Bronze Age); Kritikos and Papadaki (1963:80–150) surveyed the archaeological and historical evidence concerning the early use of opium in Greece and its colonies; and Krikorian (1975:95–114) critically evaluated the popular belief that ancient peoples in the Near East utilized the opium poppy. In chapters 7, 8, and 9, I have updated and synthesized the studies mentioned above, adding new information, and thereby present a more comprehensive discussion of man's association with the opium poppy during the late prehistoric period.

It should be noted here that commercial trade in the ancient Near East and Mediterranean regions seems to have played a very important part in the development of urbanization (Lamberg-Karlovsky 1972:22–30). Raw materials for buildings were the chief items of this early trade, but luxury goods including medicinals and precious oils were also important (Crawford 1973:237; see also Durrani 1964:51–96). Relatively long-distance trade became quite significant by the late Bronze Age (Muhly 1973). I believe that opium was one of the items traded within parts of Europe and the eastern Mediterranean regions during the Late Bronze Age. If this assumption can be substantiated, then my study should provide additional information relevant to the larger question of how and why the Old World civilizations developed.

In addition, the possible use of opium in the context of ritual worship deserves attention. Some have already suggested that opium drugs may have played a role in religion and cult worship in the ancient civilizations of Crete (Cook 1964:1165), Greece (Marinatos 1960:150; Kritikos and Papadaki 1967:1–10; and Wood 1971:232), Cyprus (Karageorghis 1976:125–29), Egypt (Merrillees 1962:287–89; 1968), and in the Levant (Merrillees 1968, 1974). In the last part of my study, I have investigated this type of application over a broader geographic area and through a longer period of time. A series of intriguing clay and ivory artifacts which may have been used for ritualistic opium intoxication in the eastern Mediterranean region during the Late Bronze Age are discussed in detail.

An historical biogeographic study can provide a unique perspective on man's experience with plants. I hope that this investigation of man's ancient relationships with the opium poppy and its drug products offers some useful insights into the motivations and patterns of past drug use and thereby furnishes valuable information concerning contemporary drug use and its economic exploitation (Dai 1970; Furst 1972: vii–xvi).

Methodological Problems and Procedures

My primary research method in this study has been an extensive literature survey of library and archival resources in domestic and foreign institutions. During the past five years, I sought information from various fields including geography, botany, genetics, archaeology, anthropology, history, mythology, religion, linguistics, philosophy, and medicine. Although data from a diversity of disciplines were utilized, the main source of evidence indicating the existence and use of the opium poppy in the prehistoric period came from archaeological reports. Recovery of the actual opium poppy itself or its products such as seed, capsules, or opium drugs is the most convincing proof of its presence. Regrettably, this has been a relatively rare occurrence. In fact, the heavily concentrated distribution of the limited fossil evidence may be misleading.

Regional Variation in Scholarly Interest

> One reason for Europe's importance is its long history of a high standard of archaeological excavation and reporting; another is the longstanding cooperation of archaeologists and scientists in prehistoric research, particularly in Scandinavia and Atlantic Europe. (Rodden 1975: 49)

As I noted previously, almost all of the early archaeological evidence indicating the presence of the opium poppy in prehistoric human settlement sites prior to the Late Bronze Age comes from locations within a relatively small area of west central Europe including parts of southwest Germany, northwest Switzerland, eastern France, and northern Italy. However, this striking concentration of the archaeobotanical evidence may be more a reflection of regional enthusiasm for the study of prehistoric plant deposits and differential preservation of the macro- and microfossil plant remains than a valid indicator of the cultural prehistory of the opium poppy.

Although the agricultural activities of some prehistoric western European culture groups are relatively well known, many are still poorly understood: "Despite the large amount of research devoted to studying the development and dispersal of major food crops in antiquity, knowledge is often sketchy" (Dennell 1978: 8). In the case of the Neolithic cultures of west central Europe, the exceptional wealth of direct evidence has probably helped define their economy and diet to a greater degree of sophistication than virtually any other in prehistoric Europe (Chard 1975: 347). In Switzerland, the structure and functioning of the federalist system has played a significant role in stimulating enthusiasm and systematic research. Archaeological, as well as other sciences, are sponsored by the individual cantons: "The advantage of this system has been to increase archaeological research and also the regional museums in addition to the Swiss National Museum" (Sauter 1976: 14).

In contrast, we have only a paucity of information about food production for the prehistoric farming groups of southwest Europe. According to Murray (1970: 51), we should recognize the fact that the lack of fossil evidence in this region "is not due to the absence of domesticated animals and plants but rather to the opinion of some excavators that finds of animal bones or carbonized plants were irrelevant and unimportant."

Differential Preservation of Macrofossil Plant Remains

In addition to this history of variation in regional attitudes among scientific investigators, differential preservation of organic materials has also been an important variable affecting research. For example, it is well known that some ecological situations such as dry caves, ash or tuff deposits, and waterlogged anaerobic environments inhibit organic decomposition and thus retard the disappearance of potential archaeobotanical evidence.

Dry cave environments have long been a major source of archaeological evidence. However, only one such environment, near the Mediterranean coast of southern Spain, has provided evidence for a prehistoric association between humans and the opium poppy (Gongora 1868). A poppy capsule excavated from a tuff deposit in southern France during the nineteenth century is of ethnobotanical interest but its age is uncertain (Fritel 1920; Neuweiler 1935; Netolitzky 1931). Unfortunately, with the exception of the data from the bog and lakeshore sites in west central Europe, the capsules from the Spanish cave (and perhaps the fossil capsule from the tuff deposit in France) are the only direct evidence we have at the moment that suggests human use of the opium poppy during the Neolithic and Early Bronze Age periods.

Although we do have considerable macrofossil evidence suggesting the use of the opium poppy in prehistoric west central Europe, other regions might yield similar data if equivalent deposits were studied as intensely:

> The concentration of *Papaver* which you can observe [in the fossil record from west central Europe] is probably artificial. Most of the sites are bog or lakeborder sites with excellent conditions for the conservation of non-carbonized seeds. Uncarbonized seeds are not apt to be present in dry soil sites and thus the concentration of *Papaver* seeds may be more closely linked to conservation conditions than to their real distribution during prehistoric times. (Lundstrom Baudais, personal communication, May 30, 1978)

Fossil Evidence

Pollen analysis is still of little help in the determination of opium poppy use in the prehistoric period because confirmed identification of *Papaver somniferum* L. pollen fossils has not yet been achieved. Although van Zeist and Casparie (1974:426) recently reported that they had identified "relatively high pollen values for *Papaver*" around the Neolithic bogside settlement at Niederwil, Godwin (personal communication, 1976), Greig (personal communication, 1976) and van Zeist (personal communication, 1980) have referred to the unresolved problem of identifying *Papaver* pollen to a specific level. Van Zeist has informed me that *Papaver* is, in general, rather poorly represented in the pollen record and may be impossible to identify beyond the genus level. In any case, we cannot be sure that fossil *Papaver* pollen indicates the past presence, let alone use, of the opium poppy.

Without agreement regarding the use of pollen analysis, we must turn to the macrofossils even though this form of evidence (i.e., fruits and seeds) is becoming both qualitatively and quantitatively less important in archaeobotanical research. As Villaret-von Rochow (1967:22) has pointed out, our knowledge of ancient plant use is so rudimentary that an attempt to develop a reasonable synthesis merely on the basis of fruit and seed finds has many limitations. In addition, new archaeological discoveries and research techniques can modify understanding of the prehistoric transmission and distribution of several species (see Bertsch and Bertsch 1949, Crabtree 1975:266–83, and Dimbleby 1975:265–67).

In any case, aside from the very infrequent recovery of fossil capsules, we must rely on opium poppy seeds as evidence of the presence and possible use of this species in the prehistoric period. The relatively small seeds (approximately one millimeter or less in length) may show up in the archaeological record, partially decomposed in petrified human feces (coproliths), as uncharred specimens (usually from waterlogged sites), or as carbonized fossils in middens, storage jars, or ovens.

> As pollen in lake sediments partly originates from distant source areas pollen analyses give only a simplified picture of the very local succession. Contrary to this, the bulk of seeds in natural sediments are of local origin, and macro fossil analyses give a more precise impression of the local vegetation. Besides, more seeds than pollen can be determined to the species level. (Fredskild 1978:197)

The relative frequencies of plant macrofossil remains may be misleading as an indicator of the actual importance of the species or its possible associations with the human groups who occupied the sites from which the fossil material was excavated:

> Just as for pollen, the seed production shows large differences among the various plant species. Moreover, the seeds or fruits of some wild species were collected intentionally, in consequence of which they are found in large numbers in the settlement samples. The seeds of many other species, on the other hand, must have arrived in the settlement by accident. (van Zeist and Casparie 1974:421)

The cautious approach suggested above was part of van Zeist and Casparie's report on the Neolithic lake shore settlement of Niederwil in northeastern Switzerland. This is one of a number of early farming sites located along the ancient lake and bog margins in west central Europe from which we have direct macrofossil evidence of the opium poppy's presence and use in the prehistoric period. These sites and the opium poppy fossils found in them are discussed in chapter 4.

In recent times, scientific inquiry in the field of archaeobotany has been enhanced by a number of technological innovations. For example, the development of several types of sophisticated flotation devices now make it possible to test extensive soil samples in a relatively systematic manner. This, in turn, has helped open up new avenues of research and improved our understanding of how settlements functioned internally.

> Instead of being concerned only with the types of plant food used by a community, researchers also hope to discover which plants were intentionally cultivated, which were the most important and how plant foods were prepared for consumption and storage. (Dennell 1978:8)

According to Dennell (1978:9), the development and implementation of the large-scale flotation devices is as significant an advance in archaeobotany as the invention and use of the telescope was in astronomy: "The most important result is that one can now obtain a far more accurate indication of the importance of each plant food in an ancient community."

Unfortunately, much of the information we now have regarding the existence and possible prehistoric use of the opium poppy comes from research reports published many years ago. Reconsideration of ancient food-processing methods and differential preservation of the plant remains, together with the future use of the new techniques and devices such as those suggested by Dennell (1972, 1974, 1976) and Hubbard (1976) may bring to light information concerning the economic status and degree of domestication of the opium poppy.

In any case, it is important to remember that what we are learning today about how we should attempt to recover and interpret ancient plant remains underscores the need for a careful appraisal of the methodological shortcomings of earlier archaeobotanical research. In the past, the techniques used to recover ancient plant remains were usually primitive and the search unsystematic:

> Food remains were usually found either by accident—when, for instance, a large deposit of carbonized (burnt) material was easily visible in the deposits—or by using crude flotation techniques. These often consisted of merely immersing small amounts of soil in water in the hope that some seeds or grains would float to the top. (Dennell 1978:8)

Contamination of Archaeobotanical Evidence

Those who deal with data on ancient plant remains (either recently discovered or based on older reports from the literature) should be aware of certain control conditions that must be maintained in order to prevent contamination of an archaeological deposit by seeds of modern origin. Keepax (1977:225) has listed a number of possible sources of contamination:

1) "Dirty" excavation and careless collection of samples.

2) Aerial contamination (seeds blowing onto the sample while it is exposed to the atmosphere).
3) Cross-contamination (seeds blowing from one sample to another while they are exposed to the atmosphere).
4) Contamination of the archaeological deposits before excavation. (Keepax 1977:225)

The problem of aerial contamination can be especially acute when a flotation machine is used; during some seasons there may be numerous wind-carried propagules moving around in the lower atmosphere. In fact, even a few grams of topsoil may contain a rather large amount of modern seeds. Furthermore, when separating the seeds from the soil samples with a flotation machine it is quite possible that disseminules from modern specimens may be blown into, or fall on, the recovery apparatus if it is not well covered and guarded carefully against contamination.

In reference to this contamination problem, Villaret-von Rochow criticized the recovery procedure used by a scientist in the early part of this centry at an archaeological site in which fossil evidence of the opium poppy was reported:

> For instance, the method described by H. Messikommer (1913: 10 & 96) for gathering seeds by washing in ditches of peat or by simply gathering off the water surface in lake bays, in Robenhausen and other places, has led to the fact, obviously, that plant varieties of the recent moor and lakeshore vegetation got into the flora lists. (Villaret-von Rochow 1967:22, trans. Inge Fischer 1976)

Even with effective action to avoid the first three sources of contamination suggested by Keepax, seeds of modern origin may have already adulterated the archaeological deposit. Keepax listed four ways in which prehistoric plant remains may become contaminated prior to any excavation activity:

a) *Ploughing* can incorporate modern seeds lying on the surface of the soil into the plough soil. This would only be of significance on very shallow sites where the plough has disturbed archaeological levels.
b) *Root holes and drying cracks* (particularly in clay soils) may penetrate into archaeological levels. They may be expected to fill up with soil washed down from the surface layers.
c) *Downwashing*. A very open textured deposit (e.g. chalk rubble) may allow modern soil containing seeds to move down through it.
d) *Earthworms* are active on many archaeological sites. They are probably the single most important factor in the introduction of contamination to archaeological levels. (Keepax 1977:225)

Although earthworms may be found in large numbers under many soil conditions, they are usually not found in extremely acid, peaty, or waterlogged soils (Keepax 1977:225). Fortunately, the prehistoric opium poppy seeds from the west central European sites have been recovered from waterlogged or peaty deposits.

Keepax suggests a number of methods that can be used to identify modern contamination of the archaeological deposits in which plant remains have been found: (1) The *radiocarbon dating technique* can be used where an adequate sample is available; however, any single date can be considered worthless since it may represent the age of either an ancient seed or a more recently arriving contaminant. (2) The *concentration of seeds* may indicate their antiquity; theoretically, the greater the concentration, the higher the probability that the sample is an ancient one. (3) A comparative survey of the *species present* can be useful in establishing the age of the seed samples, especially if the modern flora of the area is sufficiently dissimilar to that of the archaeological sample; however, here one should carefully assess the possibility that the vegetation might have changed recently. (4) The *conditions of preservation* are often a useful indication of the antiquity of the sample. Except under waterlogged conditions, it is quite improbable that uncharred seeds will be preserved for very long; therefore, those seeds excavated from waterlogged sites or charred specimens from non-waterlogged sites are the types of samples most likely to be truly ancient in origin. However, as Keepax has pointed out, special care must be taken before discounting the antiquity of a sample simply because it consists of uncharred seeds or did not come from a waterlogged site:

> Conditions on each part of a site must be carefully considered before any such generalizations are made, for example under Silbury Hill organic material was very well preserved . . . in non-waterlogged conditions. Urban sites may sometimes produce well preserved seeds from highly organic deposits although these are not always completely waterlogged.

Keepax (1977:228) also points out that charred seeds, although usually assumed to be of ancient origin when excavated from an archaeological site, may in fact be of more recent derivation, especially if the charred seed sample is small. This possibility can occur with recent burnings of agricultural fields.

Artifact Analysis

The Late Bronze Age was an important period for the Mediterranean and parts of Europe to the north. Trade between these regions developed into a flourishing exchange of goods and ideas. I believe that the opium poppy was introduced into the eastern Mediterranean region during the Late Bronze Age, probably from somewhere in central Europe or the southwestern Mediterranean area. Although presently there is very little, if any, direct evidence to support this hypothesis, I believe that there is a good deal of indirect evidence which can be used in defense of my view. Problems involving regional variation of scholarly interest and the differential preservation of macrofossil plant remains have already been discussed. In lieu of direct archaeological evidence, I have relied on artifacts formed out of clay, stone, metal, glass, and ivory which bear what I believe to be artistic representations of the opium poppy. These relics include statues, figurines, vases, jars, jugs, plates, coins, wall paintings, jewelry, amulets, pendants, and pins.

A fundamental problem in my analysis of the artifacts has been the subjective aspects of interpretation. Representations of pomegranate fruits and opium poppy capsules have been confused; both the pomegranate and the opium poppy have been referred to as the organic inspiration of the same artifact design. Since the artifacts I studied did form a significant basis of my information, in most cases, I evaluated the primary archaeological reports in which the artifacts I considered to be representations of the opium poppy were first described. In addition, in some cases, I made personal contact with museum curators and respected students in the field. I was able to observe a number of the artifacts in person. However, in many cases, I had to solicit and collect photographs of the artifacts for scrutiny and reference. Archaeological contexts which contained one or more indications of the presence of the opium poppy, or its symbolic representation are described as fully as the needs of the analysis demanded. This included a special effort to establish absolute or relative chronology.

Radiocarbon Dating

Throughout chapters 5, 6, 7, 8, and 9 there are numerous references to dates indicating the approximate age of the opium poppy fossils. The dates given may, in many cases, be not only approximate in accuracy, they may well be wrong by several hundred years!

Since the early implementation of the revolutionary radiocarbon dating technique in the early 1950s, more than 35,000 dates have been established for a wide variety of materials including many which have been recovered from archaeological excavations; yet of all the assumptions inherent in the theoretical basis of this dating technique developed by Libby (1952), only one has needed modification (Burleigh 1974:69). However, this single correction has not only affected the specific chronologies of many sites but it has also required revision of theories regarding the history of civilization in Europe and the spread of cultural innovations out of the Old World (Renfrew 1973).

Libby's original radiocarbon dating technique assumed that the proportion of carbon 14 in the atmosphere, and therefore all living things, has remained constant. However, by the early 1960s, fluctuations in the amount of carbon 14 in the atmosphere had been demonstrated and it thus became clear that these variations must be taken into account when determinations of absolute dates are made (J. A. S. Evans 1977:76–77).

Fortunately, recent dendrochronological research involving the very long-living bristlecone pine *(Pinus aristata)* of the White Mountains in California has helped both illuminate the problem and rectify the chronological discrepancies:

> Radiocarbon analysis of ring samples of known age identified the inaccuracy of the radiocarbon dates; and, with the accumulation of sufficient analyses, was able to supply curves and tables of conversion which permit correction of radiocarbon dates to approximate true age. (Gimbutas 1974:14)

Dates falling between the third and fifth millennia B.C. in the uncorrected chronology require recalibration of an increasing amount with increasing age (Gimbutas 1974:14). The necessary redating of prehistory affects this study in a number of ways, including specific reconsideration of many Neolithic sites from which opium poppy fossils have been recovered. In many cases, corrected dates have been given; otherwise the chronological reference is referred to as "uncorrected."

Another recent technological innovation in the field of radiocarbon dating should also be taken into account here because it may affect the size of known and future discoveries of opium poppy fossils. In May of 1977, it was first reported that electrostatic accelerators could be used to detect carbon-14 ions in recent samples of carbon (Bennett et al. 1978:345). Later in 1977, preliminary reports of the initial attempts to date milligram samples were published (H. E. Gove 1977, cited in Bennett et al. 1978:345). In 1978, radiocarbon dating experiments comparing the use of the conventional beta-ray counting method and the new ion counting technique were reported and discussed: "The dates determined by the two methods are consistent with each other" (Bennett et al. 1978:345).

In a more recent report, Stuiver (1978:881–83) evaluated the relative benefits of the beta and ion counting methods and pointed out that the primary advantage of ion counting is that it can be used to date very small (milligram) samples. The effective application of the ion counting method is therefore relevant to the study of the opium poppy during the prehistoric period because fossil discoveries of this species have often been quite small in both number and weight. In the future, the ion counting method may be used to determine the age of these fossils, especially when only one or a few of the tiny seeds are found. If the sample size of the fossils is not limited by weight, then the new ion counting technique" complements beta counting for the more routine carbon-14 determinations" (Stuiver 1978:881).

Utilization of Research Resources

In addition to the extensive use of the University of Hawaii Library resources, numerous interlibrary loan services, and wide-ranging correspondence with many authorities and students in several related disciplines (see Acknowledgments), I also conducted some research at relevant archaeological sites, libraries, and museums in the general study region. Research institutions and archaeological sites visited and utilized include the libraries of the University of California at Berkeley and Los Angeles, the Bodleian Library in Oxford, the British Museum Library in London, Bibliothèque Royale de Belgique in Brussels, Bayerische Staatsbibliothek in Munich, the National Archaeological Museum of Greece in Athens, the Heraklion Museum on Crete, and other regional Greek museums and archaeological sites at Mycenae, Sparta, Corinth, Delphi, Epidauros, Olympia, Knossos, Phaistos, Mallia, and Gournia.

2 • Taxonomy of the Opium Poppy, *Papaver somniferum* L.

The opium poppy, *Papaver somniferum* L., belongs to the order of Papaverales. Among the orders of flowering plants, Papaverales is generally considered to be of relatively primitive evolutionary development. According to C. L. Porter (1967 : 266), this order is composed of four families including Papaveraceae, Fumariaceae, Brassicaceae (Cruciferae), and Capparaceae. The opium poppy is classified as a member of Papaveraceae.

The Family: Papaveraceae

Papaveraceae is the family which typifies the order of Papaverales and is commonly known as the poppy family. Although the classification of this family itself is not controversial, its internal taxonomy is disputed. Consequently the quantity of genera and number of species within each genus varies according to the individual taxonomist who works with the group. For example, Lawrence (1951 : 516) recognizes 28 genera and 250 species, C. L. Porter (1967 : 267) recognizes 24 genera and 450 species, Stebbins (1974 : 350) recognizes 27 genera and 470 species, and Duke (1974 : 2) recognizes 700 taxa.

Most of the species in this family are annual or perennial herbs native to the temperate and subtropical regions of the Northern Hemisphere. The regions of greatest taxonomic diversity within the poppy family are Western North America which contains twelve genera and Eastern Asia which contains nine genera (Lawrence 1951 : 516).

Members of the poppy family produce milky or colored juice. Their leaves are arranged alternately, or the uppermost opposite or whorled. The leaves are without stipules and the margins are commonly much divided but sometimes entire. The flowers are usually solitary, perfect, regular, hypogynous, and showy. They usually have two or three deciduous sepals which fall off as the flowers open. The petals are separate and 4 to 6 or 8 to 12 in number. They are usually arranged in 1 or 2 whorls and crumpled in the bud. The stamens are numerous and free. The ovary is superior, usually comprises two or more loosely joined carpels with parietal placentation, and produces many ovules. The stigmas can be simple or divided, often taking the form of a pronged disk on top of the ovary, or a short style. The fruit is a capsule with valves or pores under the stigmas through which a multitude of seeds are dispersed.

The Genus: *Papaver*

The opium poppy belongs to *Papaver*, the genus which typifies the family, Papaveraceae. In 1753, Carl von Linné (Linnaeus) classified only eight species within this genus (1753: vol. 1, p. 508). Today *Papaver* is treated as a genus containing approximately one hundred species (Porter 1967:267 and Good 1974:88), most of which are found in the Mediterranean region (Lawrence 1951:616).

Papaver species are characterized by lobed or toothed leaves, flowers whose buds nod, petals which are commonly red and four in number or doubled, and stigmas which are united in a pronged disk. Fedde (1909) recognized ninety-nine species in the genus *Papaver*, twenty-nine of which he himself had systematically described. Fedde divided the genus into nine sections. Table 1 lists these sections of *Papaver* and the species classified within each one of them.

Table 1

Sections of the Genus *Papaver* (according to Fedde in 1909, from Stermitz 1968: 163–64)

Section 1. *Orthorhoeades*

P. rhoeas L.	*P. thaumasiosepalum* Fedde
P. rapiferum Fedde	*P. ameristophyllum* Fedde
P. caespitosum Fedde	*P. trilobum* Wallr.
P. tenerifae Fedde	*P. strigosum* (Bönningh.) Schur.
P. subpiriforme Fedde	*P. chelidoniifolium* Boiss. & Buhse
P. hirto-dubium Fedde	*P. commutatum* Fisch. & Mey
P. rhopalothece Stapf.	*P. tenuissimum* (Heldr.) Fedde
P. bipinnatum C. A. Mey.	*P. californicum* A. Gray
P. polytrichum Boiss. & Kotschy.	*P. lemmonii* Greene
P. pseudo-haussknechlii Fedde	*P. dubium* L.
P. integrifolium Vig.	*P. apicigemmatum* Fedde
P. syriacum Boiss. & Blanche	*P. arenarium* Marsch.-Bieb.
P. humile Fedde	*P. pinnatifidum* Moris
P. roubiaei Vig.	*P. lunetanum* Fedde
P. subadpressiusculo-selosum Fedde	*P. simoni* Foucaud
P. schweinfurthii Fedde	*P. mosestum* Jord.

Table 1 (continued)

Sections of the Genus *Papaver* (according to Fedde in 1909, from Stermitz 1968: 163–64)

P. erosulum Jord.
P. stipitatum Fedde
P. tenuifolium Boiss. & Hohen.
P. robertianella Fedde
P. posti Fedde
P. humifusum Fedde
P. stylatum Boiss. & Bal.
P. subumbilicatum Fedde
P. clavatum Boiss. & Hausskn.
P. gürlekense Stapf
P. umbonatum Boiss.
P. cassandrinum Charrel
P. exspectatum Fedde

Section 2. *Argemonorhoeades*

P. argemone L.
P. virchowii Aschers. & Sint.
P. belangeri Boiss.
P. apulum Ten.
P. hybridum L.
P. pavonium Fisch. & Mey.

Section 3. *Carinatae*

P. macrostomum Boiss. & Huet
P. piptostigma Bienert
P. tubuliferum Fedde
P. dalechianum Fedde
P. kurdistanicum Fedde
P. divergens Fedde & Bornmüller
P. bornmüller Fedde

Section 4. *Mecones*

P. somniferum L.
P. setigerum DC.
P. glaucum Boiss. & Hausskn.
P. gracile Auch.
P. decaisnei Hochst. & Steud.

Section 5. *Miltantha*

P. tauricolum Boiss.
P. bartuschianum Fedde
P. flahaultii Fedde
P. persicum Lindl.
P. caucasicum Marsch.-Bieb.
P. hyoscyamifolium Boiss. & Hausskn.
P. floribundum Desf.
P. acrochaetum Bornmüller
P. fugax Poir.
P. urbanianum Fedde
P. armeniacum (L.) DC.
P. triniaefolium Boiss.
P. libanoticum Boiss.
P. polychaetum Schott & Kotschy.

Section 6. *Pilosa*

P. spicatum Boiss. & Bal.
P. pilosum Sibth. & Smith
P. apokrinomenon Fedde
P. heldreichii Boiss.
P. strictum Boiss. & Bal.
P. pseudostrictum Fedde
P. lateritium C. Koch
P. oreophilum F. J. Rupr.
P. monanthum Trautv.
P. ramosissimum Fedde
P. rupifragum Boiss. & Reut.
P. atlanticum Ball

Section 7. *Macrantha*

P. orientale L.
P. bracteatum Lindl.
P. lasiothrix Fedde
P. paucifoliatum (Trautv.) Fedde

Section 8. *Scapiflora*

P. suaveolens Lap.
P. pyrenaicum (L.) A. Kerner
P. alpinum L.
P. nudicaule L.
P. anomalum Fedde

Section 9. *Horrida*

P. aculeatum Thunb.

Some genera in Fedde's classification of the whole poppy family (1936) have needed considerable revision (Lawrence 1951:516). However, *Papaver* has been one genus together with its sections which has required relatively little change (Stermitz

1968: 172–74). Nevertheless, recently the composition of some of Fedde's sections within *Papaver* have also been criticized as unnatural (Goldblatt 1974; Stermitz 1968). One of these controversial sections is *Mecones* (equals *Papaver:* see Goldblatt 1974: 265), which contains the opium poppy, *Papaver somniferum* L.

The Section: *Mecones*

According to Fedde's taxonomic classification, section *Mecones* contains *Papaver somniferum* L., *P. setigerum* DC, *P. glaucum* Boiss. and Hausskn., *P. gracile* Auch., and *P. decaisnei* Hochst. and Steud. (Fedde 1909).

Although most botanists seemed to have accepted this classification during the early part of this century (Veselovskaya 1976), more recently a number of authors such as Kuzmina (1937), Fulton (1944), Santavy (1962), Stermitz (1968), Vent (1970), and Reckin (1973) have criticized Fedde's grouping of taxa for *Mecones*. In fact, recent biochemical and cytological observations have raised some doubt about the composition and the relationships of the individual members of the genus *Papaver*. Alkaloidal aspects of the problem are discussed below and summarized for reference in table 2.

The Alkaloidal Data

Two members of section *Mecones*, the opium poppy, *P. somniferum*, and the setaceous poppy, *P. setigerum*, have been the focus of a great deal of biochemical research because these taxa were throught to be "the only plant species so far found to contain morphine" (Stermitz 1968: 166). However, in his review of the literature, Gibbs (1974: 262) listed the occurrence of the morphine alkaloid itself in the hops plant, *Humulus lupulus* L., of the mulberry family, Moraceae, as well as in five species of Papaveraceae including *Argemone mexicana* L., *Eschscholtzia mexicana* Greene, *Papaver rhoeas* L., *P. setigerum* (fruit), and *P. somniferum* (leaves and fruits).

On the other hand, James Duke (1974b: 3) has suggested that the reports of alkaloid occurrence are not always reliable because they "are not based on voucher material; hence, species determinations are suspect." Although voucher specimens should certainly be made available for reference and confirmation in such cases, it is hard to imagine that *Argemone*, *Eschscholtzia*, and especially *Humulus* would be mistaken for *Papaver*. According to Popov (1970: 486), the dried flowers of *P. rhoeas* of the section *Orthorhoeades* "contain rhoeadine and a slight quantity of morphine, often used as a soporific." However, in a personal communication Duke (August 1976) referred to *P. somniferum* and *P. setigerum* as the only two species known to contain morphine. The alkaloidal precursors of morphine are isothebaine, thebaine, and codeine. all three of these alkaloids along with a variety of benzylisoquinolines accumulate in *P. somniferum* and *P. setigerum* (Stermitz 1968: 166).

According to Santavy (1962), Kühn and Pfeifer (1963), and Reckin (1973), two poppy species, *P. glaucum* and *P. gracile*,which Fedde treated as belonging to section *Mecones*, should be separated from the other members of this section and placed in section *Orthorhoeades* because they do not accumulate morphine or benzylisoquinoline alkaloids, but do contain alkaloids of rhoeadine and papaverrubines which are

Table 2

Alkaloidal Summary for Some Species of *Papaver*

Section *Mecones* according to Danert (1958)	Section *Oxytona* according to Goldblatt (1974)	Section *Orthorhoeades* according to Fedde (1909)	Section *Mecones* according to Fedde (1909)	Section *Mecones* according to Reckin (1973)	Section *Mecones* according to Popov (1970)	*POPPY SPECIES*	Morphine	Thebaine	Isothebaine	Codeine	Benzylisoquinoline	Rhoeadine	Papaverrubine	Papaverine Alkaloids
x_b			x			*Papaver decaisnei* Hochst. and Steud.								
x_b			x	O		*P. glaucum* Boiss. and Hausskn.						x	x	
x_b			x	O		*P. gracile* Auch.						x	x	
x_a			x	x	x	*P. setigerum* DC.	x	x	x	x	x			x
x_a			x	x	x	*P. somniferum* L.	x	x	x	x	x			x
		O				*P. rhoeas* L.	x?					x		
		O				*P. strigosum* Schur.		x						
		O				*P. intermedium* (Becker) O. Ktze.		x						
		O				*P. commutatum* Fisch. and Mey.								x
	O_x					*P. bracteatum* Lindl.		x	x					
	O_x					*P. orientale* L.		x	x					
	O_x					*P. pseudo-orientale* (Fedde) Medw.			x					
				x		*P. aculeatum* Thunb.								
				x		*P. Gaubae* Cullen and Rechinger								

x = member of *Mecones*
O = member of *Orthorhoeades*
O_x = member of *Oxytona*

(Note: Danert's treatment was based on difference in filament position)

chemotaxonomically characteristic of those species classified within the section *Orthorhoeades*.

It should be emphasized here that alkaloidal relationships within the genus *Papaver* are both complex and confusing. For example, Santavy et al. (1960) reported the presence of thebaine in *Papaver strigosum* Schur. and *Papaver intermedium* (Becker) O. Ktze., whereas others have reported finding papaverine alkaloids in *Papaver commutatum* Fisch. and Mey. (Stermitz 1968: 166). The only other species known to contain papaverine alkaloids are *P. somniferum* and *P. setigerum* which, as mentioned above, also contain thebaine, a precursor of morphine (Stermitz 1968: 166–67).

Papaver strigosum, *P. intermedium*, *P. commutatum*, and *P. rhoeas* have been treated as belonging to section *Orthorhoeades* because of their close morphological relationships. However, their biochemical affinities with *P. somniferum* and *P. setigerum* of the *Mecones* have also been noted. *Papaver strigosum*, *P. intermedium*, and *P. commutatum* could be classified as intermediate between species of the section *Mecones* and the other members of section *Orthorhoeades*.

Skeptical of this conclusion, Stermitz referred to the work of Maturova et al. (1966), which reported finding traces of thebaine in *Papaver oreophilum* FJ. Rupr., a member of section *Pilosa* of the genus *Papaver*. "Perhaps the appearance of traces of thebaine will prove to be common and hence of little value systematically" (Stermitz 1968 : 167). The presence of thebaine alkaloids among members of section *Oxytona (Macrantha)* has also prompted comparisons between it and *Mecones*.

In his recent biosystematic study of section *Oxytona*, Goldblatt (1974 : 265) has treated this section of *Papaver* "as comprising three species, *P. bracteatum* Lindl., diploid (2n = 14), *P. orientale* L., tetraploid (2n = 28), and *P. pseudo-orientale* (Fedde) Medw., hexaploid (2n = 42)." Morphological characteristics of section *Oxytona* make it a clearly defined section of the genus *Papaver* (Stermitz 1968 : 169). However, the chemotaxonomy of two species, *P. bracteatum* and *P. orientale*, has been the subject of some confusion:

> These two species are closely related morphologically. Chemically, they are distinguished from the species of other sections by the accumulation of thebaine and isothebaine. The presence of thebaine indicates a closeness in alkaloid chemistry to *Mecones* as does the occurrence of oripavine in *P. orientale* L. The (presumed) demethylation of thebaine to yield oripavine is paralleled only by the similar demethylation in the section *Mecones* which produces morphine from codeine. (Stermitz 1968 : 169–70)

Stermitz made special reference to the puzzling example of *P. bracteatum*, which, according to some studies, manifested considerable chemical variation within the species:

> In two investigations (Kiselev and Konovalova, 1948; Heydenreich and Pfeifer, 1965) isothebaine was found to be the main alkaloid. However, a chemical variety of *P. bracteatum* has been developed (Neubauer and Mothes, 1963; Neubauer, 1965) in which thebaine represents 95 percent of the total alkaloid content. Although there are a number of species in the Papaveraceae that show strong tendencies toward the development of such chemical varieties, this is perhaps one of the most striking examples. Such reports emphasize the dangers inherent in basing systematics on alkaloid content alone. One may question the idea that alkaloid content could have any fundamental significance in plant biochemistry if barely distinguishable varieties of the same species are found to exhibit such marked chemical differences. However, even though the final structures of thebaine and isothebaine appear to be quite different, the basic phenol coupling reactions that lead to the elaboration of thebaine and isothebaine could easily be mediated by the same enzyme system. (Stermitz 1968 : 170)

Goldblatt (1974 : 266) reviewed the confused and conflicting literature which contains reports of the presence of morphine-related alkaloids in *P. bracteatum* and *P. orientale.* He pointed out that in the past, biochemical research dealing with these species had almost exclusively utilized cultivated material and herbarium specimens and that a satisfactory investigation of "wild plants" had not yet been undertaken. Therefore, Goldblatt (1974:275) studied the alkaloid content of naturalized populations belonging to three species in section *Oxytona.* First, he noted that all the sample populations of *P. bracteatum* did contain thebaine as a dominant alkaloid. Secondly, the sample populations of *P. orientale* yielded oripavine as the dominant alkaloid, but also contained various amounts (sometimes traces) of thebaine and:or isothebaine as well as some unidentified alkaloids. And thirdly, the sample populations of *P. pseudo-orientale* yielded isothebaine as the dominant alkaloid with other unidentified alkaloids also present (Goldblatt 1974:270).

Gibbs reported that although most alkaloids of "the morphine group" (isothebaine, thebaine, codeine, and morphine) do occur in the poppy family, Papaveraceae, they have also been located in members of Menispermaceae, Euphorbiaceae, Moraceae, Loganiaceae, and Lauraceae:

> It must be remembered that the great interest in opium has led to the most minute examination of alkaloid mixtures from the opium poppy and its relatives. Equally detailed work in the other families would without doubt lead to further discoveries in this field. (Gibbs 1974:262)

However, according to Duke, the occurrence of the morphine group alkaloids seems to be more restricted than Gibbs's research suggests:

> Analyses performed for our laboratory on several species [not listed] reports to contain codeine, morphine, and/or thebaine have been negative, except for *Papaver bracteatum*, *P. orientale*, *P. pseudo-orientale*, *P. setigerum*, and *P. somniferum*. (Duke, 1974:3)

Thus Duke's alkaloidal data for the morphine group closely parallel those of Goldblatt (1974:265–66) who found substantial amounts of thebaine in only two sections, *Mecones* and *Oxytona*, of the genus *Papaver.*

Goldblatt (1974:277) pointed out that biochemical affinities between section *Mecones* and section *Oxytona*, especially the presence of morphine-type alkaloids in both sections, "has prompted suggestions of a close relationship between these groups." For example, Jürgen Reckin (1973:250) argued that section *Mecones* and section *Macrantha* (= *Oxytona*) of the genus *Papaver* may be more closely related than had previously been suspected: "Thebaine, an alkaloid of the morphine type, occurs in both sections, however, the species of the section *Macrantha* possess plenty of rather stiff hairs whereas the *Mecones* according to Fedde's concept do not."

Although some recent chemotaxonomic reviews have stressed the value of using alkaloid content data in biosystematic research, as I noted above, Stermitz (1968:162–70) has recommended a cautious approach when considering the use of such data for taxonomic purposes. According to Goldblatt (1974:277), the biochemical similarities between a few species of *Mecones* and *Oxytona* (= *Macrantha*) can be best explained as coincidental. On the basis of this kind of data, Goldblatt thus rejects the idea of grouping these sections of *Papaver* together. A review of the chromosomal data for those species treated as belonging to section *Mecones* is presented below with the hope that it will help resolve the question of the opium poppy's relations within the genus *Papaver* and thus provide us with some additional insight into the evolutionary origins of the species.

Chromosomal Data for Members of Section Mecones

Several authors investigating biosystematic aspects of the genus *Papaver* have commented on the lack of chromosomal data for *P. gracile* and *P. decaisnei*, two of the members of section *Mecones* as treated by Fedde in 1909 (Basilevskaya 1928, Kuzmina 1937, Yasui 1940, Hrishi 1960, and Reckin 1973). In 1970, Vent presented the first published chromosome count for *P. decaisnei* (2n = 56) (Vent 1970:447–52; compare Reckin 1973:247); and in 1973, Reckin presented the first published chromosome count for *P. gracile* (2n = 28) (Reckin 1973:247). Other members of section *Mecones* as treated by Fedde have been the object of cytological research more often and earlier than *P. decaisnei* and *P. gracile*.

According to Sugiura (1931), Kuzmina (1937), and Kawatani and Ohno (1965), *P. glaucum* is diploid and has a chromosomal count of 2n = 14. According to Tahara (1915), Kuzmina (1937), Hrishi (1960), Kawatani and Ohno (1965), Reckin (1969), and Hammer and Fritsch (1977), *P. somniferum* is diploid and has a somatic chromosomal count of 2n = 22. According to Ljungdahl (1922), Kuzmina (1937), Farmilo et al. (1953), and Hrishi (1960), *P. setigerum* is tetraploid and has a somatic chromosomal count of 2n = 44. It should be noted here that in their recent study of the origins of the cultivated opium poppy, Hammer and Fritsch (1977:113–24) presented evidence indicating that *P. setigerum* produces both diploid (2n = 22) and tetraploid (2n = 44) individuals. The implications of their research will be discussed in the following chapter which deals specifically with the evolutionary origins of the opium poppy.

On the basis of comparative cytological, alkaloidal, and morphological evidence, Reckin (1973:250) suggested that *P. glaucum*, *P. gracile*, and *P. decaisnei* all belong in the section *Orthorhoeades* and that *P. somniferum* and *P. setigerum* should be retained in the section *Mecones* along with *P. aculeatum* Thunb. and *P. Gaubae* Cullen and Rechinger.

P. aculeatum is a poppy species found in South Africa and South Australia which was classified alone in the section *Horrida* by Fedde; Reckin argued that it should be treated as a member of the section *Mecones* because its diploid chromosome number (2n = 22) and morphology, as well as its general plant morphology, correspond closely with *P. somniferum* (Reckin, 1970:461–64). In addition, according to Reckin, the shoot structure and ramification pattern of *P. aculeatum* are similar to those of *P. somniferum:*

> One might object that the shape of the capsule of *Papaver aculeatum* is different from that of *P. somniferum* and *P. setigerum* respectively. The variation in the shape of the capsule of *P. aculeatum* is enormous and ranges from clavate in most Australian

> specimens to cylindrical or nearly obovate in several specimens of the African material. The variation in the size of capsules is also considerable. It appears that the capsule may be of lesser diagnostic value than supposed by Fedde. (Reckin 1973:250)

Reckin also argued that the affinity between *P. aculeatum* and *P. somniferum* is supported by the "viable hybrids" he produced when crossing these species (Reckin 1973:250).

Successful hybridization experiments have, in fact, often been used in the past to support suggestions regarding the classification of *P. somniferum* and its close relation to some members of the section *Macrantha (Oxytona);* however, according to Goldblatt (1974:277), ". . . it is now known that many species of *Papaver* with similar or different chromosome numbers can be artificially hybridized, and this fact by itself cannot be seen as any indication of taxonomic affinity." Recently, Hammer and Fritsch (1977:113) compared the chromosome morphology of *P. somniferum* and *P. aculeatum*. On the basis of their analysis of these data and the significant differences in general morphology and geographical distribution, they argued that *P. aculeatum* should not be treated as belonging to the same section (i.e., *Mecones*) as the opium poppy.

With regard to *Papaver Gaubae*, a species which was first collected in 1937 in Persia, Reckin cites Cullen's (1965) classification of the species within *Mecones:*

> It is only known from collection specimens. Unfortunately its cytology is not known yet but there is no doubt that this species must be incorporated into the section *Mecones*. (Reckin 1973:250)

Table 3 below presents Fedde's compilation of species for *Mecones* and Reckin's suggested revision. Chromosomal counts for the various species are included since they are one of the critical characters upon which Reckin's proposed reclassification is based. The data reported by Hammer and Fritsch (1977) are not included, but are discussed in chapter 3.

The Species

During the nineteenth century, several botanists including Viguier (1814), de Candolle (1824), and Elkan (1839) contributed to the intraspecific classification of the opium poppy, *Papaver somniferum* L. In the first decade of the twentieth century, Fedde (1909) reviewed the taxonomy of this species as part of his study of the poppy family, Papaveraceae.

Then in the late 1920s and early 1930s, the Russian botanist M. A. Veselovskaya organized a formidable team research effort to study the taxonomy of *Papaver somniferum* L. as well as several aspects of this plant's importance as an oilseed crop (Veselovskaya 1933, translated into English and reprinted in 1976). In total, Veselovskaya and her associates examined approximately 1,600 specimens, most of which came from Eurasian locations (see table 4).

Table 3

The Section *Mecones* of the Genus *Papaver*
(from Reckin 1973:250)

Taxon	Chromosome Number (2n)	Authors
A) According to Fedde's Treatment		
P. decaisnei Hochst. & Steud.	56	Vent 1970
P. glaucum Boiss. & Hausskn.	14	Sugiura 1931; Kuzmina 1935; Kawatani and Ohno 1965
P. gracile Auch.	28	Reckin 1972
P. setigerum DC.	44	Ljungdahl 1922; Kuzmina 1935; Hrishi 1960
P. somniferum L.	22	Tahara 1915; Kuzmina 1935; Hrishi 1960; Kawatani and Ohno 1965; Reckin 1969
B) According to Reckin's Treatment		Range
P. somniferum L.	22	Mediterranean
P. aculeatum Thunb.	22	S. Africa; S. Australia
P. setigerum DC.	44	West Mediterranean
P. Gaubae Cullen & Rech.	Unknown	Persia

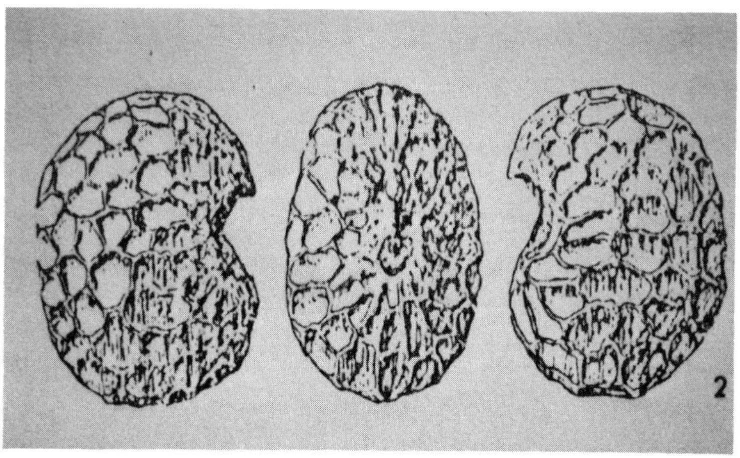

FIGURE 2. Three views of a seed of *Papaver somniferum* L., the opium poppy *(Courtesy of the Rheinisches Landesmuseum, Bonn, West Germany).*

FIGURE 3. Unripe capsule of a cultivated or semi-wild opium poppy.

FIGURE 4. Flower of cultivated opium poppy.

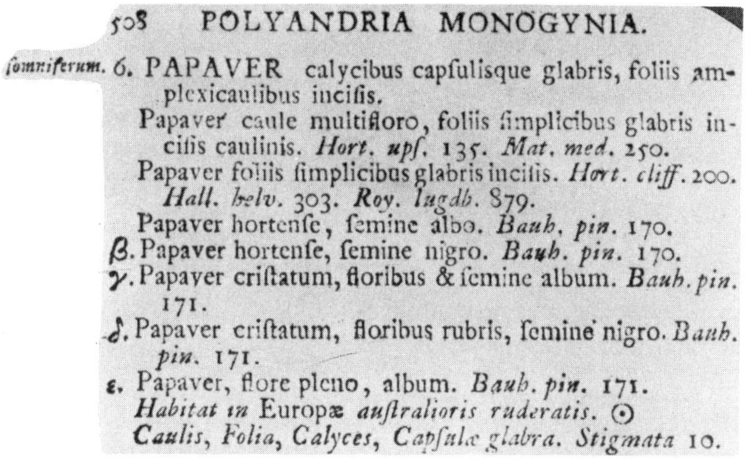

508 POLYANDRIA MONOGYNIA.

ſomniferum. 6. PAPAVER calycibus capſuliſque glabris, foliis amplexicaulibus inciſis.
Papaver caule multifloro, foliis ſimplicibus glabris inciſis caulinis. *Hort. upſ.* 135. *Mat. med.* 250.
Papaver foliis ſimplicibus glabris inciſis. *Hort. cliff.* 200. *Hall. helv.* 303. *Roy. lugdb.* 879.
Papaver hortenſe, ſemine albo. *Bauh. pin.* 170.
β. Papaver hortenſe, ſemine nigro. *Bauh. pin.* 170.
γ. Papaver criſtatum, floribus & ſemine album. *Bauh. pin.* 171.
δ. Papaver criſtatum, floribus rubris, ſemine nigro. *Bauh. pin.* 171.
ε. Papaver, flore pleno, album. *Bauh. pin.* 171.
Habitat in Europæ *auſtralioris ruderatis.* ⊙
Caulis, Folia, Calyces, Capſulæ glabra. Stigmata 10.

FIGURE 5. The early classification of the opium poppy by Carl von Linné, or Linnaeus, in *Species Plantarum* (1753). Since the publication of his binomial nomenclature, the species has been referred to as *Papaver somniferum* L.

Table 4

Geographical Distribution of Specimens Studied by Veselovskaya and Her Associates

USSR	510
Western Europe (various countries)	450
Asia Minor	470
Persia	47
Afghanistan	38
India	17
Japan	4
Mongolia	1
Manchuria	1
Palestine	1
Egypt	1

Variability Within the Species

Although we may question the small sample sizes from some areas, Veselovskaya argued that the data compiled by her group were relatively extensive and that they manifested a greater degree of ecological variability within *Papaver somniferum* L. than had been observed in earlier taxonomic studies (Veselovskaya 1976: 3). On the basis of her relatively large total sample size of specimens from a wide range of geographical regions and environmental conditions, Veselovskaya (1976: 19–77) listed the following characters as those which manifest substantial variability within *Papaver somniferum* L.:

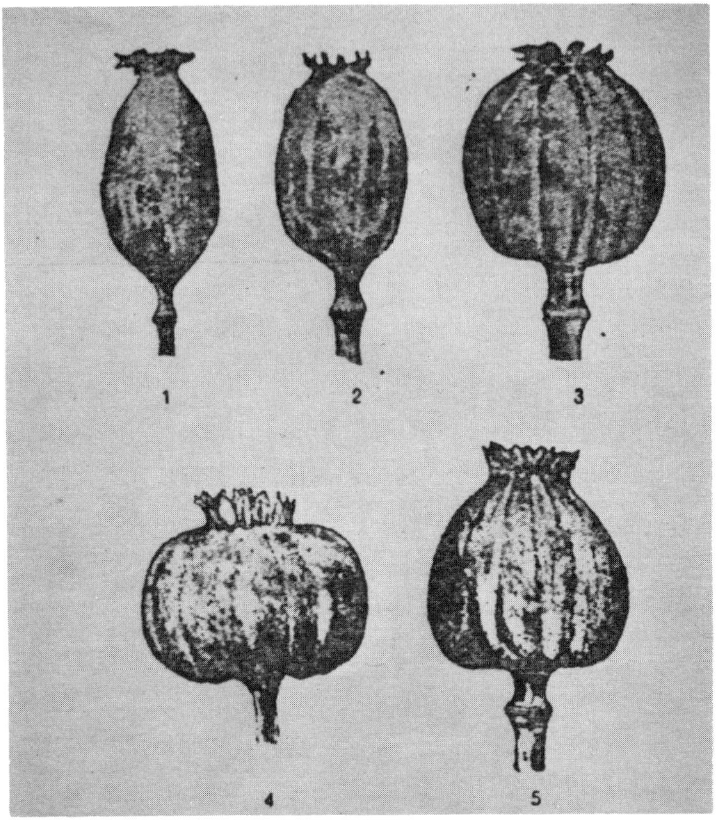

FIGURE 6. Some of the basic capsule shapes produced by the opium poppy, *Papaver somniferum* L.: (1) oval; (2) wide oval; (3) orbicular; (4) flat; (5) conical (from Veselovskaya 1976: 52; *Courtesy of Oxford & IBH Publishing Company, New Delhi, India*).

1. Seedlings
2. Stems
   A. Pubescence of peduncles
   B. Color of peduncles
   C. Height of plant
   D. Thickness of stems
   E. Branching
3. Leaves
   A. Color and coating of leaf
   B. Leaf pigments
   C. Texture of leaves
   D. Number and shape of leaves
   E. Leaf surface
4. Buds
   A. Shape of bud
   B. Color of bud
5. Flower
   A. Color of flower
   B. Size of color spot
   C. Size and shape of flower
   D. Color of the anthers and filaments
   E. Incision of the petals
6. Capsule
   A. Shape of capsule
   B. Size of the capsule
   C. Color and coating of the capsule
   D. Surface and thickness of capsule wall

E. Anatomy of capsule
F. Opening of capsule valves
G. Number of stigmatic rays of lobes
H. Shape of stigma
I. Structure of stigmatic rays

7. Seeds
   A. Seed size
   B. Seed color
8. Vegetative Period

Two Basic Ecotypes

While examining the variability of different ecotypes found in the Soviet Union, it was observed that each ecotype has had climatic requirements distinct from the

FIGURE 7. Cultivated opium poppies in northeastern France.

> other. This is why great differences are observed in ecological requirements between the cultivated poppies of Europe and Southwest Asia. (Veselovskaya 1976:78)

On the basis of her research, Veselovskaya divided the species *P. somniferum* L. into two ecological groups:

> 1) cultivated ecotypes, which are developed under agricultural conditions and which are very specific in their requirements of soil and climate;
> 2) uncultivated ecotypes which develop as a result of a natural selection and are recognized by their wide ecological adaptability and are not very particular in their agronomic requirements; they are self-growing and are found not only in poppy fields but also occur as weeds on fertile soils. (Veselovskaya 1976:78)

After extensive horticultural experimentation at various ecologically diverse research stations in the USSR (including sites near Minsk, Lenkoran on the Caspian Sea coast in Azerbaijan, Mardakyans near Baku, Sukhumi, Tashkent, and Detskoe Selo near Leningrad), Veselovskaya came to the tentative conclusion that the cultivated forms of *Papaver somniferum* L. should be broken down into the following seven major climatypes: (1) subtropical, (2) southern, (3) Asia Minor, (4) Tian-Shanian, (5) Mongolian, (6) Central Asia, (7) northern (Veselovskaya 1976:83).

Veselovskaya also divided the "semi-wild forms of poppy, known as self-sown or dehiscent (weed ecotype) poppies" into three climatypes: (1) the northern climatype (European poppies); (2) the intermediate climatype (poppies of Asia Minor); (3) the southern climatype (Indian poppies). Veselovskaya (1976:85) attributed the ecotype

FIGURE 8. First hybrid generation between self-growing or dehiscent poppy and poppy with large closed capsules (Veseleovskaya 1976: figure 20; *Courtesy of Oxford & IBH Publishing Company, New Delhi, India*).

similarity of self-sown and cultivated opium poppies to biological differences which have been produced by "prolonged diverse environmental influences observed throughout a vast area in Eurasia, extending from the Pacific to the Atlantic Ocean and from 61 degrees north to the River Indus."

Classification Keys for the Opium Poppy

When Veselovskaya (1976:99) developed her "Key for Identification of Groups, Subgroups and Varieties of *Papaver somniferum* L.," she put the highest priority on those characteristics associated with opium poppies in specific regions "which are not much influenced by any changes in environmental conditions." In other words, she gave precedence to those characteristics which are genetically constant regardless of ecological variety. According to Veselovskaya, these characteristics include: (1) structure of the stigmatic rays; (2) thickness of capsule walls which is associated with the degree of segmentation; (3) sharp differences in capsule shape; (4) thickness of the leaves; (5) waviness of the leaf margins.

Veselovskaya (1976:99) than gave secondary priority to some characteristics which do vary according to the ecology of the habitat but which "maintain a relative difference between geographical groups, particularly when cultivated under similar conditions." These characteristics include: (6) foliage (expressed in the number and size of the leaves); (7) well-defined vegetative period; (8) height of plants; (9) number of stigmatic rays. In addition, Veselovskaya (1976:102) took the following characteristics into consideration when classifying groups of semi-wild opium poppies; (a) sharp differences in branching and thickness of peduncles; (b) size of capsules, flowers, and buds; (c) openness of valves of the capsules.

Based on the above mentioned criteria and some additional characteristics of "minor importance," Veselovskaya divided *Papaver somniferum* L. into a descending order of geograhical groups (subspecies), subgroups (proles), varieties, and races. Her complicated classification key for this species provides us with a clear indication of the diverse nature of the opium poppy. Veselovskaya treats the species as comprising seven subspecies and thirty-six varieties.

Some twenty-five years after the appearance of Veselovskaya's treatment of *Papaver somniferum* L., Danert (1958:61–88) published his systematic revision for this species in which he divided the species into four convariations and fifty-two varieties.

Conclusion: Persistent Problems Face the Taxonomist

Even a cursory review of the intraspecific classification of *Papaver somniferum* L. indicates that many complexities exist and that a great amount of confusion still prevails. Some have been exceptionally skeptical of any attempt to produce a definitive taxonomic treatment of this species:

> . . . it is impossible to give a description of the plant on the basis of its external

> characteristics, for the following reasons: firstly, *Papaver somniferum* is a cultivated plant and secondly, it is found over a large part of the earth's surface. These two circumstances made it possible for a plant to produce almost infinite numbers of variations, according to the varied climatic, soil and other conditions in which it grows. Moreover, hybrids are developed by horticulturists; and in addition, the flower, while often self-pollinated, is sometimes cross-pollinated, with the result that in a single field spontaneous hybridization may occur; indeed, open and closed capsules have even been found together on the same plant.
>
> The practical conclusion is that it is impossible, from the botanical standpoint, to distinguish any precise varieties. (Anonymous, *Bulletin on Narcotics*, July–September 1953: 10)

I have presented this taxonomic review of the literature as an introduction into the enigmatic natural history of the opium poppy, *Papaver somniferum* L. What stands out about the classification of this species is that there are considerable differences of opinion. The opium poppy is obviously an extremely variable and complex species. Indeed, when I shift attention to the question of the opium poppy's natural origin, I am once again faced with substantial difficulties including a paucity of data, a perplexing diversity of educated speculation, and a relatively high degree of confusing intraspecific variation associated with this ancient and very widely cultivated species. Nevertheless, again it is my hope that the following review and synthesis of previous research (chapter 3) will shed some new light on the ancestry of one of the world's most important drug plants.

3 • The Natural Origins of the Opium Poppy

In the following discussion of natural origins, I first refer to the possibility that the ancestral derivation of the opium poppy is most closely, perhaps directly, associated with the semi-wild, self-sown dehiscent varieties. Can the direct ancestor(s) of the cultivated varieties of *Papaver somniferum* be found among the semi-wild varieties? If so, an attempt should be made to determine the evolutionary ancestry and native range of these varieties.

One of the major intraspecific distinctions between these two basic ecotypes is that there is a much greater morphological uniformity among these semi-wild varieties than among the cultivated kinds:

> Under natural selection, the self-growing poppies have maintained high ecological adaptability and appear to be comparatively "cosmopolitan"; whereas artificial selection has given rise to a number of endemic forms which are highly productive under a certain set of climatic conditions in a particular region, but their performance is adversely affected when they are introduced into other regions. (Veselovskaya 1976: 159)

Apparently, some of the characteristics which more or less differentiate the semi-wild varieties from the cultivated ones are those that clearly provide a better ecological adaptation to natural conditions and thus survival without human assistance (Harlan 1976:96).

Self-Sown and Cultivated Varieties of the Opium Poppy Compared

Differential Soil Requirements

Generally the opium poppy grows quite poorly in very heavy, clayish, nonaggregated soils because roots have great difficulty penetrating dense soil. Very heavy soils are excessively retentive of moisture; and because of ineffective drainage the opium poppy develops into stunted forms with narrow, pale green leaves, thin and simple stalks, and flowers that bloom prematurely and produce few capsules. The opium poppy also does quite poorly in very light, sandy soils which cannot retain sufficient water under natural, unirrigated conditions (Asthana 1954: 5).

Although it may develop only undersized individuals in extreme situations, the opium poppy does have a relatively broad soil tolerance. For example, varieties of the species have been cultivated in many soil types:

> the somewhat humid open forest tracts in Australia, the river clay along the Yellow River in China, the Ganges in India and the rivers in Iran, the "cotton" soils in Malwa, the loess in Central Asia, the slowly decaying volcanic soils in Asia Minor, the Nile islands in Egypt (consisting almost solely of pure sand), the "black earth" in Russia, the sandy soils of the lowlands and the calciferous loamy soils in Germany, the sea-clay soils in the province of Zealand and the gravel (sand and clay) soils in the province of Groningen (situated in the Netherlands). (Addens 1939: 12)

In Bulgaria, the opium poppy has been cultivated commercially in "light, black earth soil" of the lowlands as well as in the "grey mountain and brown mountain soils in the upland regions where wheat and other cereals are grown" (Dalev 1960: 26). In India, the opium poppy is cultivated in a wide variety of soil types including heavy clay *(Kali Matti)*, sandy loam *(Domat I)*, loamy sand *(Domat II)*, sandy *(Bhoor)*, and sandy clay (Asthana 1954: 5).

Although the various cultivated and semi-wild varieties of the opium poppy can complete their life-cycles in a rather broad range of soil structure conditions, according to Veselovskaya (1976: 160) ". . . experience has shown that self-sown poppies are not as specific regarding soil or agronomic requirements as the large-capsule, cultivated poppies."

The Dispersal Mechanism of the Opium Poppy

A major method by which the seeds of the opium poppy are dispersed is referred to as jactitation, (Good 1974: 399; Wulff 1943: 105; N. V. Polunin 1960: 17; Cain 1944: 158). This occurs to some degree in most plants that have dry fruits which dehisce when ripe. It is especially important as a dispersal mechanism among species which lack other means of scattering their seeds.

Jactitation involves the shaking of seeds from fruits held up on relatively long stalks. The impact of sufficient wind or the jolting effect of passing animals can induce shaking of the capsules and the subsequent release of seeds. Under such conditions, the dehiscent fruits may be pushed one way and then spring back in the opposite direction. Change in wind direction and animals which come in contact with the long-stemmed fruits tend to propel the seeds in a variety of paths away from the shadow of the parent. Plants which depend on this mechanism include *Melandrium (Lychnis) dioicum* Coss. & Germ., *Scilla nonscripta* Hoffmann & Link, *Verbascum spp.*, and many species of *Papaver* (Good 1974: 399; N. V. Polunin 1960: 103).

Cain (1944: 158) stated that under the impact of the wind, this mode of dispersal may catapult opium poppy seeds as far as fifteen meters from the parent plant. Polunin (1960: 103) says that if there is "a favourable slope down which seeds can be borne or roll, jactitation may occasionaly be quite effective." Seed-feeding bird species may tear open closed opium poppy capsules to get at their edible contents and thus also play a role in possible seed dispersal (see figure 10). However, I assume that their role in this case is somewhat limited.

The Importance of Dehiscence

Among those plants which largely depend on jactitation, the importance of fruit dehiscence at the appropriate time is obvious. If the ripe capsules do not open, the censor mechanism is useless.

It is well known that some species which produce dehiscent fruit in a wild or semi-wild condition have lost this ability under cultivation (Wulff 1943: 105).

> The most common genetic response in crops valued for their edible seed is a shift toward nonshattering, that is, a shift away from natural seed dispersal. (Harlan 1976:96)

One of the most striking differences between the cultivated and semi-wild, self-growing varieties of the opium poppy is the degree, or lack, of seed capsule dehiscence. Generally, within the species *Papaver somniferum*, the cultivated varieties have closed capsules, while the self-sown, semi-wild varieties, as well as most other species of *Papaver*, have open capsules (Wulff 1943: 105). The development of varieties which fail to produce any (or only occasional) individuals with effectively dehiscing fruit can probably be best explained as a result of artificial selection for mutated forms of the opium poppy with closed capsules to maximize seed harvests (Veselovskaya 1976: 161).

> Many cereals and legumes have lost the wild-type characters for seed dispersal and germination that allow them to compete in nature and have acquired characters, such as nondehiscent fruits and large seeds, that have made them better plants for man's purposes. (Heiser 1973: 192)

The practical advantage of this kind of artificial selection is also evident when one compares the cultivated flax species (*Linum usitatissimum* L.) with its wild relative *L. crepitans* Dum. The cultivated species has nondehiscent fruits as do the cultivated varieties of the opium poppy; the wild species of flax has dehiscent fruits as do the semi-wild, self-sown varieties of the opium poppy (Wulff 1943: 105).

Capsules per Plant, Their Maturity Rates, and Seed Germination

There are other characteristics which seem to provide the semi-wild, self-sown varieties of the opium poppy with a better adaptation to the vagaries of the natural

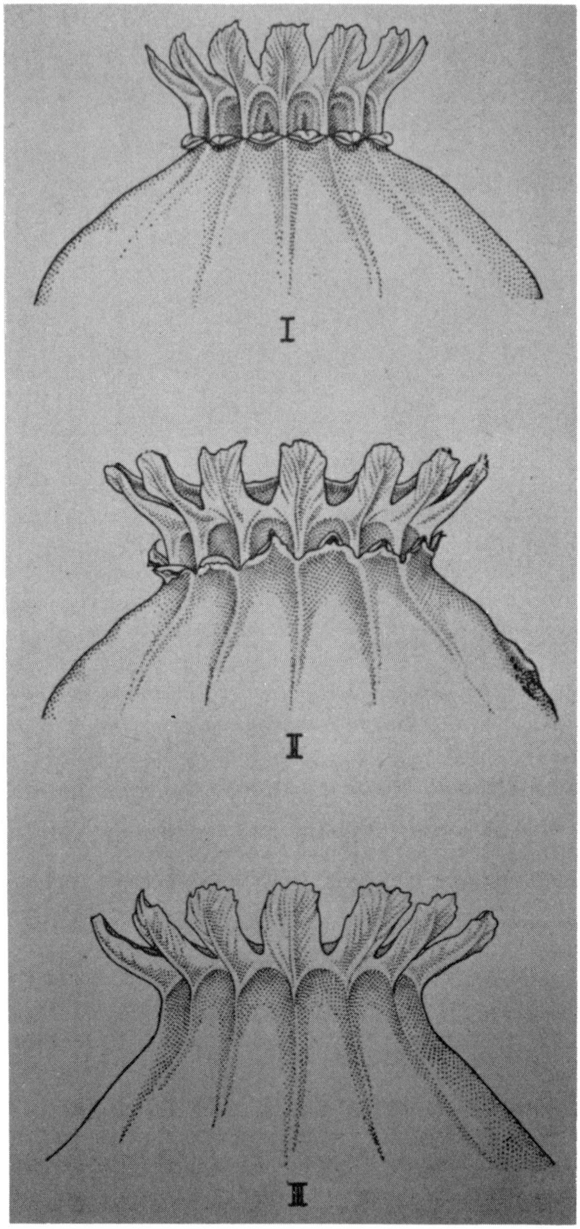

FIGURE 9. (I) semi-wild, self-sown variety of opium poppy capsule with open pores for seed dispersal involving jactitation; (II) Intermediate form; (III) closed capsule form which cannot disperse its seed through jactitation mechanism (Danert 1958:68; *Courtesy of the Zentralinstitut fur Genetik und Kulturpflanzenforschung der Akademie der Wissenschaften, Gatersleben, East Germany (DDR) and Akademie-Verlag, Berlin (DDR)*).

FIGURE 10. Semi-wild, self-sown variety of opium poppy capsules with open pores and evidence of the impact of seed-feeding birds.

environment. These include capsules which mature more gradually and, in some varieties, seeds which have variable periods of dormancy (Veselovskaya 1976:160–62). These characteristics would tend to maximize the chances for seedling survival.

> Germination represents a critical event in a plant's life cycle and its timing largely predetermines the chances of survival of a seedling to maturity. Amongst mechanisms responsible for coordinating germination with physical parameters of the environment two of the most important are the temperature response characteristics and the proportion of seeds with dormancy restrictions on growth of one kind or another. (P. A. Thompson 1973:93)

Botanical Ancestry of the Domesticated Opium Poppy

Hypothetical Evolutionary Relationships Between the Self-Sown and Cultivated Varieties of the Opium Poppy

The natural origins of the opium poppy will be best understood by looking more closely at the self-sown varieties. Indeed, Veselovskaya has suggested that the self-sown opium poppies seem to have been the "initial source of rich germ plasm from which man has isolated material for development of more productive cultivated forms" and that a lengthy history of artificial selection has thus produced varieties

FIGURE 11. Flowers and unripe capsules of a semi-wild, self-sown opium poppy growing near a cultivated field outside of Brussels, Belgium, in 1975.

with fewer and larger, closed capsules than the semi-wild, self-sown varieties (Veselovskaya 1976: 162).

In addition to the opium poppy, the progressive development of increased fruit size under artificial selection has been described for numerous other cultigens such as maize (*Zea mays* L.) and red peppers (*Capsicum* spp.) (Heiser 1973: 108, 193).

There are also other factors along with differential capsule development which have influenced Veselovskaya's conclusion regarding the ancestral origins of the domesticated opium poppy:

FIGURE 12. Comparative size of opium poppy capsules. On the left is a cultivated specimen; On the right is a semi-wild, self-sown specimen (from Veselovskaya 1976: 53; *Courtesy of Oxford & IBH Publishing Company, New Delhi, India*).

> The occurrence of transitional forms and the presence of a rich diversity of flower colors are found both in self-growing and cultivated poppies which suggests that the latter owe their origin to the self-growing poppies, selected for centuries by man. (Veselovskaya 1976:161)

The simple flow chart below represents the evolutionary development of *Papaver somniferum* L. according to the hypothesis of Veselovskaya.

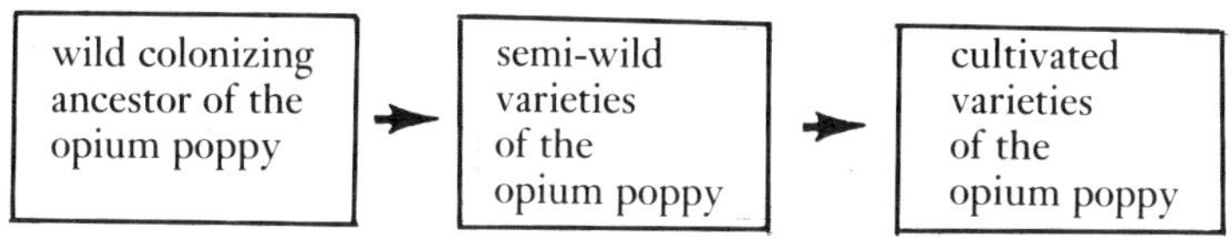

Other hypotheses might also explain the evolutionary development of both the semi-wild and domesticated varieties of the opium poppy, *Papaver somniferum*. For example, de Wet and Harlan have suggested at least three general, alternative possibilities which indicate that the semi-wild, self-sown varieties of the opium poppy may not be the direct ancestors of the cultivated forms:

> From wild to seed to cultivated may have been one route towards domestication. For most domestic species, however, cultivated and weed races developed independently from wild colonizers under the influence of different selection pressures within the man-made habitat. (de Wet and Harlan 1975:102)

In one alternative, some seeds derived from the cultivated wild plant escaped and colonized the disturbed waste areas adjacent to, or near, the cultivated fields. Progressively, through natural selection, these escapees and their descendants would have adapted to continuous habitat disturbance arising from cultivation or human settlement. This hypothesis suggests that coevolution from a wild progenitor has resulted in the development both of domesticated varieties produced under the influence of artificial selection and of semi-wild varieties produced under the influence of natural selection (see diagram below).

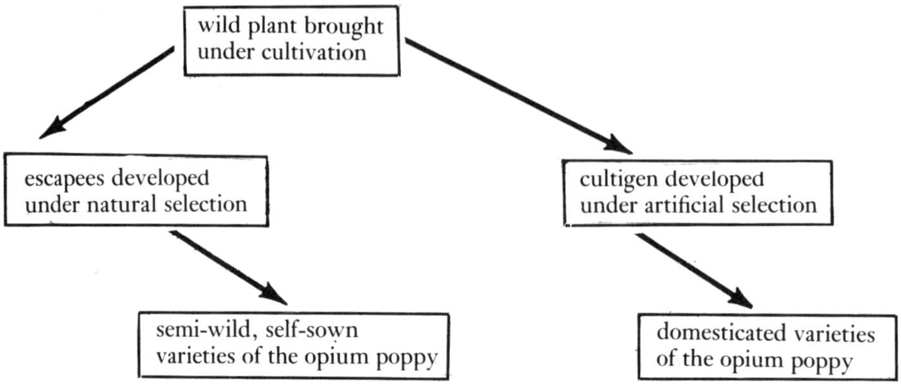

A second possibility is that the semi-wild, self-sown varieties of the opium poppy are derivaties of hybridization between the "truly wild" and domesticated forms (see diagram below).

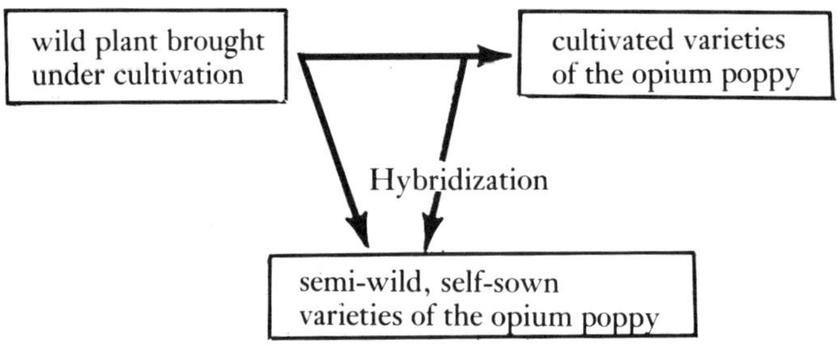

> Hybrids between wild and cultivated kinds, and their derivatives rarely successfully invade the natural habitat of the species. (de Wet and Harlan 1975: 101)

A third possibility is that the semi-wild, self-sown varieties have evolved from abandoned domesticates by reestablishing their natural seed dispersal mechanisms through natural selection (see below)

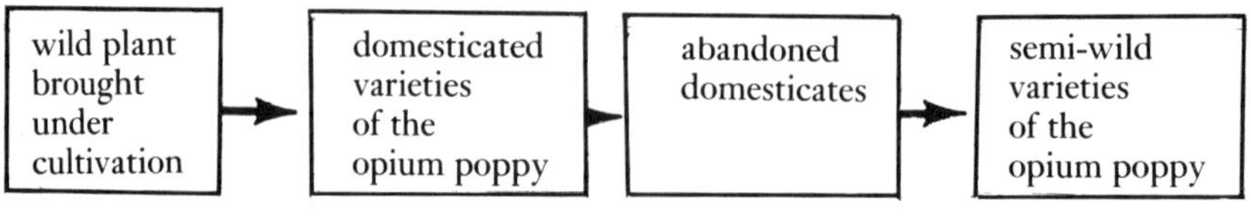

In 1976, I wrote to Professor J. M. de Wet at the Crop Evolution Laboratory of the University of Illinois at Champaign-Urbana and asked him some specific questions regarding the phylogeny of *Papaver somniferum*. In his reply, de Wet offered the following suggestion:

> If you could demonstrate truly wild kinds, natural colonizers most probably, but not weeds in the sense that they are spontaneous only in permanently disturbed manmade habitats, you should be close to the area of original domestication. (de Wet, personal communication, June 1, 1976)

The Absence of Truly Wild Populations of the Opium Poppy

The discovery of "truly wild kinds" of *P. somniferum* would amount to a major breakthrough in understanding the opium poppy's natural origins. Unfortunately, a truly wild population has not been found despite considerable efforts to do so.

Viguier (1814: 13) suggested that the red-petalled poppy with black seeds is "le type naturel de l'espèce appelée Somnifère." According to Dammel (1911: 481), the opium poppy, *P. somniferum*, can be found growing in a "naturalized" condition in Europe. However, this assertion has not been confirmed.

In fact, during the last one hundred years, most botanists have regarded the opium poppy, *Papaver somniferum* L., as an ancient cultivated plant with semi-wild, self-sown varieties, but with *no truly wild* forms known to exist in nature. For example, Bentley and Trimen (1880: Vol. I, 1) referred to the opium poppy as a species unknown "anywhere in a thoroughly wild condition."

Alphonse de Candolle (1886: 398) also claimed that truly wild forms of the opium poppy are unconfirmed: "on ne peut pas dire qu'elle *(P. somniferum)* existe à l'état vraiment sauvage, . . ." The German botanist, Georg Buschan (1895: 24), likewise described the opium poppy as a species which has not been found anywhere in the truly wild state.

In the first decade of the twentieth century, two other German botanists, Johannes Hoops (1905: 3) and Matthaus Much (1908) reiterated the common belief (perhaps without any new evidence) that *Papaver somniferum* does not occur in the wild.

Oakes Ames, a Harvard ethnobotanist, added his authority to this opinion (Ames 1939: 40). Eight years later this same conclusion could be found again in the 1949 publication of Karl and Franz Bertsch (1949: 194).

In southern England the opium poppy has long been known as "an escape" or semi-wild species (Preston 1888: 17); and recently its presence has been observed on the cliffs between Folkestone and Dover (Grieve 1959: 651). However, as the English botanist Edlin explains:

> . . . opium poppies are never found far from arable land. They claim no peace in our natural or semi-natural vegetation complexes. They would never grow, for example, in a heath moor. I think they are a good example of a plant that gets transported accidentally to a favourable spot and exists for one season or so, under a favourable summer climate. (Herbert L. Edlin, personal communication, 1976)

Apparently all varieties of the opium poppy, *Papaver somniferum*, are known only from pioneer habitats created and maintained by humans either consciously, in the form of cultivated fields, or unconsciously, in the form of "waste areas" or disturbed environments adjacent to or in the near vicinity of these fields.

> I do not know of any truly wild populations of the opium poppy today, suggesting that it, like the cocaine plants, has been highly selected. The ancestral types having since disappeared from the scene. (James Duke, personal communication, December 2, 1976)

If we accept that no populations of truly wild opium poppies exist, then *all* known varieties of this species must be dependent on man for their continued existence. The semi-wild and cultivated varieties do seem to differ only in the degree of their dependence on man.

The opium poppy, *Papaver somniferum*, is by no means the only cultivated species for which a *truly wild* population has not been found. For example, it is not known "where, if anywhere, the bottle gourd genuinely grows wild" (Harlan 1976:91). Consider also that there are over fifty species of *Nicotiana*, "yet the only one which has never been found growing wild is *Nicotiana tabacum*, the plant whose leaves have had, and continue to have, such a tremendous impact on the world in the various guises which are collectively known as tobacco" (Akehurst 1968:3). Another possible example of a cultigen without a known truly wild population is the coca plant, *Erythroxylon coca* L. which is also the source of a potent and widely used drug (R. T. Martin 1970:422).

If there are in fact no known wild populations for these domesticated species, then how can we determine the antecedent, truly wild species from which they were derived? In the case of *E. coca*, Martin has suggested that an investigation of closely related species may be helpful: "Although no examples of truly wild coca are known, the presence of several similar wild species in the Montaña region of Peru and Bolivia suggest this as the probable area of origin of the cultivated species" (Martin 1970:422). Perhaps a similar approach would be useful in the case of the opium poppy. Readers should be aware of a recent, informative botanical discussion of *E. coca* presented by Timothy Plowman (1979:103–17). He argued that *E. coca* "is now rarely found in a truly wild state"; however, he does not identify where these plants may be found. Furthermore, he also implied that we should not confuse these rare "truly wild" coca plants with the "feral" individuals which are commonly "encountered in secondary woods and thickets around coca plantations" (Plowman 1979:106).

Ancestral Relationships between the Opium Poppy, *Papaver somniferum* L., and the Setaceous Poppy, *Papaver setigerum* DC

In the past, several botanists have argued that the origins of the opium poppy, *Papaver somniferum*, could best be understood by investigating the setaceous poppy, *Papaver setigerum*. According to Veselovskaya (1976:162), some botanists including

E. Elkan (1839), Boissier (1867), K. Prantl and G. Kundig (1891), and N. Bush (1913) went so far as to suggest that *P. setigerum* is not an independent species but rather the truly wild variety or subspecies of *P. somniferum* L. (i.e., *P. somniferum* ssp. *setigerum* (DC.) Corb.) "from which all species of cultivated poppies are originated."

On the other hand, many other botanists have regarded *P. setigerum* as an *independent* species (the "mother plant") from which *P. somniferum* was derived. According to de Candolle (1883 : 398), this hypothesis was favored during his time and was inspired by the early archaeological research by the German botanist, Oswald Heer.

Early Archaeobotanical Research in Europe

Although more discussion of his archaeological evidence will be presented in the following chapter, it is appropriate here to comment on the work of Heer.

In the early 1860s Oswald Heer went to Robenhausen, Switzerland, to investigate the plant remains found among the ancient settlement sites of the central European Neolithic or Bronze Age "lake dwellers." In 1865 he published a small pamphlet entitled *Die Pflanzen der Pfahlbauten* (The Plants of the Lake Dwellers) in which he suggested that a mass of seeds and one recovered capsule (see figure 13) had a close relationship to the cultivated poppy, *Papaver somniferum:*

> In size they fall between the cultured and the wild poppy. In the wild poppy, the seeds are 0.7-0.8 mm long, and in the cultured poppy approximately 1.5 mm. The size therefore gives no definite clue as to which kind these seeds belong to; however, the sculpture indicates the cultivated poppy. This has rather large five- and six-cornered areas, which are not set in definite rows; the wild poppy has much smaller mesh, four- or five-cornered and more in rows, especially at the back of the seed. In

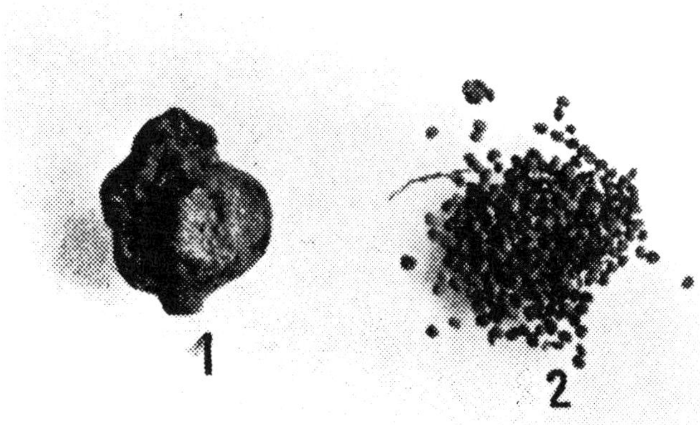

FIGURE 13. (1) The unripe poppy capsule (length 12 mm; width 10 mm) and (2) a sample of the poppy seed fossils from Robenhausen, Switzerland (from Messikommer 1913); *Courtesy of Orelli Fussli Verlag, Zurich, Switzerland*).

> this construction, the seeds of the lakedwellers correspond completely to the cultured poppy, since the mesh is larger (as in *P. rhoeas* and relatives) and forms five- to six-cornered, clearly-defined areas. Since the areas all have the same size, as we find in the cultivated poppy, and seeds are smaller, the mesh has fewer areas; this combined with the smaller size of the seeds and the eight-ray scar, shows us that the lakedweller poppy is not completely the same as the cultivated poppy of the present time. (Heer 1865 : 31, cited in Buschan 1895 : 247, translated by Inge Fischer 1976)

Heer pointed out that the unripe, charred fruit capsule found in Robenhausen corresponds in its size (12 mm length, 10 mm diameter) to that of the corn poppy *(P. rhoeas);* in terms of its overall roundish shape, it corresponds more closely to that of the cultivated poppy *(P. somniferum);* and because of the eight-ray scar of the stigmatic disc, it corresponds most closely to that of the setaceous poppy, *P. setigerum* (Buschan 1895 : 47–48; Alphonse de Candolle 1886 : 398).

FIGURE 14. Professor Oswald Heer, the nineteenth-century botanist who first identified opium poppy fossils found in the archaeological remains of Neolithic Swiss lakeshore settlements (from Messikommer, 1913; *Courtesy of Orelli Fussli Verlag, Zurich, Switzerland*).

Heer concluded that the "lake dweller poppy" was an independent "race," which he named *Papaver somniferum* var. *antiquum* (compare Bertsch and Bertsch 1947 : 196). Subsequently, Heer's analysis was used to support the contention that the opium poppy, *Papaver somniferum*, was derived from the setaceous poppy, *P. setigerum*. Although some authors such as G. Schweinfurth (1891, cited in Andrews 1952) disagreed, most of the botanists of the late nineteenth and early twentieth centuries assumed that this hypothesis was valid (de Candolle 1886 : 398).

A concise statement of George Buschan (1895 : 248) summed up the opinion at that time:

> The cultured poppy has not been found anywhere in the wild state. All botanists agree that is derived from *Papaver setigerum* DC, . . . (Translated by Inge Fischer 1976)

In 1899, Hartwich published a treatise on the "lake dwellers' poppy" *(Pfahlbautenmohn)* in which he compared the seed sizes of the prehistoric Swiss remains with recent samples of *Papaver setigerum* and *Papaver somniferum* var. *nigrum glabrum* Boiss. Hartwich's data included the following: (1) six samples of "*Papaver somniferum* L. var. *setigerum*" (about 30 seeds each) which came from botanical gardens and whose length measurements ranged in size from 0.66 to 0.97 mm; (2) "generous amounts of lake dweller poppy" seeds from archaeological sites at Moosseedorf, Niederwil, Steckborn, and Robenhausen whose length measurements ranged in size from 0.75 to 1.0 mm; (3) four samples of seeds from the "black sleepy poppy," *Papaver somniferum* var. *nigrum* DC., a semi-wild form whose length measurements ranged in size from 0.88 to 1.0 mm; (4) a seed sample from an artificially selected, nondehiscent form of *Papaver somniferum* var. *album* DC. whose length measurements ranged in size from 1.17 to 1.29 or 1.41 mm (Neuweiler 1905 : 73; see table 5).

Table 5

Hartwich's Data for Poppy Seed Length Samples (Neuweiler 1905 : 73, 1935 : 111)

		Seed Size Range
Papaver somniferum var. *setigerum*		0.66–0.97 mm
Lake dwellers' Poppy		
a) Moosseedorf	0.75–0.95	
b) Steckborn	0.83–0.97	0.75–1.00 mm
c) Robenhausen	0.83–1.00	
d) Niederwil	0.80–0.95	
Papaver somniferum var. *nigrum* DC.		0.88–1.00 mm
Papaver somniferum var. *album* DC.		1.17–1.29–1.41 mm

Based on his measurements and calculations, Hartwich concluded that the "lake dwellers poppy" represents a transitional link between the "wild form," *P. setigerum*, and the "semi-wild form," *P. somniferum* var. *nigrum glabrum* (although perhaps closer to *P. setigerum*) (Neuweiler 1935:110 and Villaret-von Rochow 1967:43–44).

The work and conclusions of Heer, de Candolle, and especially Hartwich seem to have had a significant influence. For example, consider the following translated paragraph written by the German Matthaus Much in 1908:

> In its present cultivated form, the poppy *(Papaver somniferum)* does not occur in the wild anywhere; however, a form of poppy halfway between the cultivated poppy and the wild setaceous poppy *(Papaver setigerum)*was found in the prehistoric settlements of upper Italy and in many stone-age lake-dwellings of Switzerland. The fact that this is really a primitive culture of the plant is plain, considering the many finds and the considerable quantity of found remains in individual settlements. For this reason, scientists like to presume that even the slightly more perfected form of the lake-dwellers' poppy was derived from the setaceous poppy. (Much 1908:217–18 trans. by Inge Fischer 1976).

Similar opinions were expressed by Buschan (1895:48), Hoops (1905:334), Neuweiler (1905:73), Messikommer (1913:83), Hegi (1918:33), Ames (1939:40), and Bertsch and Bertsch (1949:19).

Truly Wild Distribution of the Setaceous Poppy?

As I noted above, *P. somniferum* is not known to exist in a *truly* wild condition. There are several reports indicating that *P. setigerum* can be found growing wild in the Mediterranean region. For example, in 1867 Edmond Boissier (1867:116) reported the "wild" distribution of *P. setigerum* DC. in the Peloponnese and on Cyprus ("Haec est Forma spontanea"). Some years later, de Candolle (1886:398) reported that this species, *P. setigerum*, "est spontané dans la région de la mer Méditerranée, notamment en Espagne, en Algérie, en Corse, en Sicile, en Grèce et dans l'isle de Chypre." Subsequently, several other authors including Buschan (1895:248), Hoops (1905:334), Bertsch (1949:195), and Andrews (1952:153) have repeated de Candolle's description of the setaceous poppy's "naturalized range."

According to Veselovskaya, in 1933 *Papaver setigerum* could still be found growing "wild" in "the west Mediterranean region (southern France, Spain, western coastal regions of Italy, Corsica, Sardinia, Cyprus, Balearic and Canary Islands, northwest Africa, Morocco and Algeria)" (Veselovskaya 1976:165–66). In 1946, Fournier claimed that within France *P. setigerum* could be found only in the southwest region of that country, including the lower Rhone Valley (Fournier 1946:372, cited in Knörzer 1971a:37). In 1953, *P. setigerum* was described as a species "which grows wild in the Mediterranean region, especially southwestern Europe" (*Bulletin on Narcotics* July–September 1953:9–10). According to Mowat and Walters (1964:247), the setaceous poppy *(P. somniferum* ssp. *setigerum)* is probably native in the west Mediterranean region. In 1968, Meikle was reported to have said that *P. setigerum* can be found growing wild on Cyprus (cited in Merrillees 1968:154). According to Oberdorfer (1970:402) it has a west Mediterranean distribution (compare with Knörzer 1971a:37). Although the references mentioned above do indicate that *P. setigerum* is naturalized in parts of the Mediterranean, they do not refer specifically to *truly* wild populations.

Morphological Differences

If the setaceous poppy, *Papaver setigerum*, actually does differ from the opium poppy, *Papaver somniferum*, in that only the former has a *truly* wild distribution, then perhaps there are other significant dissimilarities between the two species. The following three comparisons describe some of the suggested morphological differences between the setaceous and opium poppies.

Veselovskaya (1976:163):

> It is well known that *P. setigerum* has highly pubescent buds, peduncles and leaves and the leaves are distinguished by a sharp, dentate margin and the presence of bristles on the large dents; flowers and capsules are considerably smaller than in *P. somniferum*. The capsules have characteristic obovate shape which is not inherent to the capsules of *P. somniferum*.

Bertsch and Bertsch (1949:195):

> [*P. setigerum*] is much smaller and more delicate than the opium poppy; its petals are dim purple and have a dark spot at the bottom. Its leaves are divided much more richly and deeply and on their tips carry bristles. The flower stems, as well as the sepals also have bristly hairs. (Trans. from German by Inge Fischer, 1976)

Hrishi (1960:8):

> *P. somniferum* L. is characterized by lightly incised leaves with the peduncle hairy or sparsely bristled, and 7–8 stigmatic rays.

Danert (1958:71) separated *P. somniferum* and *P. setigerum* from the other three members of section *Mecones* as treated by Fedde (i.e., *P. gracile*, *P. glaucum*, and *P. decaisnei*). According to Danert's key for section *Mecones*, *P. setigerum* is characterized by bristles on the end of the leaves and the calyx, while *P. somniferum* only sometimes produces bristles on the ends of some big leaves and the calyx. Hammer and Fritsch (1977:115) have pointed out that early taxonomic treatments failed to take into account the great variability within the cultivated opium poppy, and because *P. somniferum* sometimes does develop bristles on the leaves, they argued that this qualitative morphological characteristic should not be used as a basis for separating the opium poppy from the setaceous poppy. On the other hand, Hammer and Fritsch have pointed out that there appear to be some quantitative differences between *P. somniferum* and *P. setigerum* in terms of the number of stigmatic scar rays and the size of their capsules and seeds. However, in their estimation, these differences may not be significant enough to treat them as separate species. Their recent chromosomal investigation supports this position and is discussed in detail in a subsequent section of this chapter.

Alkaloidal Similarities

I noted in chapter 1 that Gibbs (1974:262) reported the occurrence of morphine alkaloids in *Humulus lupulus*, *Argemone mexicana*, and *Eschscholtzia mexicana* and that Popov (1970:486) referred to the presence of morphine alkaloids in the dried flowers of *Papaver rhoeas*. However, I also pointed out that according to Fulton (1944:49), Stermitz (1968:166), and Duke (1976:personal communication), the only species known to produce the morphine alkaloid are *P. somniferum* and *P. setigerum*. This apparently unique biochemical similarity is obviously relevant to this research. Furthermore, it seems to suggest that these two taxa do have a close relationship.

Agricultural Similarities

It should also be noted here that both *P. somniferum* and *P. setigerum* have been cultivated. Although it is well known that *P. somniferum*, the opium poppy, has been cultivated over an immense range including West, Central, Eastern, and Southern Europe, West, Central, Eastern, South, and Southeastern Asia, North and Eastern Africa, North and South America, and parts of Australia (Krikorian 1975:101; Veselovskaya 1976:5–13), it is perhaps much less well known that *P. setigerum* DC. also has been cultivated, at least in parts of France and possibly other areas of Northwestern Europe.

> *P. setigerum* is still cultivated in the north of France, together with *P. somniferum*, for the sake of its oil. (de Candolle 1886:398)

According to Karl and Franz Bertsch, in 1947 the setaceous poppy, *P. setigerum* was still being cultivated in France (next to the opium poppy, *P. somniferum* L.). Furthermore, although their argument seems to contradict itself, these authors claimed that the cultivated setaceous poppies in this area showed some definite signs of domestication such as increased fruit size, less division of leaves, and loss of bristles, all of which they felt were the result of being grown on better soil (Bertsch and Bertsch 1947:195).

Chromosomal Comparisons

In light of the above mentioned similarities and differences, I again call attention to the following question: are *P. somniferum* and *P. setigerum* sufficiently different to be considered as independent species; and if so, on what basis can this decision be justified?

According to Hrishi (1960:3),

> . . . the west Mediterranean species *P. setigerum* DC. is so near to *P. somniferum* that it is often considered as one of its varieties. Karyologically, however, they differ in their chromosome number, the former having $2n = 44$ whilst *P. somniferum* L. has $2n = 22$.

James Duke of the United States Department of Agricultural Research Service provided this perspective:

> *Papaver setigerum* I think of as a Mediterranean species still occurring in the wild. But I know of no constant character to separate *P. setigerum* from *P. somniferum* except the chromosome count. (Duke 1976: personal communication)

In an effort to determine the evolutionary relations between *P. somniferum* and *P. setigerum*, some scientists have studied the cells of these plants to ascertain the number and nature of the chromosomes for each species. Gamete cells of *P. somniferum* were studied during the early part of this century by Tahara. In 1915, he published the first illustration of these and a meiotic count of eleven (N = 11) for the species (Hrishi 1960: 10). Farmilo et al. (1953: 27) referred to a meiotic count of twenty-two chromosomes (n = 22) made on pollen cells of *P. setigerum* examined in 1951.

Although some scholars have disagreed, investigators such as Ljungdahl (1922), Yasui (1921–41), and Kuzmina (1935) have reported the somatic (2n) chromosome number of 22 for *P. somniferum* and 44 for *P. setigerum* (Hrishi 1960: 10). More recently, the diploid chromosome counts for these species were reconfirmed by Kawatani and Ohno in 1965 (Reckin 1973: 249; see also Fernandes and Quiros 1971: 342–85; Koul et al. 1972: 118).

In a preface to his research, Hrishi (1960) referred to the earlier cytological work of Kuzmina (1937) which dealt with these species:

> It is indicated in her studies, that *P. setigerum* DC. is generally near to *P. somniferum* L., but cannot be its direct ancestor as, in comparison to the latter, it possesses a double number of chromosomes, while it has been proved experimentally that the evolution of polyploid relations proceeds from low to high numbers. Reversion according to her, would only mean return to the initial forms with few chromosomes. (Hrishi 1960: 11)

In a personal communication, Peter Goldblatt of the Missouri Botanical Garden (who has worked on some species of *Papaver*) offered the following comments:

> One thing I can tell you is that since *P. setigerum* is a polyploid it is likely to have evolved from *P. somniferum* and not vice versa. Incidentally many experts consider these two as a single species with *P. setigerum* simply a polyploid race. (Goldblatt 1976: personal communication)

In a subsequent personal communication, James Duke offered this relevant and tantalizing hypothesis:

> In fields clearly sown to *Papaver somniferum* of known parentage, there are often diminutive sports ¼ to ⅓ the size of the *Papaver somniferum*, smaller in every detail, and with dehiscent pods. If they were sent to the specialist, he would classify them as *Papaver setigerum*, and they are often proportionately hairier than the *Papaver somniferum* from the same seed lot. Unfortunately I have not had the opportunity to get chromosome counts on the dehiscent and the indehiscent types from the same field. It would be quite exciting if the smaller ones were tetraploid, the bigger ones diploid. (Duke, personal communication, December 2, 1976)

If *P. setigerum* is not an independent species, but rather a separate, smaller race or subspecies of *P. somniferum*, then we might expect hybrid crosses between them to be fertile. Indeed, although reports indicate that their chromosome numbers are different (i.e., *P. somniferum* 2n = 22 and *P. setigerum* 2n = 44), crossbreeding has been attempted several times over the past one hundred years.

Interspecific Hybridization

In 1881, Focke produced the initial report of experimental attempts to produce hybrids by crossing different species of *Papaver:*

> He mentions *P. hortense* Hussen. (apparently *P. somniferum* L.), *P. officinale* and *P. setigerum* DC. as the principal species in this genus, of which *P. setigerum* DC. is considered to be the wild type of *P. hortense* Hussen. The hybrid between *P. hortense* Hussen. and *P. setigerum* DC. has been found to resemble more like *P. setigerum* DC. whilst the F_2 progeny has given more plants resembling *P. hortense* Hussen. except for the presence of the hairs of the other parent. No report has been made about the fertility of this hybrid, even though it can be inferred that it would have been nearly sterile from his statement that he could obtain only a few seeds to raise an F_2 progeny. (Hrishi 1960: 8)

In 1925, M. Blaringhem published his study of the hybrids produced by crossing *Papaver somniferum* and *P. setigerum*. According to Blaringhem, the first generation of this interspecific cross was essentially sterile suggesting significant genetic differences between these species (Veselovskaya 1976: 162–163). However, even if the two taxa were genetically identical, the hybrid progeny could still be sterile due to the formation of trivalents during meiosis (Gerald Carr, Dept. of Botany, University of Hawaii, personal communication, May 1978).

In any case, Blaringhem

> observed that the hybrid resembled more of *P. setigerum* DC. and gave only very few seeds, quite agreeing with Focke. The hybrid has been found to be very weak, early flowering and producing a large number of flowers with less and sterile anthers. He argues that if *P. setigerum* DC. and *P. somniferum* L. are the same as recognized by Battandier et Trabut, there should be only mendelian segregation; but the fact was, it behaved in a quite different way. He further states that *P. setigerum* DC. could give seeds without pollination after emasculation. The germination of the hybrid seeds has been found to be rather poor. He attributes that this poor setting of the seed after hybridization is due to incompatibility. (Hrishi 1960: 9–10)

On the other hand, Veselovskaya (1976: 163–65) reported that Blaringhem's work was in error:

> We succeeded in obtaining normal and fully fertile hybrids which gave viable seeds in F_2. We had hybrids of *P. setigerum* with Indian and European groups of *P. somniferum*, and also made the reciprocal crosses.

The hybrids obtained from the crosses mentioned above were similar to those of *P. setigerum* in external appearance in the F_1, but were easily distinguished from *P. setigerum* by their greater height and larger size of the capsules, buds, flowers and leaves. Apart from this, the pubescence of the peduncles, buds and leaves of the hybrids was somewhat less than in *P. setigerum* but the sharp dentate leaf margin and presence of bristles on the dents were maintained almost to the same extent. The seeds were of size intermediate between the small seeds of *P. setigerum* and the much bigger seeds of *P. somniferum*.

According to Veselovskaya (1976: 165), Godron also experimented with these two species and found that crosses between them produced hybrids which appeared more like *P. setigerum* in the F_1 generation, but which yielded "normal seeds." On the basis of her work and that of some other researchers, Veselovskaya argued that *P. somniferum* and *P. setigerum* are independent species but of close genetic affinity (a model of Veselovskaya's hypothesis is presented in figure 15.

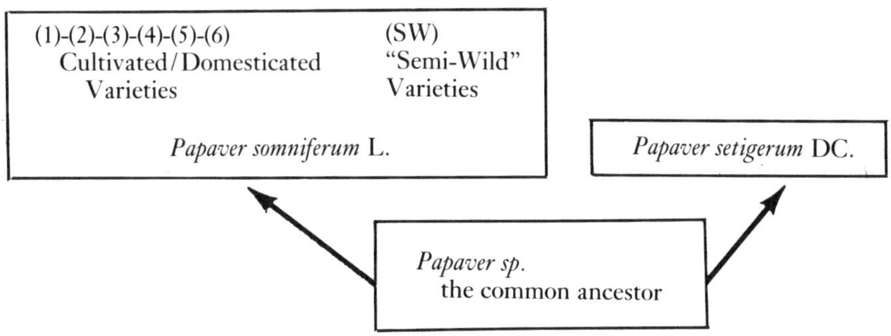

Figure 15. Veselovskaya's Hypothesis for the Inter- and Intraspecific Phylogeny of *Papaver somniferum* L.

Hrishi (1960: 10), however, failed to see any indication in Veselovskaya's work of the relationships between these two species. Hrishi's hypothetical explanation for the origin of the opium poppy is discussed below.

A Cataclysmic Evolutionary Hypothesis

In 1960, J. J. Hrishi's doctoral research on the cytology and interspecific hybridization potential of *Papaver somniferum* and *P. setigerum* was published in the journal *Genetica* (1960: 1–30). Hrishi'a major objective was to investigate the degree of relationship between these two closely related species and, if possible, to determine whether or not *P. somniferum* was derived from *P. setigerum* as many earlier taxonomists had suggested.

Hrishi reviewed previous research, presented a detailed discussion of the karyological characteristics of the two species, and developed a lengthy analysis of their interspecific hybrids and segregates in the second generation (including his own hybridization attempts). However, in summation, Hrishi only went so far as to offer the following tentative conclusion (also see figure 16):

> It has been found that *P. setigerum* DC. possesses three genomes namely genome A consisting of 11 chromosomes (which pair with the A genome of *P. somniferum* L.), genome B consisting of 3 chromosomes (which pair by autosyndesis) and genome C having 8 chromosomes (that behave as unpaired chromosomes in the interspecific cross). That the genomes B and C have been contributed perhaps by two species under Corydaleae and Chelidonieae, the former having the basic number 4 and the latter 3, has been suggested on a hypothetical basis.
>
> It has been concluded that *P. setigerum* DC. is one of the allotetraploids of *P. somniferum* L. and that the latter species perhaps has evolved from species hybrids. (Hrishi 1960:121)

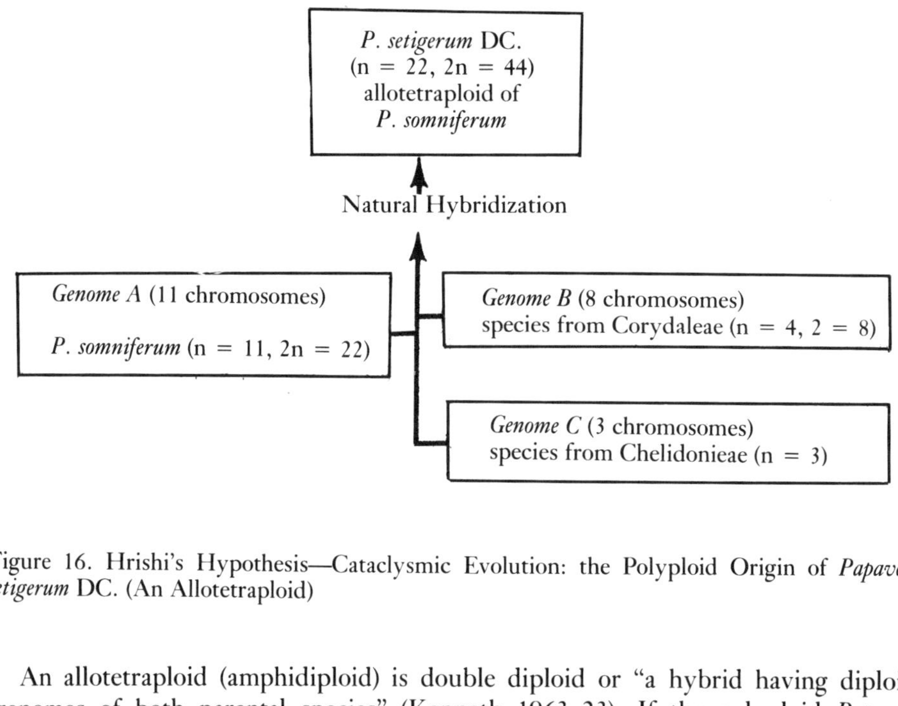

Figure 16. Hrishi's Hypothesis—Cataclysmic Evolution: the Polyploid Origin of *Papaver setigerum* DC. (An Allotetraploid)

An allotetraploid (amphidiploid) is double diploid or "a hybrid having diploid genomes of both parental species" (Kenneth 1963:23). If the polyploid *Papaver setigerum* (2n = 44) is an allotetraploid of *P. somniferum* and not the progenitor of the opium poppy, why does the archaeological evidence from the Neolithic and Bronze Age "lake-dweller" sites (according to Heer (1865) and especially Hartwich (1899), seem to indicate that possibly under the influence of artificial selection *P. somniferum* evolved from its "wild relative" *P. setigerum?* As I mentioned above, Hartwich concluded from his statistical treatment that the lake-dweller poppy is in the middle between the wild form *(P. setigerum)* and var. nigrum of the "domesticated" poppy, *P. somniferum* (Hartwich 1899:289–90; compare with Villaret-von Rochow 1967:44).

A Revaluation of the Nineteenth-Century Data

More than thirty years after publication of Hartwich's "classic treatise," Nueweiler (1935 : 111) reinterpreted the seed size and seed coat mesh-form data presented by Hartwich. On the basis of his revaluation of this data, Neuweiler argued that it demonstrated a closer relationship between the ancient lake-dweller poppy and *Papaver somniferum* than between the lake-dweller poppy and *Papaver setigerum*.

A more recent quantitative attempt to verify Hartwich's data objectively and analyze his conclusions was undertaken by the late Swiss archaeologist Margita Villaret-von Rochow (1967 : 44–45). This author began her discussion of Hartwich's "classic treatise" by raising what seems to be an important methodological consideration; what about the question of possible seed size changes over time due to shrinkage or swelling? Since poppy seeds are so small and the dimensional differences between the fossil remains and recent material is relatively minute (see table 7), it is possible that inconsistent or inappropriate preservation of the archaeological finds during the last one hundred years may have distorted the true dimensions of these seeds.

Hartwich did refer to the fact that the measured seeds from the lake-dweller sites were not charred and therefore had not shrivelled up. Unfortunately, as Villaret-von Rochow (1967 : 44) pointed out, it is not clear in Hartwich's treatise whether the fossil seeds were being kept damp constantly or whether this material was completely or temporarily dried up. Furthermore, according to Villaret-von Rochow, additional methodological problems in Hartwich's work should be considered. For example, a part of the specimens Hartwich used in 1899 came from older collections of a sample exhibit. In addition, archaeologists during the nineteenth century were probably less likely to conserve the botanical fossils in a damp condition similar to that from which they were originally recovered.

A Recent Examination of a Swiss Foreland Fossil Sample

Assuming that Hartwich's seed size frequency distribution may have been skewed by swelling or shrinkage after removal from the sediments, Villaret-von Rochow decided to compare the older data of Hartwich with newer measurements she made on the large Neolithic fossil poppy seed remains uncovered at Seeberg, Burgaschisee-Sud (Switzerland). In addition, she compared these data with recent samples of the two species in question, including a hybrid cross (i.e., *P. setigerum* × *somniferum*). Where possible, Villaret-von Rochow also randomly treated some seeds from each sample with a glycerin-water-alcohol-thymol mixture (KOH-Behandlung). The graphs below illustrate her comparative study of the variability of seed length for Neolithic *P. somniferum* samples and recent material from *P. setigerum* and *P. somniferum* (see figure 18).

Furthermore, Villaret-von Rochow made an original comparative analysis of the variation of another morphological seed character—the surface mesh sculpture (Netzmaschenazhl auf der Samenschale). Figure 19 shows the sample sizes and the minimum, mean, and maximum values of her measurements of surface mesh links for various seed populations investigated.

FIGURE 17. Uncharred opium poppy seed fossils from Neolithic Seeberg Station, Burgaschisee-Sud, Switzerland (from Villaret-von Rochow 1967: plate V, figures 15, 16; *Courtesy of the Bernisches Historisches Museum, Bern, Switzerland*).

In figure 18, the breadth of variation of the seed length in recent and ancient samples are contrasted. The solid lines represent the air-dried samples; the dot-dash, broken lines plot the samples treated with the blycerin-water-alcohol-thymol mixture which was intentionally added in order to induce some swelling and thus perhaps simulate experimentally what had occurred naturally.

In the lower part of figure 18, the seed length dimensions of spontaneous and cultivated samples of *Papaver setigerum* DC. are illustrated. Above that, the seed length dimensions of a sample from an artificial hybrid cross of *P. somniferum* × *P. setigerum* are shown.

In the middle part of figure 18, the data from the Neolithic station at Burgaschisee-

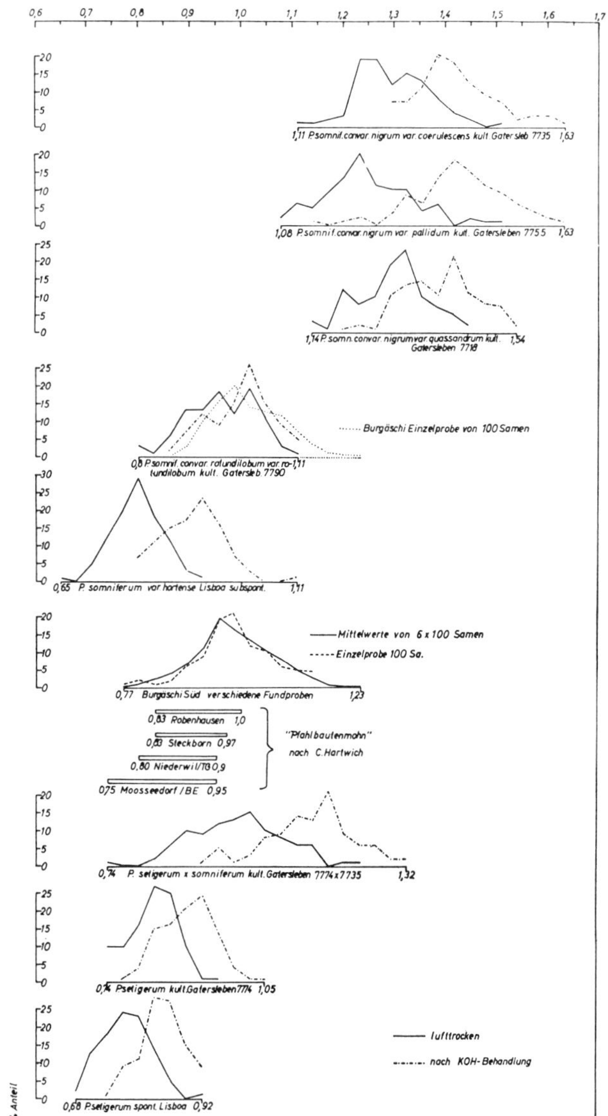

FIGURE 18. Variability of seed length for Neolithic *P. somniferum* samples and recent material from *P. setigerum* and *P. somniferum* (from Villaret-von Rochow 1967:45; *Courtesy of the Bernisches Historisches Museum, Bern, Switzerland*).

Sud are presented (the sample was taken from various finds within the site excavated) with the "pfahlbautenmohn" (lake-dweller) sample data from Hartwich placed appropriately underneath. Hartwich apparently only reported the maximum and minimum seed length dimensions and thus a statistical analysis of these data is of more limited value than had a fuller description of the size distribution been available.

Although Villaret-von Rochow did present a more complete statistical treatment than Hartwich, as she herself pointed out, we are here dealing with the variation of but a single morphological character, and thus comparative conclusions must take this limitation into account.

Finally, in the upper part of the diagram, the author has graphed the seed length variation for five different varieties of *P. somniferum*, all of which, according to her, belong to the semi-wild, naturally dehiscent or open, "shedding poppy."

In her discussion of the comparative seed length data, Villaret-von Rochow pointed out that while the dimensions Hartwich reported for *P. setigerum* (see bar graph at bottom of figure 18 under "*P. setigerum* spont. Lisboa") correspond quite closely with her more recent measurements for this species, the amplitude of seed length data from the fossil Burgaschisee-Sud sample (0.77–1.23 mm) is roughly double the average range of size distribution of Hartwich's "lake-dweller" seed samples (0.44 mm vs. 0.17 mm) (Villaret-von Rochow 1967:44; see also Neuweiler 1935:111). The difference between Villaret-von Rochow's recently compiled data and that of Hartwich is especially striking in terms of the upper limits of the fossil seed lengths.

Villaret-von Rochow argued that on the basis of these differences, her data for the Burgaschisee-Sud remains distinguish themselves much more sharply from the curves of *P. setigerum* than the fossil data reported by Hartwich in 1899. Villaret-von Rochow explained this discrepancy in terms of the possible methodological preservation problems associated with seed size change due to shrinking or swelling which Hartwich, according to her, seems not to have considered.

Villaret-von Rochow went on to point out that the samples in the upper part of the diagram (figure 18) for *P. somniferum* indicate an extremely great variability of seed length for this species, even among these dehiscent, semi-wild "shedding poppies" (Schutt-mohne). She argued that this relatively large amplitude has become larger among the derivative domesticated varieties as a result of artificial selection for large seed size on the lower end of this seed length range.

Villaret-von Rochow made special reference to the size curve of the especially small-seeded variety from Lisbon called "*P. somniferum* var. *hortense*," which closely matches those of *P. setigerum*. However, according to her, a culture attempt by the botanical garden in Lausanne showed that this variety is significantly different from *P. setigerum* in that *P. somniferum* var. *hortense* produced black seeds, capsules opening when ripe, and pale pink crown leaves with dark basal spots. Villaret-von Rochow (1967:44) felt that this Lisbon poppy resembled a dwarf variety of the "sleepy poppy" *(P. somniferum)*.

In contrast to Hartwich's conclusion, Villaret-von Rochow argued that the extreme variability of seed length size among the varieties of *P. somniferum* L. indicates that small-seededness alone is not sufficient reason to suggest a very close relationship between the fossil seed remains of Switzerland (or elsewhere) and the assumed wild "root plant" *P. setigerum*. Indeed, she pointed out that her data from the Neolithic Burgaschisee-Sud material show their closest relationship (in seed length amplitude and mean values) to her seed sample data for *P. somniferum* convar. *rotundilobum* Danert var. *rotundilobum* which was cultivated in Switzerland from seeds sent from Lisbon. According to Villaret-von Rochow, even a second sample of this convar. *rotundilobum* Danert, originally from a Rumanian population, showed closer seed size resemblance to the Neolithic material than did samples of the convar. *nigrum* that she studied. Villaret-von Rochow argued that her research indicates that convar. *nigrum*, which Hartwich (1899) and others (e.g., Bertsch 1947) have suggested as the form

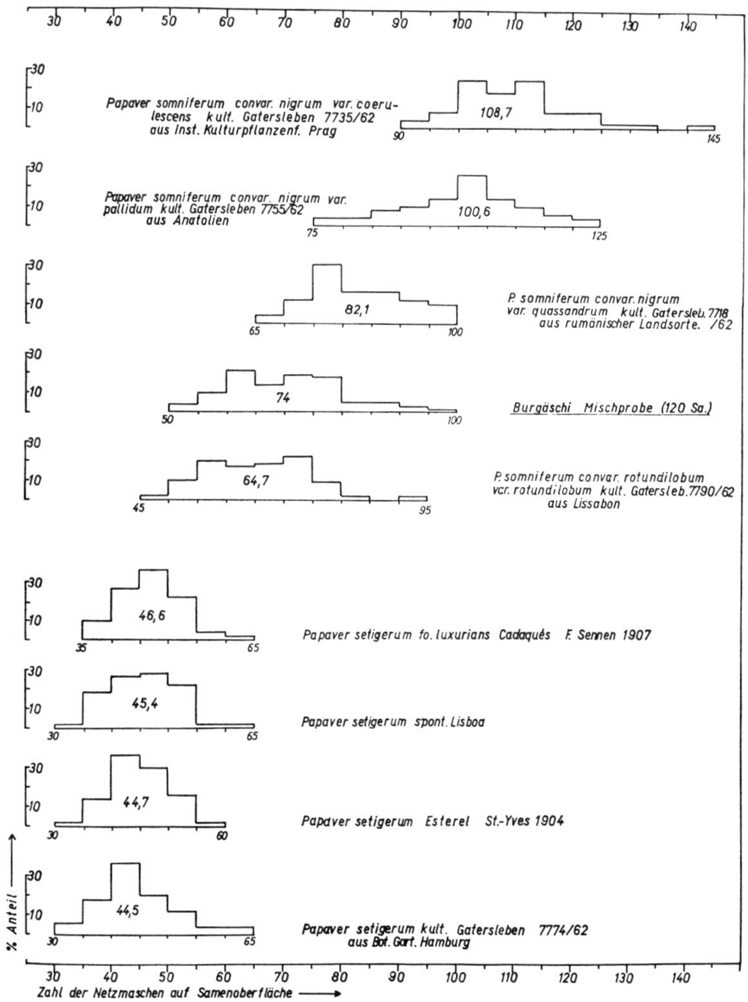

FIGURE 19. Sample sizes and minimum, mean, and maximum values for measurements of surface mesh links for Neolithic poppy seed fossils and recent *P. setigerum* and *P. somniferum* (from Villaret-von Rochow 1967:49; *Courtesy of the Bernisches Historisches Museum, Bern, Switzerland*).

closest in resemblance to the fossil remains, is not nearly as close in seed size as convar. *rotundilobum* Danert (Villaret-von Rochow 1967:48).

Villaret-von Rochow went even further in her study by statistically analyzing the seed coat sculpture of fossil and recent poppy samples (see figure 19). She assumed that a comparative quantitative treatment of this second morphological character would enhance our understanding of the origins of the opium poppy, *Papaver somniferum*, and its relationship to *P. setigerum*. However, Hartwich did not compile this type of data, and therefore, Villaret-von Rochow had to use her Burgaschisee-Sud sample when contrasting fossil material with recent samples from varieties of *P. somniferum* and *P. setigerum*.

Once again her data for this trait (figure 19) show that the varieties of *P. somniferum* L. manifest great variety (plasticity):

Range of means of mesh links on the seedcoat
of *Papaver somniferum* L. = 50.1–108.7

Range of number of mesh links on the seedcoat
of *Papaver somniferum* L. = 39–140

In contrast, the data for this same morphological seed trait for *Papaver setigerum* DC. are, according to Villaret-von Rochow's study, very limited:

Range of means of mesh links on the seedcoat
of *Papaver setigerum* DC. = 44.7–46.6

Range of number of mesh links of the seedcoat
of *Papaver setigerum* DC. = 32–64

The Neolithic Burgaschisee-Sud sample data showed greater correspondence for this trait in maximum-minimum and mean values to *P. somniferum* than to *P. setigerum*. Again the closest comparison for the fossil remains could be made with convar. *rotundilobum* Danert (from Lisbon) var. *glaucescens:*

	sample size	*min*	*max*	*mean*
fossil sample from Burgaschisee-Sud	120	50	98	74
recent sample from convar. *rotundilobum*	65	48	93	64.7

In conclusion, Villaret-von Rochow (1967:50) argued that the seed size and seedcoat mesh link quantities among the relatively large sample of Neolithic poppy seeds from Burgaschisee-Sud do not manifest a close relationship to *Papaver setigerum*, but do show a close relationship to *Papaver somniferum*. This recent thorough investigation of fossil, Neolithic poppy seeds tends to *negate* the nineteenth-century and early twentieth-century assumption that *P. somniferum* was probably derived from *P. setigerum*. In addition, as I pointed out previously, several cytological studies have provided information indicating that *P. somniferum* is diploid and *P. setigerum* is polyploid; and in accordance with the generally accepted theory that polyploid species are derived from diploid species these studies also tend to negate the long held belief that *P. setigerum* was the progenitor of *P. somniferum*.

A Recent Examination of a Rhineland Fossil Sample

Not everyone is ready to disregard the theory that *P. somniferum* was derived from *P. setigerum*. A German archaeologist, Karl-Heinz Knörzer (1971a: 34–39), recently presented data on other Neolithic poppy seed remains from the Rhine Valley which, according to him, do show a closer relationship to *P. setigerum* than to *P. somniferum* and thus suggest that *P. setigerum* may be an older form of *P. somniferum*.

Knörzer's fossil poppy seed sample consists of only eight poorly preserved specimens he recovered along with a great many charred remains of cultivated plants from five "band-ceramic" (bandkeramik) settlements near the Rhine River. These settlements include: (1) Oekoven, Kreis Grevenbroich; (2) Aldenhove, Kr. Jülich; (3) Lamersdorf, Kr. Duren; (4) Garsdorf, Kr. Bergheim; (5) Langweiler, Kr. Jülich.

According to Knörzer (1971a: 34), these fossil poppy seeds deserve special attention because they represent the oldest examples of prehistoric cultured poppy finds in central Europe. Knörzer provided three "true to form" drawings of these fossil seeds showing their size dimensions as well as the number, location, and form of the surface mesh elements.

In his discussion of the ancient Rhenish poppy seed data, Knörzer pointed out that they are dissimilar from measurements of the three "wild" poppy species (*Papaver argemone* L., *P. dubium* L., and *P. rhoeas* L.) which he claims can be found in the Rhine area today. On the other hand, Knörzer suggested that these Neolithic poppy seed remains do correlate well with *Papaver somniferum* L. and *P. setigerum* in terms of size dimensions and surface mesh characteristics (see Knörzer 1971a: 35)

However, can it be demonstrated that these are seeds of *P. somniferum* or of *P. setigerum?* Knörzer presented the following argument. First of all, when comparing the lengths of seeds one must consider that in the process of charring, seeds become more or less rounded as they shorten and thicken. However, he goes on to suggest that such shortening of the seed cannot be significant when they are, as in the case of poppy seeds, naturally almost round. According to Knörzer (1971a: 35), originally the seeds must have been shorter than 1 mm.

On the basis of seed dimensions, Knörzer saw a close correspondence between the eight charred Neolithic Rhenish seeds and recent, air-dried specimens of *P. setigerum:*

Knörzer's Data: mean and range values for length, width, height of fossil and recent samples

8 charred, band-ceramic seeds
0.78 (0.75–0.85) × 0.68 (0.6–0.7) × 0.57 (0.5–0.65) mm
10 uncharred recent seeds of Papaver setigerum DC.
0.83 (0.75–1.0) × 0.71 (0.65–0.8) × 0.68 (0.6–0.7) mm

Although Knörzer determined that the charred Neolithic Rhenish poppy seed fossils are significantly smaller than those of recent samples of *Papaver somniferum* (which according to him are over 1 mm long), he did concede that this is not enough to make an unequivocal classification. In relation to this point Knörzer referred to the work of Villaret-von Rochow (1967). However, Knörzer (1971a: 36) went on to point out that Villaret-von Rochow's study stressed the fact that there are considerably more links of mesh on the seed surfaces of *Papaver somniferum* than on those of *P. setigerum* and that the eight Neolithic Rhenish seeds (although very poorly preserved and thus difficult to measure) do show a much closer relationship in this regard to *P. setigerum* than to *P. somniferum*.

Thus, after explaining the methodological problems involved in compiling data on these damaged seeds and the unavoidable guesswork needed, Knörzer stated that five measurable seeds had, on one side, between seventeen and twenty-two mesh links; and if one doubles these values to obtain the number of links on the entire surface, they correspond fairly closely to the mean values (between 44 and 47) of the recent samples of *P. setigerum* (Villaret-von Rochow 1967:49; compare with a Knoŕrzer 1971a:37).

Furthermore, Knörzer argued that an even more convincing similarity between the ancient Rhenish finds and recent samples of *P. setigerum* (rather than *P. somniferum*) is shown in a comparison of the entire net (mesh) structure of the seed surface. Thus, in contrast to Villaret-von Rochow's conclusions regarding the classification of the seed fossils from the late Neolithic layers at Burgaschisee-Sud as probably belonging to *P. somniferum*, Knörzer suggested that in terms of seed size and the comparatively smaller number of mesh links, the "older" Rhenish seeds are much more similar to *P. setigerum* than to *P. somniferum*.

Later in his paper, Knörzer referred to two Iron Age poppy seeds, recovered from separate Rhenish archaeological excavations. According to Knörzer (1971a: 38), these two heavily damaged seeds are a little longer (0.9 mm) than the above mentioned earlier Neolithic Rhenish seeds. In addition, Knörzer argued that because these two Iron Age seeds apparently exhibit greater numbers of surface mesh links, these seeds belong to *Papaver somniferum*.

Knörzer summed up his paper with the suggestion that if one subscribes to the view that *Papaver setigerum* is the root form of *P. somniferum* (as does Oberdorfer 1970:402), one can see, in the "oldest rhenish finds," traces of the original form *(P. setigerum)* of our present-day cultivated opium poppy. According to Knörzer, this "root form" has developed in the course of thousands of years (under the selective influence of humans) towards the primitive self-sown form of the opium poppy *(P. somniferum)* which have been found in the late Neolithic Swiss archaeological sites and the early Iron Age, Rhenish remains. Knörzer suggested that these forms eventually developed into the present-day domesticated varieties of *Papaver somniferum* (the large-capsule, large-seeded, non-dehiscent forms).

Knörzer's Hypothesis

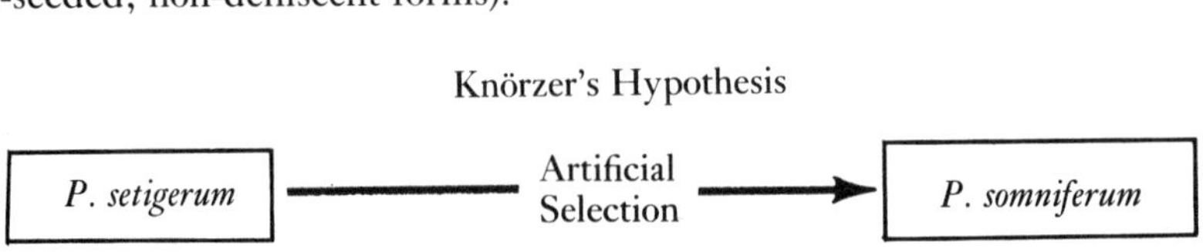

If we look critically at the research of Villaret-von Rochow and Knörzer, we cannot help but note the very small sample size of the Rhenish finds to which Knörzer refers (eight Neolithic seeds and only two early Iron Age seeds). On the other hand, Villaret-von Rochow (1967:28) utilized a fossil seed population from the late Neolithic site at Burgaschisee-Sud which was relatively large (i.e., about three thousand fossil seeds).

There is perhaps another aspect of the archaeological research referred to above that should be considered. Even if we eliminate consideration of sample size, the application of analysis of variance on one, or even two or three, morphological characteristics may not be nearly enough to use as a basis for a reliable classification of ancient seed finds.

A Recent Chromosomal Discovery

As I noted previously in this chapter, the widely accepted hypothesis of the late nineteenth and early twentieth centuries which suggested that the opium poppy *(P. somniferum)* evolved from the setaceous poppy *(P. setigerum)* has not been supported by reports of their respective chromosome numbers. Several studies have indicated that the opium poppy is diploid (2n = 22) and the setaceous poppy is tetraploid (2n = 44). Therefore, since diploids are not generally considered to have evolved from polyploids, the chances that the opium poppy evolved out of a population of setaceous poppies have been considered remote. This argument assumes, of course, that the opium poppy *(P. somniferum)* and the setaceous poppy *(P. setigerum)* are both distinct species. Furthermore, it also assumes that the opium poppy produces only diploid individuals and that the setaceous poppy produces only tetraploid individuals.

Recently, these assumptions have been challenged by Hammer and Fritsch (1977:113–24). In their investigation of the ancestral origins of the cultivated opium

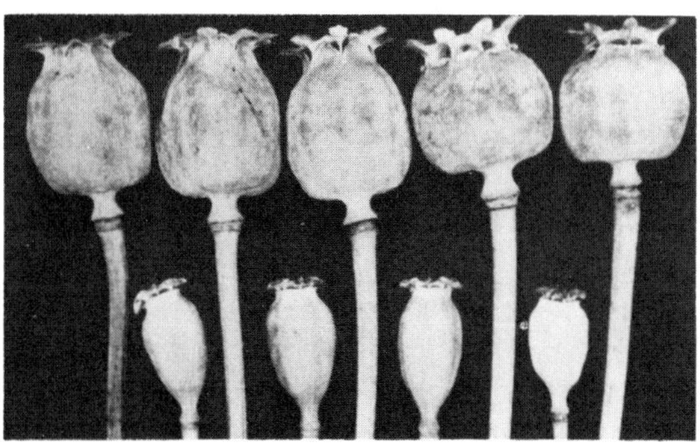

FIGURE 20. Capsules from cultivated group D 1889 (Bot. Garden Leipzig). Small capsules with characteristic *setigerum*-form and -height. Large capsules from "spontaneous" hybrid cross of *P. setigerum* (D 1889) and cultivated opium poppy, *P. somniferum;* capsules on left have closed capsule character; capsules on right have open (dehiscent) character (from Hammer and Fritsch 1977:117, figure 4; *Courtesy of Professors Karl Hammer and Reinhard Fritsch, Akademie-Verlag, Berlin, East Germany (DDR)).*

poppy, they studied the chromosome numbers and morphology of *P. somniferum* and *P. setigerum*. Although most of their sample populations for the setaceous poppy were tetraploid, others, such as "D 1889" were definitely diploid (Hammer and Fritsch 1977:114; see table 6). In addition, the chromosome morphology of the diploid population of *P. setigerum* was very similar to that of the diploid population of *P. somniferum* (see figure 21). The diploid evidence for *P. setigerum* which Hammer and Fritsch have reported is not in fact unique (e.g., see Fernandes and Queiros 1970–71). However, their strong emphasis on these data should draw attention once again to the possibility that the opium and setaceous poppies may belong to the same species (i.e., as individual subspecies of *P. somniferum*) and that the former was derived from the latter: ". . . we have found diploid strains of ssp. *setigerum* (2n = 22) which can be considered as ancestral forms for the cultivated poppy. . . ." (Hammer and Fritsch 1977:123).

It should be pointed out that Hammer and Fritsch admittedly did not succeed in

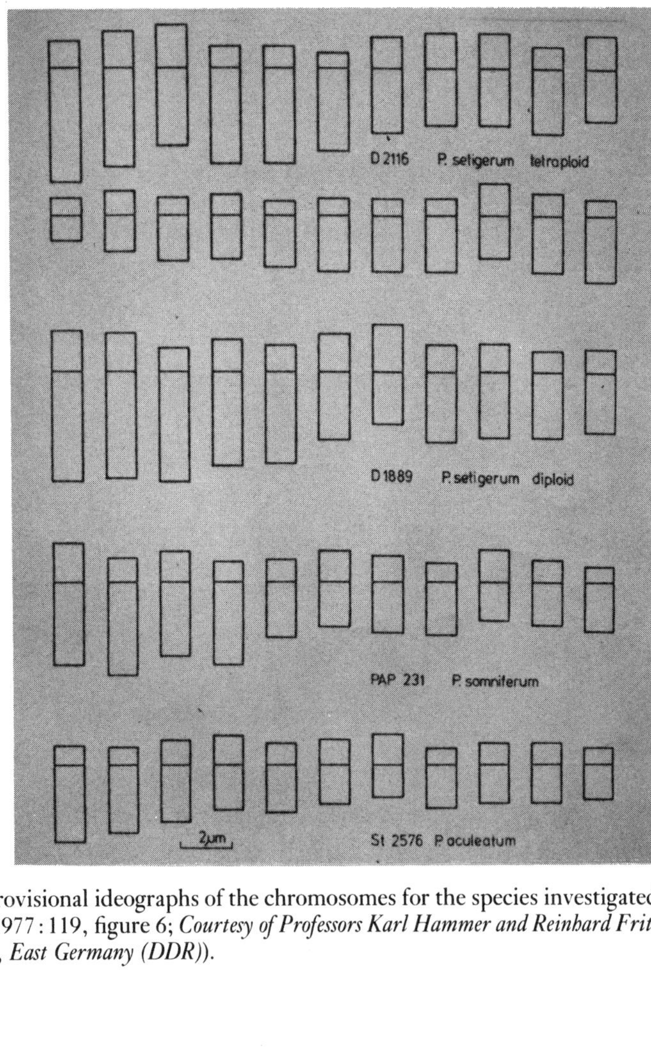

FIGURE 21. Provisional ideographs of the chromosomes for the species investigated by Hammer and Fritsch (1977:119, figure 6; *Courtesy of Professors Karl Hammer and Reinhard Fritsch, Akademie-Verlag, Berlin, East Germany (DDR)*).

identifying diploid plants among the limited self-sown ("spontan.") samples of *P. setigerum* that they investigated. In fact, they had problems even securing cultivated seed material for the setaceous poppy since it is rarely offered in international seed exchange; and when it is offered, *P. somniferum* is often sent by mistake.

According to Hammer and Fritsch (1977: 120), *P. setigerum* is a Mediterranean species with its "center of gravity" in the West. They suggested that the tetraploid individuals of *P. setigerum* are better adapted to ruderal or disturbed habitats than the diploid ones, and therefore, since such habitats are so widespread, the tetraploid plants are more common. Hammer and Fritsch also suggested that the evolution of the cultivated opium poppy may have developed before the areal spread of the tetraploid group which can now be found as far west as the Canary Islands and as far east as Cyprus. According to Hammer and Fritsch, it cannot be excluded that some cultivated opium poppies could have come from the tetraploid group of *P. setigerum*. They pointed out that one seed sample sent from Italy and designated at *P. setigerum* (i.e., PAP 231) contained tetraploid plants, but in terms of their general plant morphology, they were more closely related to *P. somniferum* (see figure 22). Hammer and Fritsch treated them as *P. somniferum* (see table 6).

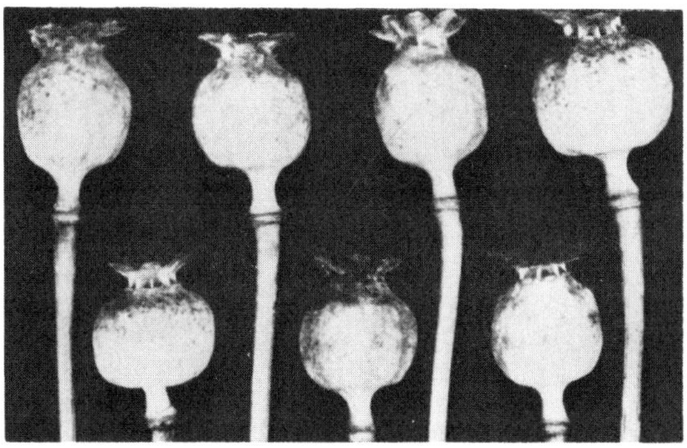

FIGURE 22. Capsules of *P. somniferum* (PAP 231: Italy, "spontan."; Hammer and Fritsch 1977: 121, figure 7; *Courtesy of Professors Karl Hammer and Reinhard Fritsch, Akademie-Verlag, Berlin, East Germany (DDR)*).

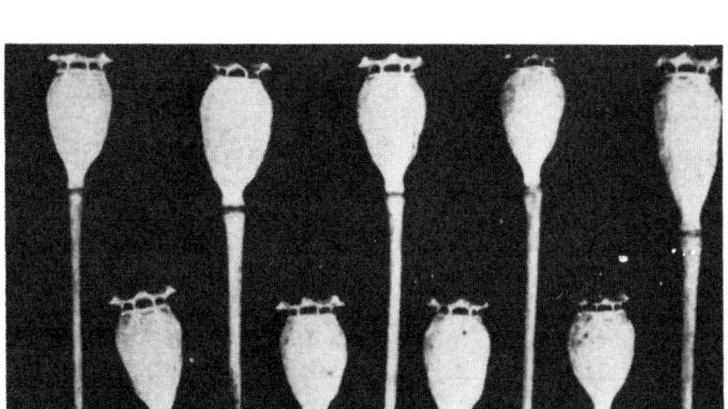

FIGURE 23. Capsules of P. setigerum (PAP 229: Canary Islands, "spontan."; Hammer and Fritsch 1977: 115, figure 1; *Courtesy of Professors Karl Hammer and Reinhard Fritsch, Akademie-Verlag, Berlin, East Germany (DDR)*).

Table 6

Plant Samples Investigated by Hammer and Fritsch (1977:114)

Cultivation Number	Origin of Material	Harvest Year	Chromosome Number	Capsule Size	Seed Size
Papaver somniferum L. subsp. *setigerum* (DC.) Corb.					
PAP 92	Bot. Garden Hamburg	1972	2n = 44	small	small
PAP 95 A	Nat. History Museum Paris, France	1972	2n = 44	small	
PAP 95 B	same as above	1976	2n = 22	large	
PAP 229	Canary Island, "spontan," via Bot. Garden Univ.	1970	2n = 44	small	large, small
	Oslo, Norway	1972	2n = 32 −36; 44	small	
PAP 230	Bot. Garden Univ. Montpellier, France	1973	2n = 44	small	small
D 1784	Portugal, "spontan," via Bot. Garden Lisbon	1975	2n = 44	small	small
D 1889 A	Bot. Garden Univ. Leipzig	1976	2n = 22	small	
D 1889 B	same as above	1976	2n = 22	large	
D 2010	Bot. Garden Univ. Lisbon, Portugal	1976	2n = 44	small	small
Papaver somniferum L. subsp. *somniferum*					
PAP 231	Italy, "spontan," via Bot. Garden Palermo		2n = 22	small	large
		1975	2n = 44	small	small

In conclusion, Hammer and Fritsch argued that *P. setigerum* and *P. somniferum* cannot be separated on the basis of chromosome number because both produce diploid and tetraploid plants. A comparison of other characters also indicated that the opium and setaceous poppies are so similar that it is very hard to separate them. Furthermore, cross breeding of diploid *P. somniferum* and *P. setigerum* produced fully fertile hybrids. Hammer and Fritsch noted that in 1839 Elkan first treated *P. setigerum* as a variety of *P. somniferum*,but they think it is more reasonable to make the differentiation at the subspecific rank. Therefore, they treat these taxa as *Papaver somniferum* L. ssp. *somniferum* and *P. somniferum* ssp. *setigerum* (DC.) Corb.

Utilizing this taxonomic treatment of the opium and setaceous poppies in a more recent publication, Fritsch (1979:217–227) described his analysis of the seed morphology of the "cultivated poppy" *(P. somniferum)*. He examined the length, width, and seed coat structure of many different races of the subsp. *somniferum* and two diploid and several tetraploid races of subsp. *setigerum*. Fritsch concluded that (1) the diploid races of subsp. *setigerum* cannot be distinguished from most races of subsp. *somniferum*, while (2) the tetraploid races of subsp. *setigerum* were difficult to distinguish from the small seeded races of subsp. *somniferum*. Fritsch pointed out that these findings should be taken into account when attempts are made to identify the archaeological remains of poppy seeds.

The Fossil Poppy Capsules

Insight into the relationship between *Papaver somniferum* and *Papaver setigerum* as well as understanding of the evolutionary and geographical origins of the opium poppy may be enhanced by a review and discussion of an interesting but meager population of fossil poppy capsules.

The Fossil Poppy Capsule from Central Germany:—in 1883, Friedrich described a relatively well-preserved fossil fruit capsule found in a lignite deposit located at Bornstedt (Province of Saxony, central Germany). According to the reported stratigraphic dating, this capsule, along with numerous other fossils, belongs to the Oligocene epoch of the Tertiary Period. If this dating is correct, the fossil capsule would be at least twenty million years old. Friedrich (1883:139) argued that the plant from which the capsule came was a member of the poppy family, Papaveraceae, and should probably be designated as a member of the genus *Papaver* (see figure 37).

The fossil's stem is furrowed lengthwise and has three distinct scar circles immediately below the egg-shaped fruit. According to Friedrich: (1) the lower impressed circle corresponds to the sepal scars—a characteristic of *Papaver;* (2) the scars forming the middle circle correspond to those of the flower crown leaves, which probably were six in number; (3) the third circle, with seven clearly visible smaller scars, represents the circle for the stamens of which there were probably fourteen; (4) the compressed fruit shows quite well the many false partitions of the *Papaver* capsule (not clearly visible in his drawings); and (5) the scar disk, which unfortunately was not preserved, seems to have had the relatively small size of that of *Papaver somniferum* (Friedrich 1883:139).

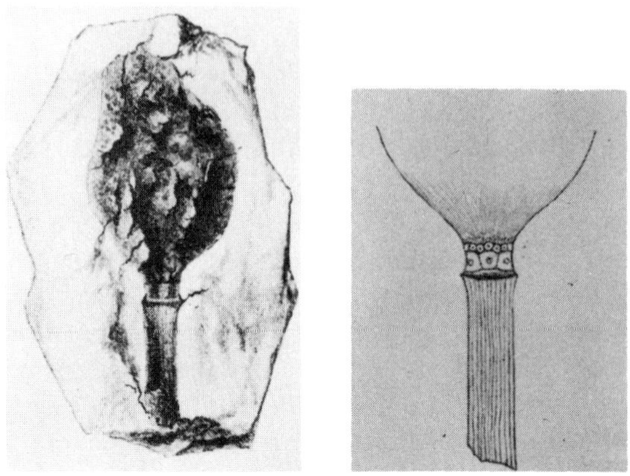

a. b.

FIGURE 24. Fossil poppy capsule from central Germany: (a) side view; (b) line drawing of the lower part of the fossil capsule (from Friedrich 1883:139, no size given).

Although Friedrich suggested that the specimen may be *Papaver somniferum*, he does interject a note of caution when he refers to the lack of other comparative examples:

> Since we have only one fossil sample we cannot be too exact, but we can state that it differs from the now existing poppy types by having only one stamen circle. (Friedrich 1883:139, trans. by Inge Fischer 1976)

Friedrich's disclaimer notwithstanding, this fossil capsule from central Germany represents one of the earliest evidences we have for *Papaver somniferum* or an ancestor of the modern opium poppy.

The Fossil Poppy Capsule from the Swiss Foreland:—In 1865, Heer described a charred, unripe poppy capsule (see figure 13) recovered from the Neolithic archaeological remains at Robenhausen, Switzerland (Heer 1865:31). As I noted previously, Oswald Heer suggested that the size of this fossil (length = 12mm; diameter = 10mm) corresponds most closely to *Papaver rhoeas;* that the shape of the capsule (round) corresponds most closely to *Papaver somniferum;* and that the number of ray scars on the stigmatic disk (eight) corresponds most closely to *Papaver setigerum*. Heer designated the capsule as a special kind: *Papaver somniferum* var. *antiquum* (Bertsch and Bertsch 1949:196). However, Neuweiler (1935:110) argued that the classification of this fossil capsule remains uncertain because of its unripe and charred condition.

Although the fossil capsule size, as reported by Heer, is small compared to most modern cultivated populations of *P. somniferum*, the self-sown, semi-wild subspecies *subspontaneum* M. Veselov. is characterized by diminutive, often roundish capsules. Table 7 presents Veselovskaya's team-collected data for the dimensions of opium poppy capsules of this group of semi-wild varieties.

Hence, the relatively small, unripe, charred fossil from Robenhausen may have come from a population of semi-wild opium poppies. On the other hand, as an unripe

Table 7

Dimensions for Poppy Capsules of *Papaver somniferum* subsp. *subspontaneum* M. Veselov.

Origin	Botanical Determination	Average Vertical (cm)	Average Horizontal (cm)
Germany	var. *sanguineum*	1.64	1.83
Russia	var. *violaceum*	1.60	1.70
Asia Minor	mixture of varieties	2.06	2.20
India	mixture of varieties	1.82	2.15

(measurements based on observation made at Lenkorn, USSR, in 1930, Veselovskaya 1976:54)

capsule, it may not have attained its fully mature size.

Furthermore, having only eight ray scars on the stigmatic disk does not necessarily preclude its classification as *Papaver somniferum*. According to Veselovskaya, the self-sown or dehiscent group of poppies that her team examined had a range of six to fourteen stigmatic lobes on the disk (Veselovskaya 1976:148).

Another point to consider is that the capsules of the self-sown and/or cultivated varieties of *Papaver somniferum* may have been smaller several thousand years ago. It is probable that a lengthy duration of artificial selection for large capsule size affected the cultivated forms directly and, perhaps, the self-sown types indirectly (i.e., spontaneous intraspecific hybridization).

The assumption that the capsule size of an ancient cultigen such as *Papaver somniferum* has remained unchanged throughout the last four thousand years or more is an improbable one. Nevertheless, we can suggest that, given Heer's data, the unripe, charred capsule from Robenhausen may well have been an early form of the opium poppy. It certainly does not resolve the questions regarding the relation between *Papaver somniferum* and *P. setigerum*.

The Fossil Poppy Capsule from Southern France:—In 1867, the Marquis de Saporta reported the discovery of numerous fossils embedded in the Quaternary tuff deposits of Provence, France (de Saporta 1867:267ff.). De Saporta left the paleobotanic collections from the Aygalades, Provence, site at the Musée National d'Histoire Naturelle in Paris. At least twenty species, including a poppy, are represented in these fossil remains. However, it was not until 1920 that another scholar, P. H. Fritel, reexamined the poppy specimen and published a short discussion of his observations and analysis (see figure 25).

According to Fritel (1920:207–8), the poppy specimen consists of a fossil capsule which is round and has its stem attached. Fritel tells us that the capsule compares in size with a small "nut" (45 mm tall) and had indentations under the stigmatic disk that were the dehiscent apertures or pores through which the presumably self-dispersing fruit could have released its seed.

The stigmatic disk on the top of the capsule was partially destroyed and encrusted. Nevertheless, Fritel suggested that a logical reconstruction of the damaged sections shows clearly that the capsule had thriteen "scar rays"5 and that the margin of the

a.

c.

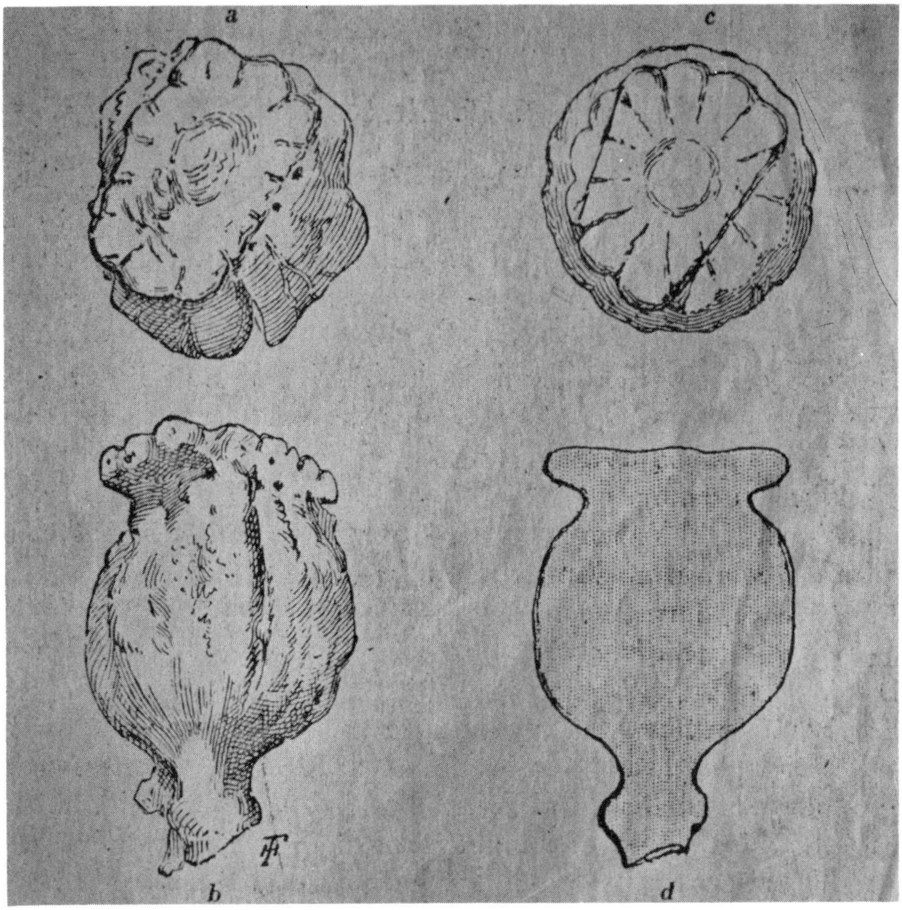

b.

d.

FIGURE 25. Poppy capsule fossil from tuff deposit in lower Rhone Valley, France: (a) top view; (b) side view; (c) top view with reconstructed stigmatic disc; (d) longitudinal profile; height of capsule = 45 mm; width of capsule = 28 mm; fruit stem = 4 mm (from Fritel 1920: 208; *Courtesy of the Société Géologigue de France, Paris, France*).

stigmatic disk is well lobed (Fritel 1920: 207).

Based on his own comparative analysis, Fritel concluded that this specimen should be classified as *P. somniferum* var. *nigrum* DC. He assumed that it dates back to the Quaternary Period and argued that this species must have been indigenous in Provence, France; but he also suggested that it disappeared from the area during the postglacial period, and was reintroduced by humans as a cultigen in more recent times (Fritel 1920: 208).

In his treatise on Papaveraceae, Fedde discussed the geographical relationships among the genera and species of this family (Fedde 1909: 65–72, cited in Stermitz 1968: 180). Fedde considered the poppy family to be of northern origin and suggested that some species spread southward prior to the climatic changes and glacial advances of the Pleistocene epoch.

> Appearance of the ice presumably eliminated all the most northern species with the exception of several members of the section Scapiflora of the genus *Papaver*, which are found native today largely in high mountains or near-Arctic regions. The widespread and numerous species of the genus *Papaver* that occur throughout central Europe today are regarded as having developed chiefly from a northward spreading of those species which had originally migrated into the Mediterranean region and had thus escaped the ice. (Stermitz 1968: 180–81)

Assuming that this theory is valid, then the opium poppy could have been one of those species of *Papaver* which in its ancestral form migrated into the Mediterranean region from a more northerly area of origin prior to the Pleistocene. Fritel's classification (*Papaver somniferum* var. *nigrum* DC.) and dating of the fossil poppy capsule found in the tuff deposits at Aygalades by de Saporta, as well as his explanation for its presence in this Mediterranean location, support this suggestion.

In 1935, Neuweiler reviewed Fritel's description, cursory analysis, and generalized conclusions concerning this single fossil capsule from southern France (Neuweiler 1935: 111). Although Neuweiler concurred with Fritel and treated the specimen as *Papaver somniferum* var. *nigrum* DC., he ended his discussion of this interesting fossil evidence with a skeptical comment. He argued that since this "large-fruited," encrusted poppy capsule is morphologically quite similar to samples of modern opium poppies *(Papaver somniferum)*, the question must be asked as to whether or not the specimen actually belongs to a more recent time (Neuweiler 1935: 112). In any case, Neuweiler called for a thorough examination of the specimen's geological origin and absolute dating.

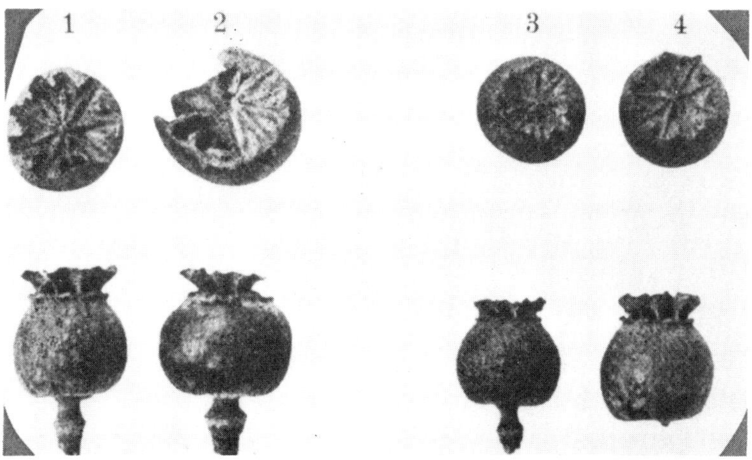

FIGURE 26. Four of the poppy capsule fossils found in Cueva de los Murciélagos, Spain. Capsule 1 = 15 mm × 13.5 mm; capsule 2 = 15 mm × 15 mm; capsule 3 = 12.5 mm × 12 mm; capsule 4 = 13.5 mm × 15 mm (from Neuweiler 1935: 109; *Courtesy of Naturforschende Gesellschaft in Zurich, Orell Fussil Graphische Betriebe AG, Zeitschriften-Abteilung, Zurich, Switzerland*).

The Fossil Poppy Capsules from Southern Spain:—In 1935, Neuweiler and Netolitzky examined four fossil poppy capsules found in a cave (Cueva de los Murciélagos) near Albuñol in the southern Spanish province of Granada (Neuweiler 1935: 108–9). Apparently, Neuweiler received these fossil capsules from Hugo Obermaier, professor of prehistoric archaeology at the University of Madrid. Obermaier notified Neuweiler that these poppy capsules and locks of hair were found in small baskets woven from "Spartogras" (*Stipa tenacissima* L.), and that these baskets were located in the cave in association with a number of naturally dessicated human skeletons, ceramic material, and tools made out of both wood and stone. According to Neuweiler, Professor Obermaier told him in writing that these bones and artifacts were about 4,500 years old (ca. 2500 B.C.) and thus "of the late younger stone age" (Neuweiler 1935: 108).

Neuweiler and Netolitzky's measurements of the four fossil capsules are presented in table 8. The capsules are divided into two groups containing two capsules in each group (Cestillo 582 and Cestillo 581).

Table 8

Measurements of the Capsules Found in Cueva de los Murciélagos (in mm)

	Cestillo 582		*Cestillo 581*	
Capsule No.	1	2	3	4
Diameter	13.5	15	12.5	13.5 mm
Height	15	15	12	15 mm
Height of scar rays	3	3	2	3 mm
Scar rays	9	(9)	9	9 mm
Width of scars	11.2	10	10	10 mm
Depth of scar sinus	1	1	1	2 mm

( ) = hypothetical reconstruction

In addition to the above listed measurements, Neuweiler (1935: 109) described other characteristics of these fossil poppy capsules. According to Neuweiler, three of the four capsules examined had stems (see figure 26). The length of these stems from the point of departure at the fruit stem (torus) to the capsule bottom is ± 3 mm. Although capsule number 2 was damaged, Neuweiler's reconstruction suggests that it had a total of nine "scar rays" (as did the other three fossil capsules in the sample).

Neuweiler (1935: 109) reported that between the scar rays there are "sinuses (grooves)." These vary in depth (horizontal length inward toward the center) from 1 to 3 mm and average 1.8 mm; thus they reach inward about ⅕ to ⅙ of the total diameter of the stigmatic disk which averaged about 10.3 mm.

Neuweiler also drew attention to their observations of "rounded holes" located under these "grooves" and next to the scar rays. These holes, according to Neuweiler, indicate that the fossil poppies were the "shedding" type (i.e., naturally dehiscent).

Neuweiler (1935: 109) described the capsules as brown, round, barrel-shaped, and comprised of nine carpels whose partitions (based on observations of the interior of the damaged capsule—number 2) reach inside from all sides up to a third of the capsule diameter.

Based on comparative morphology, Neuweiler (1935 : 110) concluded that the capsules from the southern Spanish cave should be classified as *Papaver somniferum*. According to Neuweiler, it is impossible to classify these fossil capsules as *P. setigerum* because in contrast to the fossil sample from this cave, the setaceous poppy has "longish capsules."

He argued that we should not be misled by the smallness of the fossil fruit because the size of cultivated opium poppies *(P. somniferum)* is quite variable and that to this day one can find individuals bearing capsules which are barely the size of a "hazelnut" (1935 : 110). In addition, Neuweiler reminded us that many recently observed opium poppies show nine stigmatic (scar) rays per capsule.

Neuweiler went further in his discussion and suggested an intraspecific classification for the late Neolithic fossil capsules from southern Spain. Relying on de Candolle's treatment, Neuweiler classified *Papaver somniferum* into only three varieties: (1) *P. somniferum* var. *setigerum* DC.; (2) *P. somniferum* var. *nigrum* DC.; and (3) *P. somniferum* var. *album* DC. According to Neuweiler, var. *album* produces closed, nondehiscent capsules, while var. *setigerum* and var. *nigrum* produce self-opening, dehiscent capsules.

When comparing the fossil capsules from southern Spain with recent (1935) samples of *P. somniferum* var *setigerum* DC. and *P. somniferum* var. *nigrum* DC., Neuweiler noted that the rather deep "grooves" between the "scar rays" of the fossils (depth = ⅕ to ⅙ of the diameter of the stigmatic disk) and the recession of the disk "partitions" leave approximately ⅓ of the capsule diameter free; and while these specific dimensions all correspond with the description Hartwich (1899) reported and illustrated for var. *nigrum*, the grooves of var. *setigerum* recede inward only ⅒ of the disk diameter, and the disk partitions join tightly in the center (Neuweiler 1935 : 110).

Neuweiler maintained that no comparison can be made between the Spanish fossil capsules from Cueva de los Murciélagos and the one recovered from lake sediment at Robenhausen, Switzerland. Although they are both very roughly dated as about 4,500 years old (ca. 2500 B.C.), the Spanish fossils apparently represent ripe individuals while the Swiss fossil appears to be unripe and charred. Unfortunately, there were no poppy seeds found in Cueva de los Murciélagos; therefore, no comparison could be made with the measurements of the Swiss seed fossils.

In conclusion, Neuweiler argued with self-assurance that the capsules from Murciélagos belong to *Papaver somniferum* var. *nigrum* DC. According to Neuweiler this indicates that the "prehistoric poppy" existed in the "west." On the other hand, Neuweiler also maintained that this determination does not mean that the poppy found in the Swiss "lake-dwelling" sites comes from the same species or variety. He was not sure we can deny the possibility that the Swiss fossils (thousands of seeds and a solitary capsule) are closer to, or belong to, "*setigerum*."

Nevertheless, Neuweiler (1935 : 111) suggested a tentative classification for the "lake-dweller poppy." Based on comparisons of the shape and size of the unripe capsule, form of the seedcoat sculpture, and measurements of the seed sizes, he treated the "lake-dweller poppy" as *P. somniferum* var. *nigrum* DC.

The Application of Pollen Analysis

During the last few decades, the use of pollen analysis has played an increasingly important role in paleobotany and paleoethnobotany. In fact, the natural histories of several genera and species have been greatly illuminated by the effective use of this research technique (Dimbleby 1967; J. M. Renfrew, 1973). For example, reconstruction of the early cultural history of another important and ancient drug plant, hemp (*Cannabis sativa* L.), has been aided by the application of pollen analysis (Godwin 1967).

However, students interested in the natural or cultural history of the opium poppy have not been able to utilize pollen analysis, because of unresolved technical problems. I asked Harry Godwin, the noted Cambridge botanist who successfully employed pollen analysis in a study involving hemp (Godwin 1967), about the feasibility of its use in investigations concerning the opium poppy. His response: "I have no experience that the pollen is identifiable . . ." (Godwin, personal communication, May 21, 1976). James Greig, Professor of Botany at the University of Birmingham in England, told me:

> . . . there is no evidence of opium poppy from pollen evidence, to my knowledge, because it may not be possible to identify pollen to a specific level, a common problem with pollen analysis. Another problem is that if *Papaver* does not produce much pollen, or if it does not happen to blow about or travel in streams of water, it may not be a plant about which pollen analysis can tell us much. (Greig, personal communication, April 13, 1976)

Thus, even if we found well preserved pollen fossils of a poppy species, at present we could only identify the plant to the generic level, which would not help us resolve the question of the evolutionary relationships between *Papaver somniferum* and *P. setigerum*.

A Summary of Past Research and Persistent Problems

The ancestral relationships between the opium poppy, *Papaver somniferum* L., and the setaceous poppy, *P. setigerum* DC., have long been a focus of scholarly attention and controversy. The setaceous poppy was almost always classified as a separate species during the nineteenth and early twentieth centuries (e.g., Alphonse de Candolle 1883:398; Buschan 1895:248; Hoops 1905:334; Much 1908:217; Messikomer 1913:83). However, as early as 1839, Elkan classified the setaceous poppy as a self-sown variety of the opium poppy (i.e., *P. somniferum* var. *setigerum* (DC.) Corb.; compare with Hammer and Fritsch 1977:123). In recent times, most botanists have also chosen to split the opium and setaceous poppies into separate species. Nevertheless, others still choose to lump them together into a single "natural" unit (e.g., Polunin and Smythies 1973:193; Hammer and Fritsch 1977:123). Taxonomically the opium and setaceous poppies are obviously quite close. In any case, whether one chooses to lump or split these poppies, many questions regarding the ancestral development remain unresolved.

The nineteenth-century discoveries and early analysis of fossil capsules and seeds from a number of prehistoric European sites prompted a widely held notion that the opium poppy, *Papaver somniferum*, was derived directly from its close relative the setaceous poppy, *P. setigerum*. More recently, however, some scientists have become skeptical of previous identifications of the fossil poppy seeds because of their reexamination of the nineteenth century statistical research and relatively sophisticated evaluation of new fossil evidence (e.g., Neuweiler 1935; Villaret-von Rochow 1967). Furthermore, most chromosome reports indicate that *P. somniferum* is a diploid species (2n = 22), while *P. setigerum* is a polyploid (2n = 44); and since many botanists believe that polyploids are derived from diploid species, it has been suggested that *P. setigerum* was derived from *P. somniferum* rather than vice versa (e.g., Veselovskaya 1976; Hrishi 1960).

On the other hand, a recent treatment of fossil poppy seeds found in the Rhineland suggests that the Neolithic fossils belong to *P. setigerum*, while the Iron Age fossils belong to *P. somniferum*. In fact, Knörzer (1971: 34–39), the scientist who examined the Rhineland specimens, argued that through the selective influence of humans for thousands of years, *P. somniferum* appears to have evolved from the "root form" of *P. setigerum*. In addition, the chromosomal evidence used to support the hypothesis that the opium poppy could not have evolved from the setaceous poppy has been recently challenged by Hammer and Fritsch (1977: 113–24). Their recent investigation indicates that *P. setigerum* can produce diploid as well as tetraploid strains, while *P. somniferum* can produce tetraploid as well as diploid strains. They also pointed out that the chromosome morphology of the diploid strains of *P. setigerum* is very similar to that of diploid strains of *P. somniferum*. Hammer and Fritsch argued that it is not reasonable to separate the opium and setaceous poppies into two species on the basis of chromosomal and general morphological differences. Instead, they have chosen to treat these "two taxa" as subspecies of *Papaver somniferum* (i.e., *P. somniferum* L. ssp. *somniferum*, the opium poppy, and *P. somniferum* L. ssp. *setigerum* (DC.) Corb., the setaceous poppy) and suggest that the "diploid strains of ssp. *setigerum* (2n = 22) . . . can be considered as ancestral forms for the cultivated poppy. . . ." (Hammer and Fritsch 1977: 123).

In conclusion, I think the relationship between the opium and setaceous poppies can be best summarized in the comments received from James Duke:

> I have mixed feelings about "tetraploid" *Papaver setigerum* and diploid *Papaver somniferum*, somewhat paralleled by the "hexaploid" *Papaver orientale* and the "diploid" *Papaver bracteatum*. Usually we assume that the polyploid was derived from the diploid, but I suppose it can go the other way. (James Duke, personal communication, December 2, 1976)

At present, there is no apparent consensus regarding the opium poppy's ancestry (Hrishi 1960; Reckin 1973; Knörzer 1971; Krikorian 1975; Hammer and Fritsch 1977). This situation is reflected in the diversity of suggestions concerning the "root form" of the cultivated drug plant (see table 9).

If the opium poppy were derived from the setaceous poppy, it could be assumed that the opium poppy had its origins within the native range of the setaceous poppy (i.e., *P. setigerum* or *P. somniferum* ssp. *setigerum*). A number of scholars have used this reasoning to locate the geographical origins of the opium poppy (e.g. see Schultze-

Table 9

Suggested Ancestry of the Cultivated Opium Poppy

Papaver setigerum as the direct ancestor of the opium poppy.	
de Candolle	1883:398
Buschan	1895:248
Hoops	1905:334
Much	1908:217
Messikomer	1913:83
Hegi	1918:33
Ames	1939:40
Oberdorfer	1970:402
Knörzer	1971:39
Papaver somniferum ssp. *setigerum* as the direct ancestor of the opium poppy.	
Mowat and Walters	1964:247 (from ssp. *setigerum*)
Polunin and Smythies	1973:193 (from var. *setigerum*)
Hammer and Fritsch	1977:123 (from ssp. *setigerum*)
The opium poppy, *Papaver somniferum* not derived from *P. setigerum*	
Schweinfurth	1891:661, cf. Andrews 1952:153
Bazilevskaya	1928, cf. Hrishi 1960:3
Veselovskaya	1933, cf. Vesselovskaya 1976:168
Kuzmina	1935, cf. Andrews 1952:154
Hrishi	1960:121
The opium poppy, *Papaver somniferum* derived from *P. setigerum* and *P. gracile.*	
Meikle	1968, cf. Merrillees 1968:154
The opium poppy derived from *Papaver somniferum* var. *nigrum.*	
Bertsch and Bertsch	1949:194–96

Motel 1979); this hypothesis is discussed in chapter 5. However, it is possible that, as a cultivated plant or weedy species, the setaceous poppy might have been transported (consciously or unknowingly) away from its indigenous range to a favorable environment where the opium poppy (cultivated and semi-wild varieties) could have evolved through natural hybridization or artificial selection.

If we assume that the opium poppy was not derived from the setaceous poppy, a whole range of possibilities concerning the geographical origins of the opium poppy can be suggested. During the last one hundred years, the range of speculation regarding the geographical origins of *Papaver somniferum* L. has been quite broad (see table 10). The diversity of speculation concerning the geographical origins and ancestry of the opium poppy is a manifestation of the fact that as yet no verified, truly wild population of *Papaver somniferum* L. has been located.

Unfortunately, this unsatisfying situation may not be resolved in the near future, especially if the opium poppy was in fact derived from a now limited, possibly extinct, truly wild diploid population of the so-called setaceous poppy. Perhaps the archaeological record can shed some useful light on this problem. This is one of the

Table 10

The Geographical Origin of the Opium Poppy: Suggested Locations

Southwest Europe	
Fritel 1920:687	
Kuzmina 1935, cf. Andrews 1952:154	
Bertsch and Bertsch 1947:194–96	
Polunin and Smythies 1973:193	
Schultze-Motel 1979:213	
Southeast Europe	
Bentley and Trimen 1880, Vol. I, no. 1	
Zeuner 1954:358 ("possibly Greece")	
Gimbutas 1978:personal communication	
Central Europe	
Polunin	1969:103
General European Mediterranean Region	
de Candolle	1883:398
Buschan	1895:248
Hoops	1905:334
Much	1908:217
Messikomer	1913:83
Boisacq	1916:632
Hegi	1918:33
Veselovskaya	1933; cf. Veselovskaya 1976:168
Ames	1939:40
Fulton	1944:1
Western Asia	
Lindley and Moore	1889:84 (from the "Levant")
Vavilov	1949/50:33
Zeuner	1954:358 (from the "Middle East")
Collom	1957:614 (from "Asia Minor")
Kritikos	1960:63 (from "Anatolia")
Taylor	1965:209 (from "Asia Minor")
Cohen	1969:76 (from "Asia Minor")
Hutchinson	1973:520
de Wet	1976:personal comm. (from "central or western Asia")
Duke	1976:personal comm. (from the "Middle East")
Asia (general)	
Dammel	1911:48
Sayre	1917:200
Vavilov	1949/50:25 (western China)

basic objectives of the following chapters, in which I discuss the archaeological evidence and attempt to synthesize these data into a meaningful survey of our earliest evidence of man's association with one of the world's most famous drug plants.

4 • The Prehistoric Use of the Opium Poppy

> An historical geographer must use the materials of many people and must go into many fields of study, with all of the interest and danger that such trespass entails. Although many of the following pages are concerned with materials not immediately geographical, the structure of the argument can be compared to that of an arch, for which the many voussoirs, that individually give no suggestion of a vault, must be put in place before the final form is apparent. (Stanislawski 1975:427)

In chapters 2 and 3, I surveyed the literature for research relevant to the natural history of the opium poppy, discussing the available evidence and describing many of the unanswered questions related to the ancestry of this widely known drug and food plant. In this chapter, I address myself to questions concerning the enigmatic ethnobotanical origins and possible prehistoric uses of the opium poppy. First, I discuss the multipurpose utilization of the species, drawing attention to the more important applications of the plant in the past. Then, I offer some general speculations pertaining to the initial man-opium relationships and early use of the plant.

Multipurpose Possibilities

The opium poppy is one of the most widely known herbal drug plants. However, its value as a medicinal and its notoriety as a narcotic have attracted so much attention that they have obscured the plant's history as a multipurpose species. Where the opium poppy was available during the prehistoric period, it could have provided a number of valuable resources: some economic, some therapeutic, and some perhaps ritualistic.

Human Food and Animal Fodder Resources

Although the young leaves of the opium poppy have been used as a potherb and salad vegetable, the most important edible resource of this plant has certainly been its small, oily seeds which can be eaten raw, pressed for a comestible oil, baked into poppycakes, ground into poppy flour, and sprinkled over dough for flavor. According to Jane Renfrew (1973: 194), plants which produce edible oil seeds "were recognized at an early stage of agriculture to have nutritional value, and the first crop of this type to be cultivated for food was flax" (*Linum usitatissimum* L.). Like the flax plant, the opium poppy's seeds are also very nutritious. In fact, the diminutive but plentiful seeds of the opium poppy contain relatively large percentages of carbohydrates, fats, protein, and calcium as well as small amounts of phosphorus, iron, thiamine, riboflavin, nicotinic acid, iodine, manganese, copper, magnesium, sodium, potassium, zinc, and lecithin (Duke 1973: 390–91).

The oil of the opium poppy seed is especially noteworthy. It constitutes approximately 45 percent of the total seed weight (Duke 1973: 391) and can be used as an edible oil without heating or refining. Therefore, as a vegetal fat resource, the seed oil could have been a major factor attracting early human groups to the opium poppy

FIGURE 27. Nutritious opium poppy seeds have a long history as s source of food for humans and animals.

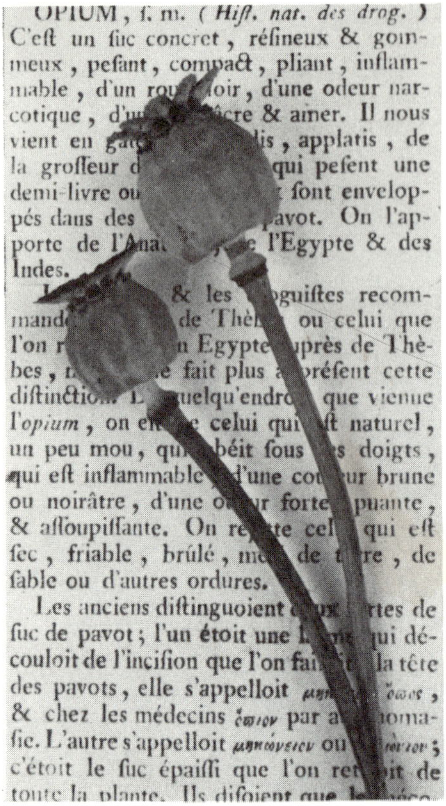

FIGURE 28. Capsules of the opium poppy can be soaked in water to produce a weak narcotic drug solution.

FIGURE 29. Opium poppy seed oil can be used for cooking and in lamps.

(Van Zeist and Casparie 1974:426). In addition, as a staple of supplementary flour-meal source, ground opium poppy seed could have provided a nutritious porridge or cake filling, a function it still serves in some European countries such as Austria, Czechoslovakia, and Germany (Hoops 1905:297, Maurizio 1915:225, Netolitzky 1930:61, and Singer 1954). Furthermore, the use of the opium poppy seeds as a reserve or famine food, or simply as a source of tasty seeds, may have played a role in the cultures of some prehistoric peoples. Relatively large, well-preserved masses of poppy seeds found among the organic refuse and remains uncovered from the Neolithic lakeshore settlements in Switzerland indicate that a relationship between man and the opium poppy did exist more than five thousand years ago. This archaeobotanical evidence and its relevance to food use are discussed in chapter 5.

Man's use of the opium poppy as a food for animals is also worthy of consideration here. Duke (1973:391) has referred to studies which indicate that poppyseed oil cakes produce enough digestible crude and true protein to make them useful animal fodder (also see Addens 1939:43). In Paleolithic or Mesolithic times, such cakes may have been used as baits for attracting wild beasts or semidomesticated animals. During the Neolithic period, they may have been used as cattle, pig, sheep, goat, or horse feed by early farmers. Some of the Neolithic archaeobotanical evidence from west central Europe supports the idea that such cakes were indeed used as animal fodder (see chapter 5). Furthermore, the oil in the opium poppy seeds "is similar to olive oil and can be used for cooking and also in lamps—in which it is said to burn for much longer than olive oil" (J. M. Renfrew 1973:162; see also Addens 1939:43). In any case, the abundant growth of the opium poppy in fallowed grain fields and adjacent waste areas would have put the seed resource in a convenient position for early exploitation.

Drug Resources: A Paradoxical Herb

The opium poppy is at the same time both a remarkably beneficial and socially dangerous drug source. On the one hand, as the foremost natural source of analgesic medicine, opium drugs can greatly dissipate the anxiety associated with pain, and when taken in moderate dosage, opium drugs can induce restful, therapeutic sleep. On the other hand, except in cases of paralytic ileus and diarrhea, opium drugs do not attack the causes of pain; they merely affect the symptoms of distress (Wright 1961:22). Unfortunately, when such drugs are used immoderately they can be viciously habit forming. And, if they are not readily available, the addict may, ironically, experience extreme pain. Even more alarming is the fact that an inadvertent or premeditated overdose may stupefy or cause death.

There are still other contradictory aspects of opium drugs, for while they are known to stimulate mild euphoria and pleasant hallucinations, in some situations opium drugs may produce illusionary perceptions of a frightening nature. As a source of powerful, tranquilizing pain-killers, the opium poppy is "a blessed thing when properly used, but when abused it is one of the cruelest enemies of mankind" (Ames 1939:40).

The paradoxical nature of opium drugs and their use is in great part due to the contradictory qualities of the various alkaloids that the opium poppy produces.

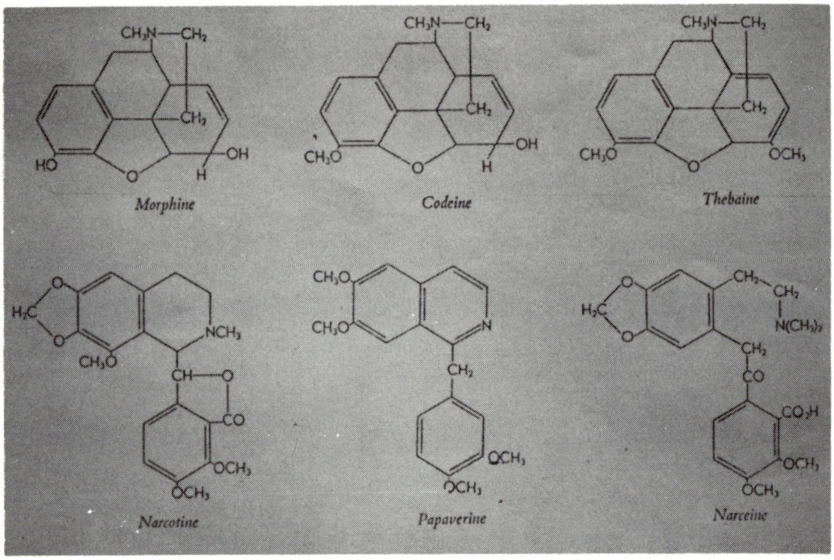

FIGURE 30. The principal opium alkaloids.

Duke's (1973 : 394–96) description of the more important alkaloids contained in opium provides us with some idea of the diversity of their potential effects:

> *Morphine* is a powerful analgesic, narcotic and stimulant. It depresses thalamus, sensory cortex, respiratory and cough centers; it stimulates the spinal cord, vagus and vomiting centers and the third nerve center; and it increases tone of involuntary muscles especially in alimentary sphincters. It reduces secretions except for the skin glands. It induces euphoria (within ½ hour after hypodermic injection) with muscular relaxation, lessened physical activity, dimness of vision, loss of pain and hunger, slowing of respiration and contraction of pupils. Larger does (15–20 mg) induce sleep and may cause nausea, delirum and convulsions. Still larger doses cause depression, unconsciousness, or even death (394).
>
> *Papaverine*, with little narcotic or analgestic action relaxes the involuntary muscles (intestinal, biliary, bronchial trees, urether, blood vessels). It slows the heart less than morphine or codeine. Intravenous injection increases cerebral blood flow but affects respiration little (396).
>
> *Codeine* is less narcotic, constipating and euphoric than morphine. It stimulates the spinal cord and lower part of the brain less than morphine and depresses higher cerebral centers less. There is less sedation on respiratory centers than with morphine. Small doses are soporific, whereas large doses cause restlessness and increase reflex excitability.
>
> *Narcotine*, the most abundant alkaloid after morphine, is a very mild narcotic and spinal stimulant. Falling between thebaine and codeine for convulsant activity it accelerates respiration and has a colchicine-like action on mitosis.

FIGURE 31. A field of cultivated opium poppies in northeastern France, 1975.

> *Thebaine* is convulsant rather than narcotic. In dogs, it acts as an anti-emetic. In large doses it may cause tetanic spasms (like strychnine) or may paralyze the peripheral motor nerves. (Duke 1973 : 394–396)

The important point to note here is that the different alkaloids contained in opium produce a variety of effects. Some of the alkaloids depress, while others stimulate the central nervous system: "Ascending from morphine to papaverine, codeine, narcotine, and thebaine, the narcotic action diminishes and powerful reflex stimulation increases until, with thebaine, the effect is a strychnine-like convulsion" (Duke 1973 : 394).

Medicinal Use:—". . . eating-opium is a household medicine par excellence and is used in case of diarrhoea, dysentery, cold shivers, fits of fever, asthma, chronic coughs, rheumatism, diabetes and pains of all kinds" (Addens 1939 : 106 n. 11).

In modern times, it is well known that the products of the opium poppy have been extremely useful in the widespread symptomatic treatment of numerous medical and psychological afflictions. This is not, however, a recent phenomenon: the medicinal history of opium drug use has ancient roots in human culture: "opium is the strongest of the drugs which numb the senses and induce a deadening sleep" (Claudius Galen 13.273, 2nd Century A.D.). Nevertheless, it is important to make a distinction here between the kinds of powerful opiate drugs that are available to modern man and those which were in the pharmacopoeia of earlier societies.

Opium is simply the thickened juice or sap exuded from the unripe capsules of the opium poppy (see figure 33). Although this sap contains the potent alkaloids described above, it was not until certain scientific and technological innovations occurred in the nineteenth and twentieth centuries that man was able to identify and extract these alkaloids from their organic sources. In other words, use and abuse of morphine or codeine alone was not possible until after the alkaloids were isolated. Incidently, the isolation of the morphine alkaloid by the German pharmacist, Friedrich Sertürner, in 1803 was the first discovery of any alkaloid, and therefore was an important milestone in the history of biochemistry itself: "To have discovered the nature of an alkaloid and to have isolated morphine in the process would have entitled him to the Nobel Prize if one had existed then" (Taylor 1965 : 212).

Once the potent alkaloids of the opium sap were isolated and extracted, the dangerous social abuse potential of the opium poppy rose significantly. During the latter part of the nineteenth century, these abuse problems were intensified by the spreading use

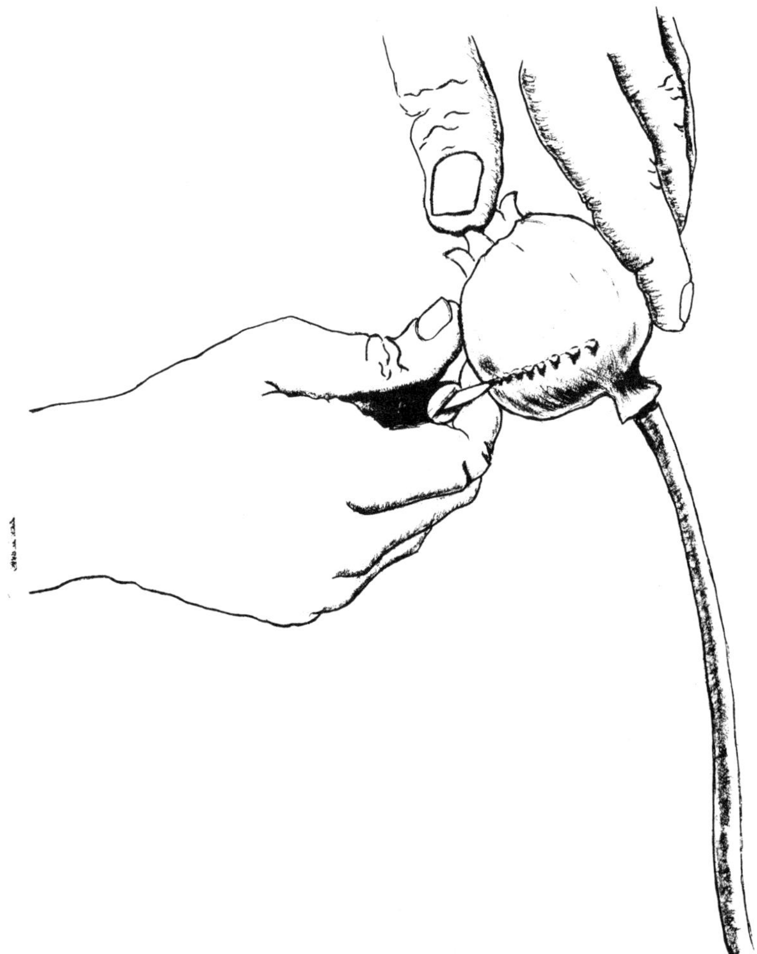

FIGURE 32. Opium poppy capsule and cutting tool *(Courtesy of Denise Peltier, Los Angeles, California).*

FIGURE 33. Opium sap seeps out of a freshly incised capsule *(Courtesy of Dr. James A. Duke, Chief of the Economic Botany Laboratory, United States Department of Agriculture, Science and Education Administration).*

of the hypodermic syringe and the discovery of heroin (diacetylmorphine, a chemically bonded synthesis of acetic anhydride, a common industrial acid, and morphine). In addition, the sophisticated study of the natural molecular structure of opium's psychoactive alkaloids has inspired the modern development of synthetic narcotics known under such trade names as "Demerol" and "Percodan." These man-made opiates have also been widely abused both in terms of application and dosage.

The powerful, modern opiate drugs and their chemical substitutes were certainly not available to early peoples. Even the alkaloid-rich sap of the opium poppy may not have been *directly* available during the prehistoric period. We must remember that although opium can be extracted rather easily with the proper cutting tool and technique (see figure 32), such an instrument and skill may not have been invented until late prehistoric or early historic times. Most, perhaps all, prehistoric people who used the opium poppy to produce drugs, probably did so by boiling, steeping, or soaking the capsules in order to extract the psychoactive ingredients which we now know are contained in the milky juice of the seed vessels. Early peoples may have eaten the capsules of the opium poppy directly. However, I think that, at least in most cases, they drank infusions or decoctions of the capsules and, therefore, *normally* consumed only relatively mild doses of the opium poppy's potent alkaloids. Historical accounts in the early Greek literature tell us that the early medical use of the opium poppy involved a decoction of capsules, stem, and leaves called *meconium* (compare Hippocrates, Theophrastus, et al.), and that this potion was used as a soporific drug mixed in with a beverage and taken internally or as a "demulcent fomentation when applied hot to inflamed parts, bruises, sprains, and other painful afflictions" (Bentley and Trimen 1880: vol. 1, no. 18).

The earliest drug use of the opium poppy may very well even reach back past the Greeks into the early Neolithic, Mesolithic or upper Paleolithic periods of cultural evolution: "Mankind has always been engaged in a search for greater comfort, and high priority has been granted to the development of drugs to relieve pain" (Cohen 1969:76). We can be sure that man has suffered from "hurtful influences" from the very beginnings of his existence:

> He could be burned by the sun and by fire; he could be injured in hunting or fighting; he could be chilled by the night and the winter; he could be hurt by infections and a thousand kinds of physical pain. We know that man very early turned to the use of the narcotic actions of such drugs as opium, alcohol, and mandragora to diminish pain but that none of these was really effective against the torment of very severe pain. (Malcolm 1971:61)

Although ancient opium drugs may not have quelled the symptoms of the most severe pain, they certainly would have reduced the intensity of unpleasant sensations and associated anxiety. Opium drugs apparently do not remove the actual pain of an injury or other afflictions. They do, however, seem to dissipate the apprehension connected with pain. This, in turn, evidently relaxes the patient to the point where he may be able to tolerate the actual sensation of pain: "Twentieth-century research, moreover, suggests that even the beneficial effect of the opiates on *physical* pain is in considerable part a tranquilizing effect" (Brecher 1972:12). In any case, even a perceived reduction of a painful condition would have been enough incentive to develop a special admiration for the opium drug and to give it an important place in the local pharmacopoeia:

> Any drug that actually alleviates severe pain must necessarily confer on its user massive relief. The gratitude that is felt by the recipient in his pain-free state will be closely remembered against the return of the hated pain. (Malcolm 1971:65)

Thus, the symptomatic pain relief that opium drugs could have provided for many afflictions might have stimulated a high degree of esteem for the plant. Perhaps it acquired a sacred status in some early societies as a pain-killer, tranquilizer, cough suppressant, or source of therapeutic oil:

> Everywhere men tasted this or that in seeking what they might with safety eat, and learned what each might do to bowel or head or heart. What cooled the fiery, painful skin they noted, too, and learning, treasured deep and long in memory. So when desire rose to alter bowel, to dull the ache of head or heart, or soothe the flushed or irritated skin, remembrance showed the way. (Leake 1947 : 7)

It is not unreasonable to suggest that the opium poppy may have been as important a medicinal drug plant in some regions during the prehistoric period as it is over a vast range of the world today (see Fulton 1944 and Duke 1973).

Psychoactive Use:—Ingestion of certain plant materials for mind-altering purposes is probably a very ancient cultural activity. Some even suggest that this type of botanical drug use is as old as mankind and "probably began when our ancestors browsed their way through the forests and found that among the foods they sampled, some produced interesting changes in how they perceived, and how they could accomondate themselves to the world" (Aaronson and Osmond 1971 : 4).

Several anthropologists, botanists, and culture historians have published reports on the use and relative importance of psychoactive plant drugs in the ethnographic present and prehistoric past. This group of scholars includes Spruce (1908), Lewin (1929), Osmond (1955), Schultes (1937, 1940, 1941, 1954, 1957, 1963, 1970, etc.), Heim (1956, 1957, 1967), Fernandez (1964, 1965, 1972), Dobkin de Rios (1968, 1969), Furst (1967, 1972), Martin (1970), Sharon (1972, 1977), Wasson (1972, 1978), Wilbert (1972), Merlin (1972), Emboden (1979a), and Stanislawski (1975). In addition, the recent formation of an academic society in the United States (The Ethnopharmacological Society' and the publication of an international scholarly periodical *(Journal of Ethnopharmacology)* are manifestations of the continued, indeed growing, interest in this subfield of ethnobotanical research.

Today there are perhaps no more than 150 botanical species used to produce a mind-altered experience; all of these plants are concentrated among the fungi and angiosperms. A great majority of them are used by subsistence-oriented, remotely located culture groups in the New World Tropics (Schultes 1961 : 141–67; 1966 : 293–308;). Although it may be more a reflection of cultural history and discovery than actual past use, only a handful of plants in the Old World are known to have been or are still used for mind-altering purposes (see Schultes and Hofmann 1980 : 19–24). The opium poppy is, by nature of its psychoactive contents, one of these very special Old World species. It has the rare quality of being able to induce strong psychological states of well-being and intense pain-relieving, hypnotic conditions. "Opium has been used as a drug at least since classical Greek times, not only, of course, because it deadens pain but also because it gives rise to euphoria" (Snyder 1977 : 44). The early use of the opium poppy may have, therefore, been for psychic or visionary purposes such as ritualistic intoxication, shamanistic divination, cult worship, or ecstatic reverie (see La Barre 1972 : 261–79).

One of the more remarkable, if often frivolous, social developments of this century has been the increasing use, abuse, and misuse of psychoactive substances of botanical or synthetic origin (Schultes 1969:245). Weil (1972:66) has suggested that the contemporary indulgence in hallucinogenic drugs is, as it was in the past, basically a response to a normal innate drive perhaps as powerful as that directed to satisfy hunger and sex: in other words, he argues that there is a genetically inherited desire to alter consciousness.

La Barre (1970:75–76), on the other hand, suggests that only some groups are "culturally motivated" to utilize mind-altering plant drugs for visionary shamanism or other ecstatic religious ritual; for example, he argues that the "basic culture" of the "pan American paleo-Indian" incorporated the use of such plant drugs as an important geographically widespread culture trait:

> All the invaders from Siberia were simple hunters and gatherers. All American agriculture and domestication (save for the mesolithic dog) were local developments wholly independent of Asia. No later animal domesticates and no cultivated plants were intercontinentally shared in pre-Columbian times [gourd? sweet potato? coconut? cotton?]—though, interestingly, the use of wild-gathered narcotic mushrooms, which . . . I believe to be authentically mesolithic in age, was perhaps even part of an incipient "narcotic complex", both in aboriginal Siberia and America. Basic religion in both Americas was the visionary shamanism of hunters (animal familiars, animal "owners") quite like and in fact culture-historically continuous-with paleo-Siberian shamanism and the shamanism of mesolithic Eurasiatic hunting peoples. As ecstatic-shamanists they valued the psychedelic state; as simple hunters and gatherers they were under pressure continuously to explore their plant environment for food and accidental new narcotics. (La Barre 1970:75–76)

Discovery of a plant substance which could produce an ecstatic-visionary state would have been highly valued by a group that held such kinds of psychological effects in high esteem:

> Primitive cultures, where sickness and death are usually ascribed to a supernatural cause, have long accorded psychoactive plants a high, even sacred, rank in their magic, medical, and religious practices, because their ethnopharmacology often values the psychic effects of "medicine" more than the physiological. (Schultes 1969:245).

Acquisition of the spirit powers inherent in the plant substance motivates the imbiber to partake. In the past, ethnocentric values such as accepted supernatural imagery and specific paradigms of spiritual reality may have been the prime mover or suppressant force affecting the use or nonuse of a known mild-altering drug such as the opium poppy. The ethnographic record provides us with some interesting examples that may have more than a little relevance to the ancient uses of the opium poppy. However, here we should make a special note concerning the use of ethnographic analogies. Although there may be a strong temptation to consider examples from the ethnographic present as proof that similar ethnobotanical relationships existed far into the past, it is prudent to retain a somewhat skeptical perspective. Although the analogy may seem quite logical, a particular use of a specific plant substance today does not necessarily confirm its application for similar purposes in antiquity:

> Inevitably theses based on the life of present-day primitive societies assume that such societies reflect the economic and social conditions of the Paleolithic or Mesolithic and ignore the possibility of cultural regression and breakdown. (Isaac 1970:19)

These cautious comments notwithstanding, let us review a few examples from the ethnographic record for the insight they may provide into the widespread cross-cultural phenomenon of psychoactive drug plant use; specifically, this perspective may shed some light on the probable early mind-altering use of the opium poppy in regions where it was available and its psychoactive qualities were recognized.

Tobacco: There are numberous examples of drug plant use in sacred magico-religious contexts; most of the documented ones come from the New World. For example, La Barre (1970:77) suggests that there is copious evidence indicating the supernatural status of the tobacco (*Nicotiana* spp.) plant among all aboriginal groups that had access to it in the New World" from middle Canada southward to Patagonia." Wilbert (1972:53–83) has documented a contemporary sacred relationship between this plant and the Warao Indians in Venezuela (See Also Furst 1976:23–32).

Coca: Martin (1970:422–23) and others have discussed the use of the coca plant *(Erythroxylon coca)* in the uplands of South America: " . . . no Andean communicant would dream of approaching the supernatural without being intoxicated with the chewed coca leaf; more exactly, the psychedelic effect *is* the supernatural" (La Barre 1970:77).

Psilocybin Mushrooms: The ancient and contemporary ritualistic use of "teonanacatl" ("flesh of the gods") mushroom (*Psilocybe* spp.) among Mexican Indians such as the Aztecan and Mazatecans is reminiscent of the paleo-Siberian cultural trait involving the ingestion of the fly-agaric *(Amanita muscaria)* fungus by the local shamans (Wasson 1972:185–200).

The Red Bean Cult: Among the native American Indians of the Great Plains region, the red bean *(Sophora secundiflora)* was the central item of a widespread aboriginal cult that celebrated the "supernatural power" of the species; this spirit force was experienced through the sacred act of ingesting the alkaloid-rich red seeds of this psychoactive leguminous plant: " . . . the evidence is archaeological" (La Barre 1970:77). In fact, the archaeological evidence for these hallucinogenic red seeds comes from several sites in Texas and northern Mexico and dates farther back in time than any other documented human association with a psychoactive plant:

> It is certainly noteworthy that apparently a single hallucinogen, the *Sophora* bean, should have enjoyed an uninterrupted reign of over 10,000 years—from the ninth millenium B.C. well into the nineteenth century and the disintegration of traditional Indian culture—as the focus of ecstatic-visionary shamanism, and for all but a few centuries of that enormous span of time in the context of the well-documented, conservative, and evidently highly successful homogeneous ecological adaptation we know as the Desert Culture of southwestern North America. (Furst 1976:9).

The Peyote Cult: The more recent expansion of the peyote cult among the Plains Indians of twentieth century America also manifests the important cultural role of a mind-altering plant: " . . . with all specific tribal cultures now largely gone, residual generic Plains culture still regards the hallucinations produced by eating peyote, *Lophophora williamsii*, as visionary proof of the presence of the supernatural" (La Barre 1970:78).

The San Pedro Cactus: Sharon (1972:114–35; 1977:374–81) has described the sacred role of the San Pedro cactus *(Trichocereus pachanoi)* in coastal Peru. Once again there is evidence of a psychoactive plant substance being used to communicate with divine spirits to transcend normal consciousness. Numerous other ethnographic examples of drug plant use in connection with religious ritual could be cited (e.g., see Furth 1972, Harner 1973, Emboden 1979a, and Schultes and Hofmann 1980). An important point to remember here is that there seems to be a consistent role which the drug plays:

> Whether shaman alone, or shaman and communicants, or communicants alone imbibe or ingest the Ilex black drink, datura infusions, tobacco in whatever forms, native beers and wines, peyote cactus, ololiu seeds, mushrooms, narcotic mint leaves or coca, the ayahausca *(Banisteriopsis caapi)* or "death vine"—or any of the vast array of Amerindian psychedelics—the principle is the same. *These plants contain spirit power.* (La Barre 1970:78)

Many ancient cults involving the ritualistic ingestion of a "divine" drug plant substance have probably disappeared as a result of basic changes in patterns of livelihood. La Barre (1970:78) suggested that farmers may be less inclined to use the psychoactive drug plants in the same way as their ancestral hunting and gathering forefathers:

> It is possible that agriculturalists have different anxieties from those of hunters. Is a "technology" of rain-making more needed than animal-helpers? Does religion itself become more bureaucratized with denser town-living? And do town-living people become more sophisticated and demanding epistemologically than individualistic hunters naively content to rely on similarly individualistic religion, the direct mystical experience?

La Barre saw the roots of our own Western culture as being embedded in this Paleolithic-Mesolithic shaman cult complex. He asks us to consider the pantheon of ancient Greece:

> Behind Olympian king of the gods himself—an image significantly consistent with the *tyrannos* of the early city-state—lurks the old Indo-European, nay ancient Eurasian shaman. Northern Zeus, who in Greece married local Hera of the "clear sky" remained the weather-shaman and rain-king.
>
> . . . even the gods most characteristically Greek often still had their shamanistic ecstatic oracle, Zeus at Olympus and Apollo at Delphi (some say intoxicated with laurel leaves). Indeed, the Mysteries and even later cults drank immortality with the blood of the vine-god and ate it in the flesh of the divine animal of the [poppy] *seed-cake of Ceres.* (La Barre 1970:78; emphasis added; see also Meuli 1935, Dodds 1951, and Butterworth 1966)

Ceres, the Roman earth mother goddess, is of course cognate with her Greek predecessor, Demeter. The cult of Demeter and many other Neolithic earth mother goddesses was essentially an agricultural fertility cult. Isaac supplies some cogent comments regarding the roots and relevance of this kind of cult:

> The religious revolution that I believe underlies agriculture supplanted the widespread Upper Paleolithic Eurasian religion, which centered as far as can be determined on mother goddesses with pronounced lunar and sometimes stellar attributes. The new world view rested on the belief that the world emerged from the death of a primeval being or god, whose slain and severed body came to constitute the existing world. In its most ancient form the myth involved the death and resurrection of a mother goddess. (Isaac 1970: 107)
>
> At the same time the myths refer predominantly not to the creation of the world, but to the emergence of plant life from the slain body. The myth has become the distinctive myth of agriculture. (Isaac 1970: 108)

Considerable indirect evidence suggests that there was a mystical afterlife cult use of the opium poppy as part of the visionary mystery rites in ancient Greece:

> All the myths of the dying fertility gods are ritual myths, which means that they express what is conceived as the key to a culture's origin, meaning, and survival. Hence they must be reenacted at critical times, such as a change of season, crisis periods of human life like death and birth, and in national crisis. (Isaac 1970: 108)

Gimbutas describes such a ritual use in ancient Greece at a "critical time":

> At threshing time another festival, called *Skirophoria*, took place in Ancient Greece. Young virgins were clad in white robes and at night given sacred objects called *skira*. These objects were figurines of suckling pigs and cakes made in the shape of serpents. After the festival they were deposited in the sanctuary of Demeter. In the goddess' sanctuary at Lykosura, offerings made to Demeter were listed in inscriptions: oil, honeycomb, parsley, figurines, poppy seed, lamps, incense. (1974: 215; compare with Nilsson 1957: 312ff.)

Although the opium poppy was used in ancient Greece (possibly for ritualistic as well as practical purposes) it may not be indigenous to the region. In fact, I believe that the opium poppy and its use were introduced during the Late Bronze Age, sometime in the second millennium B.C., possibly from the north along with the old Indo-European or ancient Eurasian shaman referred to above. This obscure aspect of the culture history of the opium poppy is discussed in detail in chapters 7 and 8. The important point to consider here is that the opium poppy was a potential psychoactive drug source which may have been incorporated into the group of plants used by some early humans, not only as a valuable source of medicine but possibly as a crucial divinatory vehicle for magico-religious experience. Such an experience could have formed the basis for a more secure psychological adaptation to the realities of life and death:

> It might even be argued that this material kind of secular adaptation to the environment accumulates and evolves like any other animal adaptation. By contrast, sacred culture is mere psychological adaptation to inner anxieties. Some of these inner anxieties (e.g., those concerning death) perhaps exist in all human beings; hence faith in an animistic, separable, and indestructible soul is well-nigh panhuman. (La Barre 1972:262)

La Barre clarifies his position in regard to the use of hallucinogenic plant substances and the origin of religion; and in doing so he sets the stage for my investigation of the early relationships between man and the opium poppy:

> . . . all our knowledge of the supernatural derives *de facto* from the statements made by religious visionaries or ecstatics (i.e., prophets and shamans)—the priests only administrate the ecclesia established on this supernatural basis—and that the nature of the shamic ecstasy may be illuminated by attention to ancient hallucinogens. (La Barre 1972:262)

La Barre has indeed assumed that the cultural motivation for ecstatic visionary experience through the use of hallucinogenic plant substances reached far back into the Bronze Age and Mesolithic period (La Barre 1972:262).

> Some individuals, "spiritually" quite ungifted in spontaneous visionary skill, are nevertheless capable of authentic hallucinations, given pharmacodynamic help. With some other anthropologists, I believe that the use of powerful botanical hallucinogens has been a real and important vehicle of shamanistic ecstasy, not only in modern ethnographic time but also in prehistoric antiquity. (La Barre 1972:270)

Ethnobotanic Clues from Shanidar Cave and Parts of Europe:—Although we have no direct evidence of medicinal or ritualistic use of plants from any Cro-Magnon sites, we do have some remarkable indirect ethnobotanical evidence that comes from the Neanderthal cave of Shanidar in the Zagros mountain highlands of Iraq. Here during the month of June about sixty thousand years ago ". . . a man was buried in unusual circumstances" (Leakey 1977:125).

Pollen analysis of the soil samples from the Shanidar cave grave site has revealed a fascinating arrangement of pollen grains surrounding the grave. Paleobotanists at the Musée de l'Homme in Paris have documented an orderly distribution of the pollen microfossils found among the human funeral remains indicating that the flowers from which the pollen were derived must have been deliberately placed in bunches surrounding the dead body in the grave site (Solecki 1971:247–49):

> It would appear that the man's family friends, and perhaps members of his tribe had gone into the fields and brought back bunches of yarrow, cornflowers, St. Barnaby's thistle, groundsel, grape-hyacinths, woody horsetail, and a kind of mallow. The branches of the woody horsetail are particularly suitable for weaving a rough bedding on which the body appears to have been laid; and the white, yellow, red, blue, and purple flowers of the other plants must have added greatly to a poignant scene. (Leakey 1977:125)

The flowers in this ancient grave can be interpreted as "condensed symbols."

> Flowers are in common use throughout the world as decorations, as religious offerings, objects of divinatory manipulation and symbols of groups and activities. in contemporary European culture flowers are appropriately present at marriages and funerals, at military ceremonies and pacifist demonstrations. Ladies are courted with flowers and the sick are cheered by them. Flowers thus have diverse and contradictory referents. (Gillen 1976:276)

According to Cirlot (1971:104–5) there are at least two special symbolic floral associations. On the one hand, flowers in their "essence" represent the transitory nature of life and beauty; and on the other, their shape can be interpreted as a symbolic center of attention or the "soul." Flowers at the Shanidar grave are especially revealing:

> The fact of deliberate burial is interesting enough for it betrays a keen self-awareness and a concern for the human spirit. And to have decorated a corpse with flowers adds enormous significance. (Leakey 1977:125)

The symbolic association of flowers with funerals has also been reported in classical culture; Cirlot (1971:104–5) referred to the use of flowers in ancient Greece where they were strewn "over the corpses as they bore them to the funeral pyre and over the graves (not so much as an offering as an analogy)."

The opium poppy, of course, is well known for its lovely, delicate flowers; but this aspect of its physical nature is overshadowed by the great therapeutic potential it could have provided early man in his search for effective painkillers and soporific tranquilizers. In this respect the pollen record from Shanidar cave is even more curious:

> . . . perhaps most intriguingly of all, it turns out that of the various species used in the Shanidar burial, several have until recently been used in local herbal medicine! That the Shanidar people were aware of at least some of the medicinal properties of these flowers is not unlikely. As they became more and more independent of the demands of the environment, early members of the human family must also have had an intimate knowledge of what Nature offers. This, indeed, is the key to success among hunting and gathering communities. (Leakey 1977:125)

One of the pollen clusters found in the Shanidar cave grave site is the so-called "woody horsetail." Actually, this species is not a true "horsetail" (i.e., *Esquisetum* sp.), but rather is a member of the genus *Ephedra*, which produces the familiar nerve stimulant ephedrine. The French paleobotanist, Mme. Arlette Lesoi-Gourhan, identified the *Ephedra* pollen from the Shanidar cave as belonging to *E. altissima* (Solecki 1971:248–49). It is important to note that although the genus *Ephedra* is known to contain medicinally valuable, physically stimulating alkaloids (see Taylor 1965:101–14), the Neanderthal people who occupied Shanidar cave may have used their *Ephedra* species for psychoactive, perhaps visionary purposes. Some have even suggested that a species of *Ephedra* (e.g., *E. pachyclada* or *E. intermedia*) was the botanical source of Soma, the ancient sacred drug plant of India referred to often in the Rig Veda (Emboden 1979a: 60). According to Schultes and Hofmann (1980: 367), *Ephedra nevadensis* allegedly produces hallucinogenic effects. "It is certainly tempting to speculate that if thse Neanderthals, whom Solecki and other scholars now believe to be in modern humanity's direct line of evolution, possessed knowledge of so many effective medicinal plants, they may likewise have been familiar with some of the psychedelic flora of the region" (Furst 1976: 5). As improbable as this hypothesis may seem to some, we should consider the following statement made by Solecki, the archaeologist who excavated Shanidar cave: "From archaeological findings elsewhere we already know that Neanderthal man seems to have had a spiritual concept, since he evidently practiced funerary rites over his dead" (Solecki 1971: 250; see also Solecki 1975: 880–81). Perhaps their "spiritual concept" was derived from a shamanistic, nonordinary altered state of consciousness brought on by the use of a powerful psychoactive substance(s): "It is probably not too much to say that mysticism, or religion, has always been a fundamental aspect of the human condition, with its beginnings reaching back perhaps to the primitive origins of self-consciousness" (Furst 1976: 3–4).

The fact that this very ancient evidence of possible ritual and medicinal knowledge is associated with Neanderthal man should not obscure the significance of this remarkable archaeological evidence:

> The findings at Shanidar are at least as important in telling us something about the minds of our true ancestors as they are direct evidence of the ill-fated Neanderthalers. Although as yet there are no signs of ritual as subtle as the flower burial for our true ancestors, we can be sure that their culture was not less developed. (Leakey 1977: 125)

First thought to be savage, brutish, and stupid, Neanderthal man is now considered to have been an intelligent primate, "almost certainly a specialized form of early man who slid into extinction" (Leakey 1977: 32).

By the time that Neanderthal types disappeared from the archaeological record approximately 30,000 years ago, they had apparently established themselves over a relatively far-reaching range. Evidence for their existence has been found in China, Java, Iraq, Zambia, Yugoslavia, Italy, Spain, France, and Germany (Leakey 1977: 125).

Whether or not the Neanderthal or the Cro-Magnon people knew of and used the opium poppy or opiate-producing ancestors of this species remains conjectural. Certainly the substantial magico-medicinal potential of the opium poppy would have made the capsule an appropriate burial gift of ritualistic or symbolic importance; but as of now there is no evidence of the opium poppy's presence in either the macro- or microfossil record from the Shanidar cave. Indeed the conspicuous absence of any Neolithic fossil remains of the opium poppy in western Asia suggests that it probably was not present in this general region until a relatively recent period of the prehistoric past (Hans Helbaek, personal communication 1976). On the other hand, two presumably Paleolithic fossil poppy capsules (one found in central Germany and the other recovered from southern France) at least suggest that the opium poppy or a close relative may have been extant in the early western or central European environments of Cro-Magnon or Neanderthal man. Some aspects of the taxonomy and morphology of these fossils are discussed in chapter 3.

Neither the specific nor generic classification of the poppy fossil from Bornstedt in central Germany has been confirmed. However, Friedrich (1883:139) noted that this fossil may have been an opium poppy *(Papaver somniferum);* and according to Fritel (1920:208) and Neuweiler (1935:111–12), the poppy capsule fossil found in the Aygalades tuff deposits near Marseilles should be treated as an opium poppy (*P. somniferum* var. *nigrum*). Fritel referred to the fossil's age as belonging to the Quaternary period. According to Lumley-Woodyear (1971:353), the Aygalades tuff deposits were laid down before 200,000 B.P. However, Neuweiler suggested a cautious approach, arguing that the close resemblance of the fossil capsule to modern specimens of the opium poppy indicates that it may have been much younger, possibly of recent origin.

If, on the other hand, the suggestion of Fritel and the concurring conclusion of later authors such as Bertsch and Bertsch (1947:196) are correct and the opium poppy fossil capsule from the Aygalades tuff deposit is an indication that this species was a member of the Pleistocene Flora of southwestern Europe, then we can assume that our Paleolithic ancestors probably came into contact with it somewhere in this region and perhaps learned early that it could provide edible leaves and seed, useful oil, medicinal substances and a whole host of other applications including euphoric intoxication.

Whether or not these kinds of ethnobotanical possibilities were recognized by our human ancestors at such an early stage of cultural development may never be known. Nevertheless, the circumstantial evidence from the Shanidar cave and the poppy capsule fossils from Europe are intriguing. I suggest that the combination of the poppy flower's attractive appearance, its edible products, and its potent psychoactive substances would have made the plant very appealing to prehistoric peoples living in environments where this species could be found. In this perspective, it is not difficult to imagine that an early relationship developed between man and this multi-purpose herbal plant.

The idea that the opium poppy was used to assist in the interpretation of the spirit world in some human groups is a challenging one; perhaps it is the most controversial and difficult speculation to confirm or negate about this plant. Yet this notion is supported by the documented examples of other psychoactive plants such as those referred to above, as well as a number of myths and relics of the ancient Greek and Roman civilizations which indirectly implicate the opium poppy (e.g., see Kritikos and Papadaki 1967). Furthermore, and probably more dramatically, this concept is supported by the remarkable discovery of numerous opium poppy capsules and seeds found as "gifts of love" in association with a large communal burial site—"asilo de la muerte"—discovered inside a southern Spanish cave near the Mediterranean coast and dated to the Late Neolithic or Early Bronze Age (Gongora 1868:35; Neuwiler 1935:108; Pellicer 1963:7). This provocative archaeological discovery is described and analyzed in chapter 5.

Initial Man-Opium Poppy Relationships

As I noted in the first chapter, the archaeobotanical record of early man is in many cases quite fragmentary. This relative lack of direct information generally becomes even more apparent as we look further and further back into our remote ancestry. In fact, the early ethnobotanical record is so full of gaps that we must often rely on reasonable speculation as a means of explaining prehistoric man-plant relationships:

> The archaeological record provides a logical beginning for any attempt to answer the where, when, and why of the domestication of plants and animals. Yet despite the dramatic achievements of twentieth century archaeology, grave doubts about the ability of archaeology to answer questions of origins remain. For although archaeology can presumably establish sequences and even cast light on the process of change, it cannot discover where the first crucial step was taken. As Carleton S. Coon has said, "It is too much to ask of the law of chance that the first archaeologist who digs a hole in the ground should locate the place where the invention took place." (Isaac 1970: 17)

According to Hawkes (1969: 17), if we want to develop an insight into the origins of the domesticated plants we must approach the inquiry from an appropriate point of view:

> . . . we must look at wild and cultivated plants associated with man as an ecological complex and view this in relation to the ecology of man himself. If we do this we can make considerable progress in understanding the mechanism of domestication, verifying and, when necessary, modifying our hypotheses by reference to exact archaeobotanical data.

I hope that the reader finds the following discussion both reasonable and stimulating. Perhaps it will illuminate new avenues of research and encourage more investigation into the puzzling prehistory of mankind's association with the opium poppy.

> The important truth that students should learn early in the game is that science deals not with absolute truth, but with successive approximations to the truth. Half a truth is better than no truth at all. (Fagan 1974: xvi)

Ecological and Speculative Considerations

When and where did humans first come into contact with the opium poppy? How and why did they first experiment with this plant? Some years ago I applied these same questions to another well-known drug plant, hemp or marijuana, *Cannabis sativa* L. (Merlin 1972: 36–45). In light of the ecological and ethnobotanical information available for that species, I formulated some hypotheses which may also apply to the opium poppy. In fact, hemp and the opium poppy have much in common. Both species are herbaceous annuals, have been used to produce oil from their seeds, have also been used to produce psychoactive drugs, and from an ecological perspective, both can be appropriately described as "weedy" species:

> The common factor in cultivated and in closely related wild species is their "weedy tendency," their ecological adaptation to "open," disturbed, or unstable habitats with bare soil and less competition from other plants. This makes use of an ecological, rather than a horticultural definition of weeds, as Bunting and Harlan and de Wet have pointed out, and defines a weed, not in terms of whether it is an unwanted plant competing with those we are trying to cultivate, but one which is adapted to disturbed or open habitats, often requiring high soil nitrogen and able to grow only in those areas where climax forest and grassland have been destroyed. (Hawkes 1969: 18)

Man has had very long and curious associations with both hemp and the opium poppy. In different places, and often for different reasons, humans have found important as well as frivolous uses for these unusual plants. In some regions these famous drug plants may have played significant, sometimes crucial roles in the fundamental cultural changes that occurred after the end of the latest ice age. My discussion of the ethnobotanical origins and early cultivation of the opium poppy is a slightly modified version of that which I suggested for the hemp plant (Merlin 1972: 36–45). It draws heavily upon the theories of Sauer (1952) and Anderson (1967).

Most of man's existence has been spent roving in small groups, hunting and gathering wild food in many varied kinds of environments. In Europe, after the end of the last ice age, human settlement and subsistence patterns began to change. As global temperature increased, massive continental ice sheets and alpine glaciers receded. In addition, the higher latitudes, formerly covered by large masses of ice, opened up to possible colonization. Furthermore, the melted waters of the receding glaciers swelled existing rivers, enlarged some lakes, and created new water bodies. These environmental changes in turn invited man's ingenuity: "It was above all a rarely favorable time for man to test out the possibilities of water-side life, and especially of living along fresh water" (Sauer 1948, cited in Sauer 1963: 258).

Some groups of men began to take up more permanent residency along seacoasts, in benign river valleys, or beside the relatively open margins of freshwater lakes and bogs. And with stationary settlement patterns, the pace of man's cultural evolution accelerated significantly. He domesticated dogs, possibly at first for pets, and produced such cultural innovations as bows and arrows, newly styled chisel-edged axes and adzes, crude pottery, and elaborate fishing gear such as nets, lines, and poisons necessary for the development of fishing (Sauer 1952).

Indeed, the arts of fishing played an important part in the cultural development of man. Without agriculture or domestication, and with rather simple technologies, hunting and gathering groups led nomadic ways of life required by a foraging economy. But the development of ways to trap and collect fish and waterfowl often provided man with a relatively stable and stationary protein food supply. These fixed and abundant food resources allowed some groups of early fishermen to become more sedentary. A more settled life probably allowed for more free time to experiment with a great diversity of plants and animals. Eventually, under more settled conditions, in some cases stimulated by the development of fishing, man experimented with the growing of plants.

Those innovative bands of men who took up sedentary residence along streams, rivers, and lakes during the climatically more favorable, postglacial period were perhaps some of the earlier incipient cultivators of plants. Intensified use of inland and coastal water resources:

> allowed a considerably higher degree of sedentary occupation, freeing man of the necessity of continuous wandering. One precondition for the acceptance of farming—residential stability—was thus automatically fulfilled. (Waterbolk 1968: 1096)

In a recent study dealing with modern attempts by Australian Aborigines to cultivate crops, Kimber argued that residential stability was the motivating force which stimulated the beginning of farming among the people he described. Furthermore, he suggested that his argument may apply as well to other human groups during earlier periods of time:

> The one really great change in the life-style of each of the men involved has been that of a nomadic person to one living in a fixed locality, and recognizing that change to be permanent. In this context these first ever attempts at cultivation of native-plants are very interesting. It is my belief that, despite differences in man-plant-animal relationships that must have existed in the various prehistoric societies, these . . . examples by a traditional modern hunter-gatherer people hint at the driving forces behind the beginnings of farming. (Kimber 1976: 149).

Through a long, slow, and very important process referred to as the Neolithic revolution, farming was invented. Man thus became an active producer of his own food supply. This gradually changed his subsistence patterns from hunting and gathering to a new mode of life based on the economy of agriculture.

A Dump Heap Theory for the Origins of a Man-Opium Poppy Relationship

When the early fishing groups settled down on stream-plowed banks, gravel bars, or muddy lake shores, they cleared off more of the land and constructed shelters. In the process of clearing, they helped expand the number and size of the disturbed or open habitats available. This type of newly opened habitat is quickly invaded by a series of successional plants which we may refer to as "weeds." One so-called weed, the opium poppy, is a sun-loving plant that thrives in light soils which are rich in nitrogen compounds. This kind is often found in and around the open dump heaps or waste areas associated with human disturbance and settlement.

Perhaps an early people, consciously or inadvertently, collected some opium poppy seeds along with other useful plant materials gathered from the surrounding environment and brought them back to the settlement area. Subsequently, some of the tiny opium poppy seeds may have been dumped unknowingly into the waste heaps. If so, the poppy seeds would have had an adequate micro-environment in which to grow near the human settlement area. This would have resulted in the development of an intimate association between the opium poppy and man. Eventually, man could have learned to utilize this plant (now found growing in his dump heaps) for a number of purposes including food, oil, medicine, animal fodder, euphoric intoxication, and ritualistic ceremony. Ultimately, man would have encouraged and enlarged the distribution of the opium poppy more directly by protecting or cultivating the plant.

According to Sauer (1948, cited in Sauer 1963:257–58), "well-drained alluvial lands" are quite favorable for plant growth, especially "useful mesophytic species" which are well adapted to such environments. Therefore, Sauer (1952:27) assumed that some of the oldest cultivated plants were versatile species grown around fishing settlements for starch food, tough water-resistant fibers, euphoric and medicinal drugs, and poisons. The opium poppy does not produce usable fibers nor are there indications that it is one of the oldest cultivated plants. However, as we have pointed out above, the opium poppy is a quite versatile species that can provide human food, animal fodder, psychoactive drugs, and even heating fuel from the clean-burning dried stalks. In addition, we know that this species was utilized by a number of Neolithic lake shore and bogside dwellers in west central Europe during the prehistoric period more than 5,000 years ago. These early farmers and their use of the opium poppy are the subject of the following chapter.

5 • The Archaeological Record for the Neolithic and Early Bronze Age Periods in West Central Europe

During the past five years, I have reviewed and analyzed archeobotanical data containing indications that prehistoric human communities have used, or lived in an association with, the opium poppy. The great bulk of this evidence, especially for the Neolithic period, comes from the Swiss Foreland of west central Europe. In this chapter, I provide some background information regarding the early farmers in west central Europe. This is then followed by a survey of the numerous prehistoric agricultural settlements in this region from which fossils of the opium poppy have been recovered (see figure 34). Finally, I discuss the opium poppy's possible associations with the early farmers of the region.

The Prehistoric Settlements of the Swiss Foreland

Three of the four geographic zones in Switzerland should be briefly noted here. First there is the permeable limestone Jura Mountain Range which had a southwest-northeast orientation and which had long presented a formidable barrier to human movement in the region. Second and most relevant to this study is the Swiss Foreland, a plateau "characterized from the Neolithic to the end of the Bronze Age by numerous lakes of all sizes; we shall see the important part they played in agricultural colonization" (Sauter 1976: 191).

The Foreland is constricted to a relatively narrow passageway in the southwest

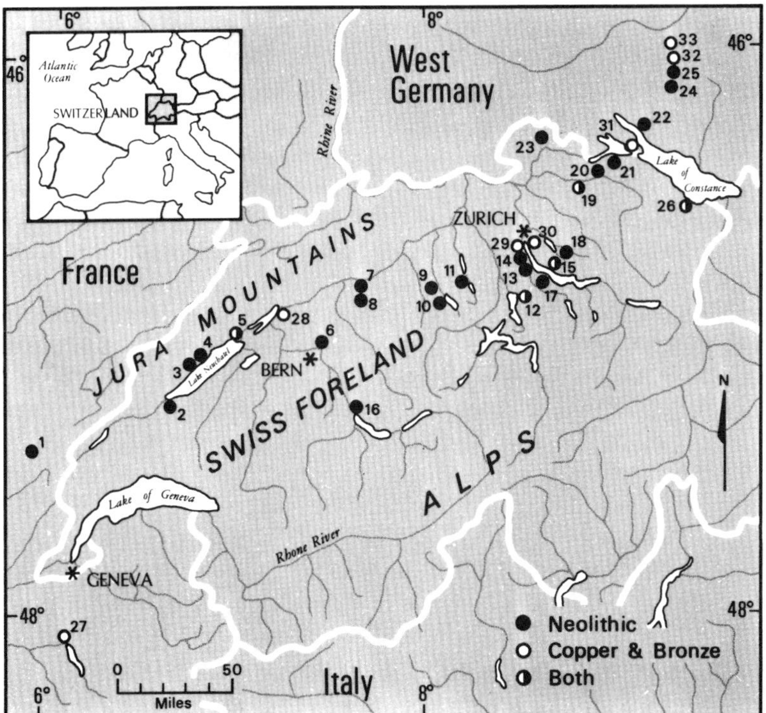

FOSSIL EVIDENCE FOR THE OPIUM POPPY

IN NEOLITHIC, COPPER, AND BRONZE AGE SITES OF THE SWISS FORELAND

1 Lake Chalain	10 Oberkirch	19 Niederwil	27 Lake Bourget
2 Yverdon	11 Lake Baldegger	20 Pfyn	28 Morigen
3 Cortaillod	12 Zug-Sumpf	21 Steckborn	29 Wollinshofen
4 Auvenier	13 Utoquai	22 Sipplingen	30 Alpenquai
5 St. Blaise	14 Mannedorf	23 Weier	31 Langrain
6 Moosseedorf	15 Obermeilen	24 Reute (Waldsee)	32 Buchau
7 Solothurn	16 Thun	25 Riedschachen-siedlung	33 Dullenried
8 Seeberg (Burgaschi)	17 Horgen	26 Arbon	
9 Egozwil	18 Robenhausen		

FIGURE 34. Fossil evidence for the opium poppy.

where it is bordered by Lake Geneva. In the north, the Foreland opens widely, offering a relatively large corridor which in effect links the Rhone, Rhine, and Danube river basins. The third zone to be noted here lies south of the Swiss Foreland and is, of course, the massive Alps. Although impressive in its grandeur, this alpine mountain range is not impenetrable. Some mountain passes such as the Saint Bernard have actually played an important cultural role since the prehistoric period, allowing people, materials, and ideas to move north and south. The corridors of the Foreland and transalpine passes have thus been instrumental in facilitating the cultural flow to and from the Mediterranean by way of the valleys of the Po and Rhone, to the northen European Plain up and down the Rhine Basin, and to and from central and eastern Europe by way of the Danube River Valley (Sauter 1976: 20).

As I point out in the following section, it has been in the Alpine Foreland region that the great bulk of evidence for man's use of the opium poppy has been found. I survey this evidence below and suggest possible explanations for the geographical distributions, chronological relationships, and cultural contexts of some of the data. It is hoped that this will serve as a useful reference for historians and ethnobotanists as well as archaeobotanists specifically interested in the ancient relationships between man and one of the world's most widely known, revered, and feared drug plants. Interpretations and speculations are proposed as tentative explanations which in many cases will have to be altered or perhaps discarded entirely as new evidence becomes available and more reasonable reconstructions are presented: "Archaeological facts are often dull, but they have the compensating merit of possessing a permanent value; archaeologist's inferences, on the other hand, are often interesting, but are apt to be precarious and ephemeral" (O'Rahilly 1946: 440).

During the winter of 1853–54, the Alpine Foreland of west central Europe experienced a remarkably long drought. So prolonged was the dry, cold spell that rivers dwindled and lakes retreated to degrees never before recorded in the area; and although these conditions adversely affected farmers and fishermen, the attention of the world's scientific community was attracted to the extraordinary prehistoric remains that were uncovered along the exposed lake margins (Keller 1878: 11). The first discovery took place at Obermeilen on Lake Zurich, "thanks to the initiative of a schoolmaster, Johannes Aeppli" (Sauter 1976: 16).

The wealth of archaeological information that these remains have provided in the years since their discovery was first brought to notice by Ferdinand Keller, the president of the Antiquarian Society of Zurich, who went to Obermeilen in 1854 to investigate the original discovery of a so-called "lake-dwelling" settlement (Keller 1878: 11). In the years immediately following the exciting excavations at Meilen, a

FIGURE 35. The peatbog of Robenhausen, where Neolithic lakeshore dwellings were uncovered in the nineteenth century (Messikommer 1913, plate IV; *Courtesy of Orelli Fussli Verlag, Zurich, Switzerland*).

number of additional lake-shore sites were uncovered in northern Switzerland at Moosseedorf, Robenhausen, Niederwil, Wauwil, and other locations.

Among the early scientists who came to excavate and examine these prehistoric remains was Oswald Heer who worked at the Robenhausen site located along Lake Pfaffiker near Zurich in 1865; Heer studied the plant remains recovered from several of the lake shore sites, including Robenhausen, and eventually published a treatise on the subject in which he described the excellent preservation of fruits and seeds (Heer 1866). Heer classified the residue of about seventy species in what may be considered the initial systematic documentation of postglacial archaeobotany (Villaret-von-Rochow 1967:21). Among the flora established for the prehistoric Robenhausen site, Heer identified a "whole cake of the seeds of the garden or opium poppy" (1878:523) and one unripe capsule of this species (1878:531; see figure 13); Heer treated these specimens as *Papaver somniferum* L. var. *antiquum* (see chapter 3).

Since the initial discoveries were made over a century ago, approximately thirty prehistoric lake shore settlements in the alpine foreland or nearby regions of west central Europe have yielded fossil opium poppy seeds in either charred or unburnt condition (see figure 34). Obviously this species was well known; perhaps it was

FIGURE 36. Jakob Messikommer (a) a nineteenth century archaeologist, is shown (b) recovering specimens from a flooded lake shore settlement site in the Swiss Foreland at Robenhausen (from Messikommer 1913; *Courtesy of Orelli Fussli Verlag, Zurich, Switzerland*).

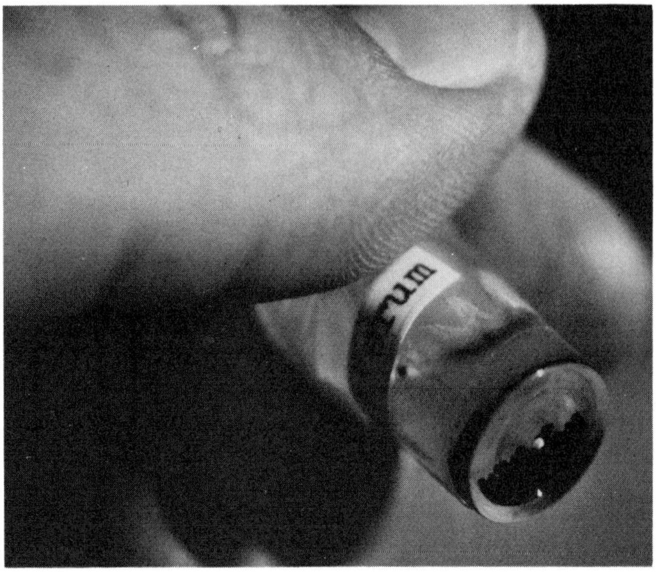

FIGURE 37. The small vial holds Neolithic fossil opium poppy seeds sent to me by Dr. Schweingruber. The fossils were recovered from a Cortaillod settlement at Seeufersiedlung Twann on the shores of Lake Bieler in 1975.

FIGURE 38. Nineteenth-century excavations of the lake shore settlements in the Swiss Foreland uncovered a number of Neolithic and Bronze Age artifacts (from Messikommer 1913, figure 2; *Courtesy of Orelli Fussli Verlag, Zurich, Switzerland*).

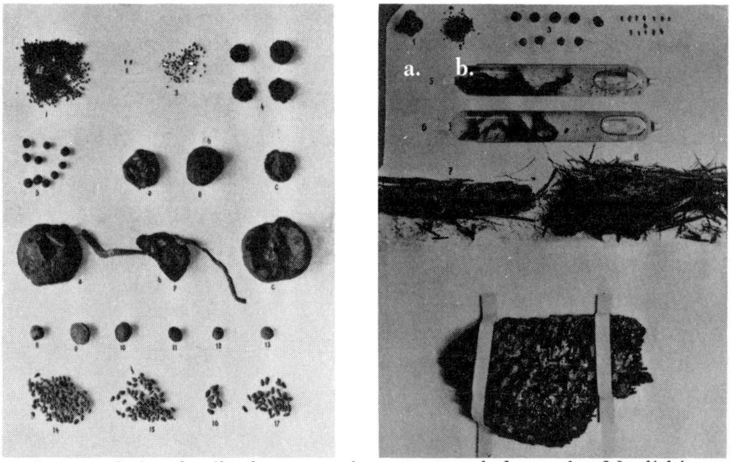

FIGURE 39. Some of the fossil plant remains recovered from the Neolithic settlement at Robenhausen. The fossil poppy capsule (a) and poppy seeds (b) were found among these archaeobotanical remains (from Messikommer 1913, plates XXXV and XXXVI; *Courtesy of Orelli Fussli Verlag, Zurich, Switzerland*).

FIGURE 40. Jakob Messikommer sorting ancient seed fossils found in Swiss peatbog (Messikommer 1913; *Courtesy of Orelli Fussli Verlag, Zurich, Switzerland*).

cultivated by the early farming people who occupied these sites.

First I discuss some of the basic aspects of these sites and the cultures that they represent; this, in turn, is followed by a review of the earliest lake shore sites from which we have evidence of the opium poppy and a consideration of the possible geographical origins of man's association with this species in west central Europe or elsewhere. Subsequently I present a more detailed survey of remaining prehistoric west central European sites that have yielded fossil evidence of the opium poppy during the Neolithic and Early Bronze Age.

The Lake Shore Dwellers of West Central Europe

> The first agricultural groups in the alpine foreland preferred lakeshores or bogs as settlement sites. (Waterbolk 1968: 1099)

After the initial discovery of the "lake dweller" settlement at Obermeilen in 1853, a number of exploratory expeditions were carried out in the surrounding region. Several extraordinary settlement remains were uncovered; and because most were found within the geographical borders of Switzerland, they were referred to as the "Swiss Lake-dwellings." In actuality this type of Neolithic settlement was also found in Austria, Bavaria, eastern France, and the Italian Alps. During the prehistoric period, the margins of almost all of the lakes and bogs of Switzerland and many of those in southwest Germany, northern Italy, and southwest France were eventually occupied by early farming groups. These village settlements were once considered to have been constructed on piles over lake waters for protection (see figure 41).

FIGURE 41. Artist's reconstruction of Neolithic and Bronze Age dwellings built on piles over the lakes and bogs of west central Europe. For approximately one hundred years, scholars believed that such settlements were constructed over water for protection (from Messikommer 1913, figure 3, first published by J. Heierli, 1901; *Courtesy of Orelli Fussli Verlag, Zurich, Switzerland*).

> It is hardly necessary here to state, what in the year 1855 was new to antiquarians, that from a series of discoveries the fact is made manifest, that in the very earliest times groups of families, or probably whole tribes, subsisting by hunting and fishing, with some knowledge of agriculture, lived on the borders of the Swiss lakes, in huts built not on dry ground, but on a series of piles in the shallows near the shores. (Keller 1878:2)

However, it is now assumed that these village settlements were located on exposed margins of bogs or lake shorelines (Waterbolk 1968:1099, see figure 42):

FIGURE 42. Artist's reconstruction of Neolithic dwelling built on the shores of the lakes and bogs of the Swiss Foreland. Archaeological evidence now indicates that these prehistoric settlements in west central Europe were located on the margins of the lakes and bogs, not over water as many scholars previously believed *(Courtesy of the Bernisches Historisches Museum, Bern, Switzerland).*

> . . . in most cases the lake- and moor-dwellings of the Neolithic and Bronze Age were not built over the water but on the shore which had been recently uncovered as a result of a fall in the water-level. Since the ground was very damp protection was needed; this was secured—very unsatisfactorily—by a bed of oak bark laid on the earth, or interlaced branches or floors made of closely-set poles that were partly or entirely covered with a coat of clay. (Sauter 1976:77)

The prehistoric lake shore dwellers of west central Europe resided in rectangular houses "built on the damp ground between the lakes' reed beds and scrub brush of the valley behind" (Fagan 1974:219).

It may be asked why these settlements were constructed on the damp ground when more stable, firmer terrain lay behind the remains of the ancient domicile structures:

> The answer may depend on the vegetational conditions of the time: the primary deciduous forest covered most of Switzerland during the Neolithic period. The Neolithic and Bronze Age farmers needed fields to grow corn, barley, and millet. Since such fields could not be located on the shoreline where there was no arable land they had to use clearings, or create them by burning. They would therefore have built their dwellings in the spaces between the forest and the water. (Sauter 1976:77)

In some cases, numerous lake shore sites have been located along the same lake bed or peat moor. For example, at least fifty have been found around Lake Neuchatel; thirty-two have been uncovered along the margins of Lake Constance; at least twenty-four have been located bordering Lake Geneva; and about twenty have been found along Lake Bienne:

> Some villages grew to include between twenty-four and seventy-five houses clustered on the lake shore, a density of population considerably above the average for the hamlets of Norman Britain, for example, which usually contained about thirty households. (Fagan 1974:219)

In some of the lake shore sites no metal implements, ornaments, or weapons have been found; these sites have been classified as Neolithic or New Stone Age settlements. Other lake shore sites are characterized by the presence of copper and bronze artifacts as well as some made of stone, wood, and bone; these sites have been classified as Bronze Age settlements. Other sites such as that at Niederwil are characterized by the appearance of stone, wood, and bone artifacts in the lower cultural layers of peat, mud, and clay, while the upper layers have also yielded metal materials such as copper and bronze along with artifacts fabricated out of stone, wood, and bone.

Lüdi (1955, cited in Villaret-von Rochow 1967:22) has pointed out that a separation of the Neolithic from the Bronze Age flora is in many cases not possible due to lack of, or poor, control exercised in the excavation and recovery of the fossil remains.

The chronological problems notwithstanding, we will now survey the series of Neolithic lake or bog shore cultures that occupied sections of west central Europe during parts of the fourth and third millennium B.C. and which in many cases had an apparently important relationship with the opium poppy.

Table 11 summarizes the major phases of Late Glacial and Early Post-Glacial Europe. The Middle Neolithic settlements along numerous lake shores and bogside environments in the Alpine Foreland occurred during the Atlantic phase (ca. 5500–3000 B.C., see Rodden 1975:55; 6200–3300 B.C., see Butzer 1971:571). Indeed, the initial spread of farming activity into Europe took place during this relatively warm, moist phase:

> With the Boreal-Atlantic transition, about 5500 B.C., there was much environmental change north of the Alpine belt. Pine and hazel lost their prominent place to a mixed deciduous forest.
>
> Increase in precipitation promoted the vegetation on lake-shores and the filling up of basins. Raised bogs appeared in former lakes or marshes, . . .

Table 11

Major Time Periods and Associated Biota in Europe (after Rodden 1975:56)

	TIME SCALE	---	*Key BIOTA*
Late Würm	14,000 B.C.		
	11,500 B.C.	Bölling	Reindeer, mammoth, woolly rhinoceros, musk ox, giant elk, bison, horse predominate.
	10,500		
	10,000 B.C.		
		Allerød	All above (except reindeer) extinct; reindeer approaching total extinction.
	8,800 B.C.		
	8,300 B.C.		
	7,500 B.C.	pre-Boreal	Red deer, roe deer, elk, aurochs, boar dominate; open woodland.
post-Glacial		Boreal	Appearance of hazel.
	5,600 B.C.		
	5,500 B.C.		
		Atlantic	Period of warmer temperatures and increased precipitation ("climatic optimum"); appearance of elm, lime, oak; aurochs and red deer populations decline.
	3,000 B.C.		

> The mixed oak forest was more shady than the fairly open pine woodland, and there was less grazing for large animals. Big game such as aurochs and red deer decreased in number. All these changes required readaptation by man, who now faced a rapid decrease of the natural resource he had been relying on for two millennia. (Waterbolk 1968:1096)

As a result of adaptation to roughly the same ecological conditions, the Middle Neolithic lake shore and bogside settlements of the Alpine Foreland had similar artifacts, architectural design, and subsistence economy. These resemblances encouraged early scholars to lump all of the settlements into one group, the "Lake-dwelling culture"; however, more recent research clearly demonstrates the diverse origins and development of at least three separate culture groups that are now known to have occupied the Neolithic lake shore communities of west central Europe (Hibben 1958:151). Probably the earliest of these culture groups is the most famous one; extremely well documented, it is referred to as the "Cortaillod culture" of Switzerland (Murray 1970:51; Chard 1975:347).

The Neolithic Cortaillod Culture

As early as 1929 Paul Vouga identified in his stratigraphic excavations of the Neuchatel lake-dwellings a lower layer to which he gave the name "Néolithique lacustrae ancien" in his classification of the lake dwelling Neolithic. Emil Vogt was

> to widen this concept to the whole of Switzerland, giving it in 1934 the name "Cortaillodkultur." (Sauter 1976:47)

Like other ancient lake shore dwellers of west central Europe, the people of the Cortaillod culture(s) manufactured simple pottery, produced some axes and adzes, and combined both farming and hunting as parts of their subsistence economy (Hibbens 1958:150–54): "The communities which gathered in these different types of settlements were all farmers— . . . with several varieties of wheat, barley and millet—and shepherds (with dogs)" (Sauter 1976:49).

Progressively, the levels of the adjacent water bodies rose and many of the pile-and-beam supported lake shore dwellings of the Cortaillod people were drowned. In fact, it was the eventual submergence of their residences and the waterlogging of their kitchen middens which has helped preserve so much information concerning their ancient culture (Murray 1970:51; Chard 1975:347); ". . . the gently rising waters of a prehistoric lake slowly covered the living floors and preserved them for posterity with tools lying where they dropped" (Fagan 1974:17).

The Cortaillod people raised a number of domestic animals and hunted several others in the environments surrounding their settlements. They also utilized scores of plants; some, such as hazel nuts *(Corylus avellana)*, acorns (*Quercus* spp.), raspberries *(Rubus idaeus)*, and sloes *(Prunus spinosa)*, were sought out for their edible wild fruits, while others, including several cereals, peas *(Pisum sativum)* and flax *(Linum usitatissimum)*, were cultivated (Villaret-von Rochow 1969:205). In addition, the opium poppy (*P. somniferum* and/or *P. setigerum*) appears to have been an important, possibly domesticated, plant in the Cortaillod culture:

> Poppy was also a popular plant, but its exact status does not appear to have been thoroughly investigated. These seeds must have been of great significance to the Cortaillod people, judging by their frequency at some sites, as might be expected from parallels of certain modern primitive tribes. (Murray 1970:54)

Among the prehistoric sites classified as belonging to the Cortaillod culture group, there are a number which have yielded evidence of opium use (see figure 34). These include the lake shore settlements at Moosseedorf near Münchenbuchsee in Canton Bern (Hartwich 1899), St. Blaise on the northern shore of Lake Neuchatel (Guyan 1955), Egozwil on Lake Wauwiler (Neuweiler 1924), Thun on Lake Thuner (Rytz 1930), and at least three sites on Lake Burgaschi (Neuweiler 1905; Villaret-von Rochow 1967; Murray 1970). According to Murray (1970:51), the Cortaillod culture developed between 3200 and 2600 B.C. (uncorrected).

Evidence from Lake Burgaschi

Evidence of opium poppy use at a Neolithic Cortaillod lake shore settlement along Lake Burgaschi was first reported from a site near Solothurn by Neuweiler in 1905. More recently, excavations at Seeberg Station on the southern shores of this Swiss lake have uncovered evidence of the prehistoric association of man and the opium poppy. Villaret-von Rochow (1967 : 22) has described and analyzed a large quantity of approximately 3,000 opium poppy seeds extracted in "veritable masses" from the site of Seeberg. She noted that the small seeds of this species were found in every sediment sample examined from this site and that almost all of the seeds were uncharred: "The material . . . examined consisted of a great number of loose plant remains which had been found during the digging, and which were spread out across many sectors of the digging area and mostly come from the culture areas" (Villaret-von Rochow 1967 : 23; trans. Inge Fischer 1976).

Villaret-von Rochow investigated the frequency distribution of the seed size of her sample from Seeberg and discussed her analysis of variance. She came to the conclusion that the prehistoric opium poppy seeds from south Burgaschisee show a closer relationship to *Papaver somniferum* L. than to *P. setigerum* DC. (see chapter 2); Villaret-von Rochow (1967:28; 1969:205) also considers the opium poppy to have been a "crop" of the Swiss Neolithic peoples.

In a recent personal communication with Dr. Schweingruber of the Swiss Federal Institute of Forestry Research in Zurich (December 12, 1977), I learned that fossil seeds of the opium poppy, "*Papaver somniferum*," were found in the excavations of the Cortaillod cultural settlement at Seeufersiedlung Twann on the shores of Lake Bieler in 1975. Dr. Schweingruber graciously sent me a sample of the seeds (see figure 37).

Evidence from the Chassey-Chalain Culture Group

The Chassey-Chalain culture has, along with the Cortaillod group, been classified as a "Middle Neolithic" culture group (Sauter 1976:44). It is known from sites found in most of the fertile regions of France (Murray 1970:57). These sites have been dated to between 4800 B.C. and 3200 B.C. (Whitehouse 1975 : 142).

Mesolithic influences have been suggested for both the Chassey-Chalain and Cortaillod culture groups; in addition, amber has been found in some sites of the former group, indicating early trade (Murray 1970:59). The major crops of the Chassey-Chalain culture appear to have been einkorn wheat, emmer, bread wheat, spelt, and barley: "Broomcorn, flax, poppy seeds and legumes are also recorded at one or two sites, but there is no indication that they were represented in large amounts" (Murray 1970:59). Poppy seeds were found along with flax seeds, einkorn, emmer, and barley at the Chalain lakeside Neolithic site (Lac de Chalain) in the Jura Mountains of northeast France (Neuweiler 1935 : 121). The waterlogged conditions and proximity to the Cortaillod settlements may help explain the existence and preservation of the fossils of this species in this location. In fact, Sauter (1976:50) groups the Chalain and Cortaillod settlements together as sites of a "kindred culture."

The Neolithic Pfyn Culture

Although the pottery styles of the Pfyn group differ from those of the Cortaillod group, there are a number of striking similarities between them as well. Both groups placed their settlements in similar environmental locations near the shores of lakes or bogs; both groups fabricated their tools, weapons, and ornaments out of flint, bone, and wood in distinctly similar fashion (Waterbolk and van Zeist 1966: 579). The Pfyn sites are considered "to be contemporary with the late Cortaillod phase" (Murray 1970: 51).

Fossil opium poppy seeds have been found in several Pfyn culture sites of northeast Switzerland, including those at Niederwil (Hartwich 1899; Waterbolk and van Zeist 1966; van Zeist and Casparie 1974), Pfyn-Breitenloo (Neuweiler 1946: 129), Obermeilen (Neuweiler 1946: 124), Steckborn (Hartwich 1899: 85–87), and Weier (Fredskild 1978: 186). In addition, fossil seeds of this species have been recovered from the Pfyn group settlement at Sipplingen in southwest Germany (Bertsch and Bertsch 1949: 198). The cultural context of the opium poppy in two of these Pfyn group sites, Niederwil and Weier, is discussed in the following sections. A close relationship between the human settlement phase and the presence of the opium poppy seems apparent.

Evidence from Niederwil: A Case Study

Nineteenth-century peat cuttings first brought to light the Neolithic Swiss bog settlement of Niederwil near West Germany in northern Switzerland (see figure 34). In 1862, Messikommer excavated a number of superimposed timber layers upon

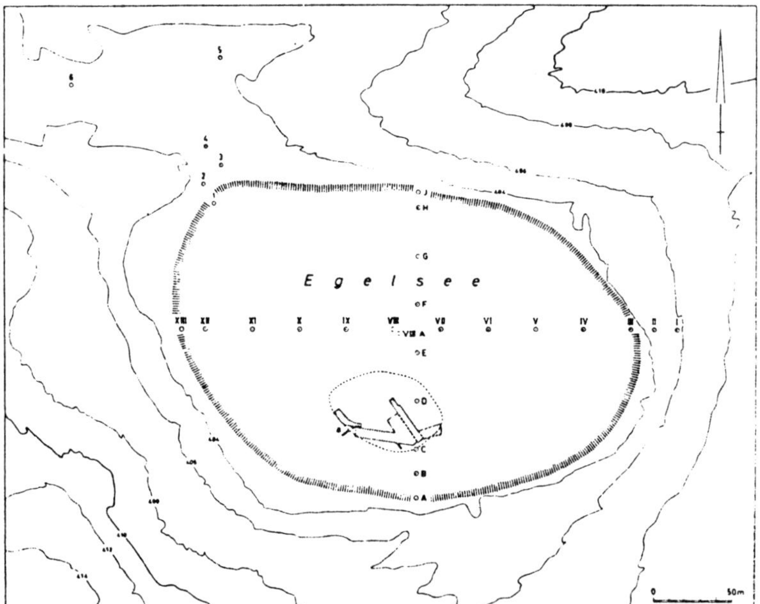

FIGURE 43. The location of the Niederwil bogside settlement, now located within the Egelsee Basin of the Swiss Foreland (from van Zeist and Casparie 1974:416. *Courtesy of the Royal Geological and Mining Society of the Netherlands, Amsterdam, the Netherlands*).

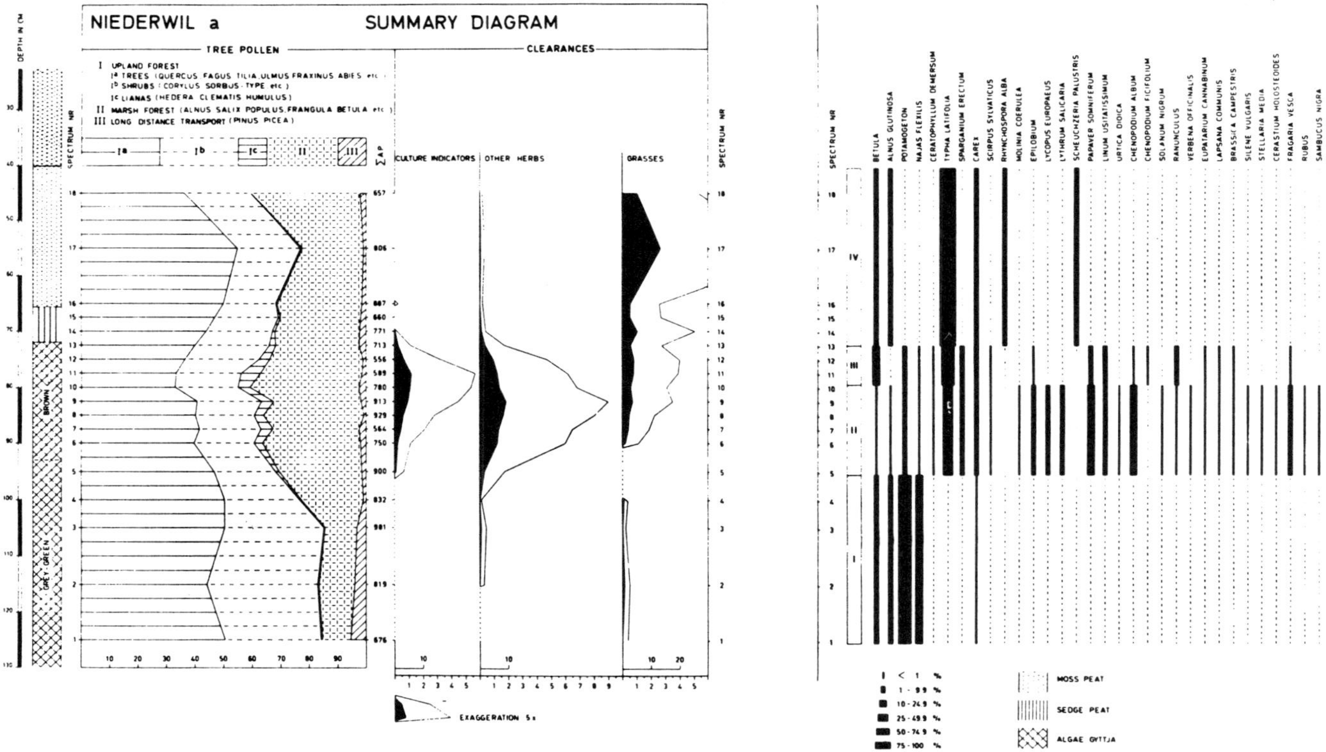

FIGURE 44. Seed diagram from Niederwil bogside site (from van Zeist and Casparie 1974:417, figure 5; *Courtesy of the Royal Geological and Mining Society of the Netherlands, Amsterdam, the Netherlands*).

which prehistoric buildings were constructed near the margins of the bog. Following Messikommer's early archaeological work at Niederwil, intermittent exploitation of the peat formation continued for several decades. More recently, in 1962–63, prior to the initial use of the bog as a sedimentation basin for anew sugarbeet factory, the Thurgau Museum sponsored a substantial excavation in order to recover information and material from the prehistoric settlement (Waterbolk and van Zeist 1966:559).

The bog at Niederwil covers about 60,000 square meters and is located within an oval depression on a moraine ridge between the valleys of the Thur and Tegelbach rivers. The prehistoric settlement area itself was about 2,800 square meters in size and rested on a higher, oval region approximately thirty meters from the bog's edge (see figure 43). By the time the 1962–63 excavations took place, most of the top lying peat layer had been artificially removed. This resulted in the exposure of various mud and clay deposits as well as some extraordinary prehistoric cultural remains:

> After the settlement was abandoned it was overgrown by peat. These circumstances were favourable for the preservation of organic remains. In addition to the timber used for the construction of the houses, animal bones, antler and bone tools, fruits and seeds, textiles, and wooden objects were preserved. (van Zeist and Casparie 1974:416).

Although few or no metal objects were found in the Niederwil bog site, great quantities of plant remains have been uncovered. The micro- and macrofossil remains of several domesticated species found in this site have been identified; the most important of these are bread wheat *(Triticum aestivum)*, naked barley) *Hordeum vulgare*

FIGURE 45. Prehistoric remains of houses at Niederwil, Switzerland. House floor on the left of the passage is made of planks; house floor on the right is made of closely laid round timbers (from Waterbolk and van Zeist 1978, figure 54; *Courtesy of Professor W. van Zeist, Biologisch-Archaeologisch Instituut, Rijksuniversiteit, Gronigen, the Netherlands*).

FIGURE 46. House floor made of planks with rectangular wooden substructure of hearth at Niederwil, Switzerland. In the background a passage and wall posts can be seen (from Waterbolk and van Zeist 1978, figure 59; *Courtesy of Professor W. van Zeist, Biologisch-Archaeologisch Instituut, Rijksuniversiteit, Gronigen, the Netherlands*).

FIGURE 47. House floor made of planks and lower part of house wall of split planks at Niederwil, Switzerland (from Waterbolk and van Zeist 1978, figure 90; *Courtesy of Professor W. van Zeist, Biologisch-Archaeologisch Instituut, Rijksuniversiteit, Groningen, the Netherlands*).

var. *nudum*), flax)(*Linum usitatissimum*), and the opium poppy (*Papaver somniferum*) (van Zeist and Casparie 1974:415). In addition, many wild, edible species including hazel nuts (*Corylus avellana*), wild apple (*Malus sylvestria*), strawberry (*Fragaria vesca*), blackberry (*Rubus fruticosus*), and raspberry (*Rubus idaeus*) as well as several ruderal plants have been identified (Waterbolk and van Zeist 1966:576).

The presence of one "weed" species among the archaeobotanical remains recovered from the Niederwil site may be of special interest to this research project; this plant is "fat hen" (*Chenopodium album*). Its tiny, nutritious seeds (see figure 49) have been found in at least twelve of the prehistoric lake and bog shore sites of west central Europe (Bertsch and Bertsch 1949:190). At Niederwil, the presence of fossil fat hen seeds in the cultural layers of the soil samples is quite common; moreover in one archaeological context seeds of this species were found together with fragments of a ceramic container in which they apparently were held in store (Waterbolk and van Zeist 1966:576). Neuweiler (1905) pointed out that before the turn of this century fat hen seeds were eaten during times of famine. This adds a special significance to the finds at Niederwil:

> The fact that the seeds of this weed could be collected in large quantities suggests that the farmers exercised the practice of fallowing. On fallow fields this weed could expand so considerably that the seeds could be harvested on a comparatively large scale. (Waterbolk and van Zeist 1966:576)

Perhaps the opium poppy was also an abundant volunteer in the fallow fields (or cultivated fields?) of the Neolithic lake shore dwellers of west central Europe. Today, as Godwin (personal communication 1976) points out, the opium poppy frequently appears as a weed in English-grain fields because its small seed is often mixed in with the preferred grain seed. The fat hen and the opium poppy would have been among

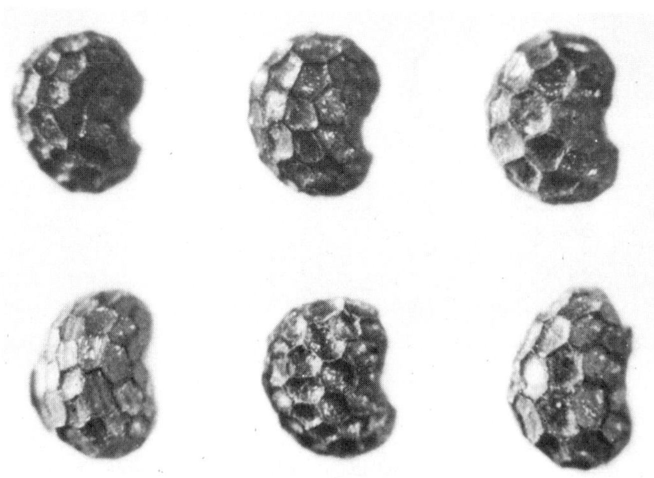

FIGURE 48. Opium poppy seed fossils from the prehistoric Niederwil settlement in Switzerland, magnification ca. 35x (*Courtesy of Professor W. van Zeist, Biologisch-Archaeologisch Instituut, Rijksuniversiteit, Groningen, the Netherlands*).

FIGURE 49. Fat hen, *Chenopodium album* L., is an annual herb. It is often found growing as a weed in cultivated fields, gardens, rubbish heaps, fallow plots, and compost piles; therefore, it shares a common ecological niche with the opium poppy. In fact, fossil seeds of both the fat hen and the opium poppy have been found in several of the shore and bogside settlements of Neolithic and Early Bronze Age west central Europe. Both species could have provided edible seeds (from Slavik 1974:156).

the pioneer (or volunteer) species that successfully invaded those environments near the lake and bog shores opened up by timber exploitation and used for shifting cultivation and grazing. Indeed, many of the wild edible species reported from the prehistoric lake shore farm sites of west central Europe such as raspberries, blackberries, hazel nuts, fat hen, and the opium poppy are successional plants adapted to disturbed environments.

At the Niederwil bogside settlement, the evidence of human activity in and around the vicinity of the site is concentrated in the sediment samples at depths between 95 and 72 centimeters below the surface:

> The pollen of crop plants, weeds from fields and ruderal places, and of other upland herbs show high percentages in this section. As for the macrofossil remains, *Papaver*, *Linum*, *Chenopodium album*, *Lapsana communis*, *Stellaria media* and other

> species are only represented in the section between 95 and 72 cm. (van Zeist and Casparie 1974:417)

Whether or not the opium poppy fossils found in the lake shore or bogside settlements represent specimens of a domesticated species or a weedy plant has not been fully resolved (Hans Helbaek, personal communication 1976). However, it has been assumed that this multi-purpose plant was cultivated at Niederwil and used for medicine and vegetal fat:

> Another crop plant of the Niederwil farmers was *Papaver somniferum* (opium poppy). It is likely that the cultivators of opium poppy were acquainted with the narcotic properties of the milky juice of this species. Whether the narcotic properties were also used to forget the hardships of life from time to time or whether they were only reserved for medicinal purposes, must remain an unanswered question. The relatively high pollen values for *Papaver* suggest that this plant was also grown around the depression. Together with those of *Papaver*, the seeds of flax would have met the demand for vegetal fat by the inhabitants of the site. (van Zeist and Casparie 1974:426)

Evidence from Weier: Another Case Study

The Weier Valley in northeastern Switzerland is relatively small, approximately one mile in length. It is oblong in shape and encircled by small hills. The bog that now occupies part of the valley bottom is a successional landform that eventually replaced the small (ca. 50,000 square meters), shallow lake which was contained in the same place during most of the Atlantic and Sub-Boreal periods. Radiocarbon dating indicates that the human settlement on the moist exposed margins of the ancient lake shore occurred about 2800 B.C. (uncorrected; Fredskild 1978:197, citing his personal communication with J. Troels-Smith). In a recent personal communication (April 19, 1978), Bent Fredskild informed me that the radiocarbon date referred to above is an average of some five to seven datings; ". . . in calendar years this approximates 3600 B.C."

Prehistoric occupancy of the lake shore site probably took place during part of a relatively brief dry spell that resulted in a lowered lake level and consequent exposure of the damp mudflats. A return of more humid climatic conditions caused the lake level to rise once again, which in turn resulted in the eventual submergence of the prehistoric settlement site. This cyclical sequence of humid-dry-humid conditions and the resultant effect on terrestrial surface waterbodies apparently manifested themselves in several west central European lakes. In many situations, prehistoric communities took advantage of the ephemeral exposure of these open habitats along the lake and bog margins.

However, at least in the case of Weier, the rising lake level and its return to a more normal areal extent may not have forced the farmers from their homes along the open mudflats of the lake shore. The early cultivators at Weier probably had to move on before their wooden houses were submerged. According to Fredskild (1978:200), it was soil exhaustion and the diminishing productivity associated with an unsustainable slash-and-burn agriculture that seems to have precipitated the demise of the Weier settlement. Interpretation of the Weier site has largely been developed by Troels-Smith (1954: 1–64) and more recently by Fredskild (1978: 189–201). The work of the latter has been based primarily on archaeobotanical evidence.

Excluding the charred grains of the cereals from the Weier soil samples that were examined by someone else, Fredskild classified more than 11,000 macrofossil seeds and fruits recovered from the Weier site. In the process, he established a score of species that had previously not been found among the Neolithic flora of Switzerland:

> A transect perpendicular to the shore of the former lake was investigated in 1958. In a sequence of layers of quite natural lake sediments, some layers, rich in charcoal, charred grains, sand and clay were interpreted as having been washed down from a contemporary field on the surrounding slopes, and nine macrofossil samples

Table 12

Successive Layered Sediments Examined by Fredskild (1978) for Archaeobotanical Remains, Excavated from Weier, Switzerland

Frequency of seeds and fruits from the different plant groups in the layers. + denotes frequencies less than 1 percent.

Layer No.	19		18		17		16		
Group of plants	No.	%	No.	%	No.	%	No.	%	Total
1. Hydrophytes	120	10							120
2. Plants from swamp etc.	429	36	17	3	1053	11	66	22	1565
3. Trees and shrubs	327	27	38	6	35	+	14	5	414
4. Plants from forest-floor			2	+	26	+	3	1	31
5. Apophytes	189	16	24	4	3512	38	93	31	3818
6. Weeds	5	+	27	5	84	1	20	7	136
7. Gathered or cultiv. plants	115	10	496	80	4615	49	100	34	5326
8. Varia	19	2	15	2	21	+	2	+	57
Total	1204		619		9346		298		11467

In the above table, the plants have been arranged in the following groups:

1. Plants from the lake proper.
2. Plants from the edge of the lake, mainly from the reed, but also from moist, peaty meadows.
3. Trees and tall shrubs.
4. Herbs from the forest-floor.
5. Apophytes, i.e., species favored by the activity of man; by forest clearance, by trampling down of the natural lake shore vegetation (e.g., *Ranunculus sceleratus*) and by enrichment of the soil by dung (e.g., *Urtica dioeca*).
6. Weeds, almost only annuals or biennials, bound to cultivated soil or ruderals.
7. Plants collected for food supply, or cultivated *(Linum)*.
8. Varia.

Layer No. 16	Few Feral Plants, Settlement Abandoned
Layer No. 17	Pronounced Human Disturbance
Layer No. 18	Forest Cutting and Farming
Layer No. 19	Alder, Birch, Reed Swamp Herbs, Pristine Vegetation

Table 13

The Content of "Seeds etc." in the Nine Samples from Weier (from Fredskild 1978: 198, table 4)

	Layer No.	19	19	18	18	17	17	17b	17a	16	
	Series No.	P20	P18	P20	P18	P18	P20	P20	P20	P20	Total
Hydrophytes	1.										
	Najas flexilis	61	21								82
	Characeae, oospores	2									2
	Potamogeton cf. natana	1	3								4
	Potamogeton obtusifolius	14	6								20
	Potamogeton cf. praelongus	2	2								4
	Potamogeton sp.	5	2								7
	Hippuris vulgaris	1									1
Plants from the swamp etc.	2.										
	Caltha palustris	22									22
	Carex elata	47	4					1			52
	Carex pseudocyperus	46	6	1			1	5	2		61
	Carex paradoxa/diandra	174	16	1		4	1	15	5	1	217
	Cicuta virosa	6	2		1 '			1			10
	Peucedanum palustre	4									4
	Carex vesicaria	1									1
	Menyanthes trifoliata	3	6	1							10
	Lythrum salicaria	2		1				2	1		6
	Epilobium palustre	1									1
	Filipendula ulmaria	2				1	1			1	5
	Prunella vulgaris	1									1
	Carex cf. distans	13								1	14
	Lastraea thelypteris, leaf fragm.				90						90
	Stellaria aquatica	10		1	5	349	1	1	56	19	442
	Luzula campestris, multiflora					1					1
	Luzula sp.					2					2
	Scutellaria galericulata						18	1		21	40
	Mentha aquatica/arvensis			1				5			6
	Scirpus silvaticus	1		1	2			10	322	6	342
	Lycopus europaeus	15	16		1	1		17	182	16	248
	Solanum dulcamara	6	13			2		5	30		56
	Carex flacca	12			1				8		21
	Potentilla erecta								1		1
	Stachys palustris								1		1
	Comarum palustre									1	1
Trees and tall shrubs	3.										
	Frangula alnus	9									9
	Prunus spinosa	1	2								3
	Fraxinus excelsior	1									1
	Alnus glutinosa	79	21			1		2	5	1	109
	Betula sp., fruits	172	35	1	36	14	3	6	2	13	282
	Betula sp., catkin scales	49	1		8					3	61
	Cornus sanguinea	3	2						1		6
	Corylus avellana		2			1					3
	Viburnum opulus				1						1
Herbs of the forest floor	4.										
	Mochringia trinervia			2		8	3	1		1	15
	Laserpitium latifolium					3					3
	Agrimonia eupatoria							1			1
	Voila riviniana silvestris							1	4		5
	Ajuga reptans								5		5
	Carex silvatica									1	1
	Polygonum dumetorum									1	1

Table 13 (continued)

The Content of "Seeds etc." in the Nine Samples from Weier (from Fredskild 1978: 198, table 4

Species of cleared spaces	5.										
	Cirsium vulgare	6									6
	Arctium sp.		1								1
	Campanula rapunculoides			2	1	1				1	5
	Silene alba				1	2					3
	Lapsana communis				6	1			5	5	17
	Carduus sp.					1	1				2
	Ranunculus repens					2	3	2	9	1	17
	Galeopsis subgen. Galeopsis	1		3		1		1	11	7	24
	Hypericum perforatum					1		17	14	3	35
	Ranunculus sceleratus	171	3	4	2	19	5	900	1494	52	2650
	Urtica dioeca	4	2	1	4	11	6	25	977	21	1051
	Carex Sectio Muhlenbergiana		1						2	2	5
	Polygonum minus								1		1
	cf. Carum carvi									1	1
Weeds, Ruderals	6.										
	Brassica sp.			2	8	7	5		1	2	25
	Polygonum aviculare			1	3	6	2	2	4	3	21
	Polygonum persicaria							18			18
	Chenopodium album				2	6	1		3	3	15
	Chenopodium sp.	1		2	1				3	4	11
	Chenopodium polyspermum				3			1	3	4	11
	Sonchus asper			2		3		1	3	1	10
	Polygonum convolvulus				1		5			2	8
	Anagallis arvensis	3									3
	Aethusa cynapium							1	1	1	3
	Solanum nigrum							2			2
	Valerianella dentata					1					1
	Vicia hirsuta						1				1
	Chenopodium glaucum rubrum	1									1
	Lamium cf. hybridum					1					1
	Vicia sativa						1				1
	Arenaria serpyllifolia					1					1
	Verbena officinalis								1		1
	Cirsium arvense				2						2
Gathered or Cultivated Plants	7.										
	Fragaria sp.	10	5		9	2187	490	454	241	59	3455
	Papaver sp.	1	1	137	161	530	30	17	8	19	904
	Rubus caesius/fruticosus				141	6		11	27	2	187
	Rubus idaeus				3	349	4	55	54	13	478
	Rubus caesius/fruticosus/idaeus	26	16								42
	Linum usitatissimum	9	34	7	26	28	38	29		7	178
	Linum usit., capsule fragments					1	1			1	3
	Sambucus sp.	7	3	1		3		14	15		43
	Pyrus malus	3		1	10	17	8				39
Varia	8.										
	Cenococcum geophilum, scleroties	4			4	16	3	4	33	5	69
	Carex sp.	4	12			2	1	4	3	1	27
	Gramincae			1	14		4				19
	Potentilla sp.	2							3		5
	Stachys sp.		1					2			3
	Labiatae sp.								1		1
	Galium sp.					1					1
	Viola sp.									1	1
	Total seeds (excl. scales etc.)	966	238	174	445	3574	633	1630	3509	298	11467

> of unknown but different volume were handed over to me for a closer examination. Four subsequent layers: Nos. 16–19 (No. 19 being the deepest one) were sampled in two series (P 18 and P 20) with only half a metre's distance in the transect. In one of the series three samples originate from layer No. 17, viz. one sample covering the whole layer, and two, Nos. 17a and 17b, covering the upper and lower part of the layer. (Fredskild 1978: 189; see table 12)

Fredskild's ability to analyze and explain the fossil samples has been obstructed to a significant degree by methodological limitations beyond his control:

> Unfortunately most of the macrofossil samples at hand cover the entire layers. Without drawings of the transects, descriptions of the sediments, knowledge of the precise position of the samples and of the remnants of the houses, it is not possible to go into a detailed discussion. However, from the determinations the main features may be deduced in the development of the local vegetation during the one or two centuries which saw three periods of habitation around 4,800 radiocarbon years ago. (J. Troels-Smith, personal communication; Fredskild 1978: 197)

According to Fredskild's reconstruction (1978: 197), the pristine vegetation surrounding Lake Weier (layer 19 of his sample) was "an alder-birch swamp mixed with a species-rich reed swamp." He argues that the ". . . very few real culture indicators *(Chenopodium, Anagallis arvensis, Linum, Papaver)* most likely originate from the very top of the sample at the transition to the covering layers, or they have been trampled down in the soft sediment during the habitation period" (Fredskild 1978: 197). It should be noted here that only two poppy seeds were found in the sample from layer 19, which yielded a total of 1,204 fossil seeds for all the species combined (see table 12).

In strong contrast, the subsequent layer (no. 18) in Fredskild's samples produced little or no evidence of any original swamp vegetation, but did yield a relatively high percentage of potentially valuable food plants—especially for the poppy, "*Papaver sp.*" (i.e., 298 poppy seeds out of a total of 619 fossil seeds, or 48 percent of the samples for layer 18). The abrupt change in vegetation at the Weier settlement site is probably a reflection of forest clearance and farming activities in or near the prehistoric habitation area. According to Fredskild (1978: 197), the inhabitants of the Weier settlement practiced slash-and-burn agriculture: "*Triticum aestivum*, *T. monococcum*, and *Hordeum vulgare* (hulled as well as naked barley) were the main crops (Jørgensen 1975), but also *Linum usitatissimum* seems to have been cultivated."

In layer 17, the fossil flora largely consists of species associated with pronounced human disturbance. However a comparison of the lower (i.e., older, No. 17) and upper (i.e., younger, No.17a and No. 17b) sediments within this layer reveals a significant drop in the percentage of species such as the poppy, *Papaver sp.*, and strawberry, *Fragaria sp.*, which are dependent upon human activity for their survival, either as cultivated or ruderal plants. Although there are a number of other possible explanations, according to Fredskild (1978: 197–200), this evidence indicates that resource depletion may have played a major role in the culture history of the Weier settlement.

The uppermost layer of Fredskild's samples (No. 16) contained even fewer species intimately connected with human activity and in fact shows a return of more plants native to the local swamp vegetation. However, there is no evidence from this layer of an increasing water level. As suggested previously, this apparent reversion toward a more native, truly wild flora was not, at least initially, influenced by a rising lake level which eventually did occur to such an extent that the pile dwellings were ultimately submerged and covered over by heavy deposits of lake sediments. Apparently rejecting the possible influence of such conditions as war, pestilence, and anarchy, Fredskild (1978:200) suggested that the settlement at Weier was abandoned because of poor land use management:

> One of the problems which contemporary agriculture has to struggle with is that of soil erosion, i.e., the washing off of the upper soil of the fields when the natural vegetation cover, usually a forest, is replaced by crops. The present investigation has shown that on a smaller scale the phenomenon, caused by a primitive stone age agriculture occurred already some 5,000 years ago.

Fredskild (1978:196) examined the size of 114 fossil *Papaver* (poppy) seeds found in the Pfyn culture layers (No. 17) at Weier, Switzerland. He compared the specimens with dry reference materials of *P. dubium* (three collections from Denmark), *P. rhoeas* (two collections from Denmark), *P. hybridum* (two collections from Spain and England), *P. argemone* (two collections from Denmark), and *P. somniferum*, the opium poppy (three collections from Denmark). Although the fossil seeds are intermediate in size between *P. somniferum* and the other species, Fredskild argued that the fossils are probably examples of the opium poppy (see figure 50).

Although Fredskild suggested that the *Papaver* fossils from Weier should probably be treated as "small-seeded" *P. somniferum*, he did not consider the possibility that the opium poppy was used by the Neolithic inhabitants of this settlement.

Evidence from Other Pfyn Sites

At the Pfyn-Breitenloo lake shore settlement, Neuweiler (1946:129) found fossil opium poppy seeds among the plant remains of approximately thirty species; the mean length size for this poppy seed sample (0.9 mm) is very close to those established for fossil poppy seed samples recovered from other Middle Neolithic Swiss lake or bog shore sites including Steckborn, Niederwil, Robenhausen, Moosseedorf, and Weier.

Fossil evidence of the opium poppy has also been reported from a Middle Neolithic Pfyn culture site in southwest Germany. In fact, according to Bertsch and Bertsch (1949:195), they found fossil seeds of this species "down to the lowest layer of the finds" in the lake shore settlement at Sipplingen along the Bodensee.

Additional Neolithic Evidence from West Central Europe

According to L. Lundstrom Baudais (personal communication, May 5, 1978), fossil

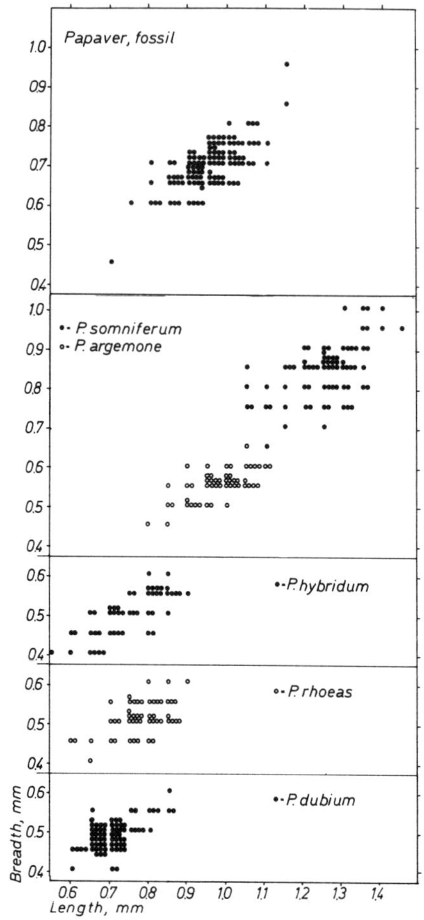

FIGURE 50. Size measurements of fossil and recent *Papaver* (Fredskild 1978: 194, figure 6). "The fossil seeds are evidently swollen and in many cases even broken. The average size of the fossil seeds is 0.93 × 0.77 mm, of *P. rhoeas* is 0.77 × 0.50, of *P. dubium* 0.69 × 0.48, and of *P. hybridum* 0.74 × 0.51 mm. Thus recent seeds of these three species should swell 22–36% in length and 51–62% in breadth to obtain the size of the fossil seeds from Weier. Villaret-von Rochow (1967) showed by experiment that a short treatment of recent seeds with a thin potassium hydroxide dilution caused an increase of approximately 10% (0.1–0.2 mm) in length of *P. somniferum* seeds. If an increase by ¼–⅓ in length in unrealistic, the seeds must originate from a small-seeded *P. somniferum*, as concluded for the material from Burgaschisee-Sud. The length of this material (1967, Abb. 11) is almost identical to that from Weier" (Fredskild 1978: 196; *Courtesy of Professor Bent Fredskild, Botanical Museum, University of Copenhagen, Copenhagen, Denmark*).

seeds of the opium poppy *(P. somniferum)* have been recently reported from two additional Neolithic sites near Lake Neuchatel; these are Auvernier on Lake Neuenburger (also see Villaret-von Rochow 1971: 206–8) and Yverdon: Avenue des Sports.

In a number of cases the archaeological reports failed to indicate the exact context in which the fossil evidence was found. Thus in these situations we cannot be sure if the fossils represent cultivated plants or weedy species. Some examples are presented below.

Neuweiler (1905, 1925; compare with Bertsch and Bertsch 1949:198) reported the discovery of opium poppy seed fossils among the Neolithic plant remains found in the "Lake dweller" site on Lake Baldegger and at Oberkirch on Lake Sempacher in Canton Luzern, at Storen and Furren on Lake Greifen, and later (Neuweiler 1946:129, 134) at Uerikon on the eastern shore of Lake Zurich.

Bertsch and Bertsch (1949:198) found opium poppy seed fossils at "Reute bei Waldsee" and "Riedschachensiedlung im Federsee," two Neolithic sites in southwest Germany. Bertsch (1950:95) also reported the unusual recovery of a Neolithic poppy capsule from Wasserburg Buchau on the Federsee in southwest Germany:

> I found a single capsule of poppy *(Papaver somniferum)* but considering the small size of the capsule and the dark, even color of seed and peat, the discovery of one of these small capsules is almost a matter of accident, even if one searches a large area of peat with a magnifier. According to its size and the network of its surface we are dealing with the Lake-dwellers' poppy. (Bertsch 1950:95, trans. Inge Fischer 1977)

Among the "Single Grave Complex" group of northern Switzerland, Murray (1970:380) refers to two sites as having yielded fossils of the opium poppy; these are the lake shore dwellings at Robenhausen and Meyer. As noted previously, Oswald Heer (1866:32f; 1878:523, 531) recovered a poppy seed "cake" and one unripe capsule from the Robenhausen site on Lake Pfaffiker (see figure 13). Heer treated the poppy specimens as *Papaver somniferum* L. var. *antiquum* (see chapter 3).

The Neolithic Michelsburg culture (2800–2500 B.C. uncorrected) was located in bogside settlements in northeast Switzerland and Württemberg and "hillcamps in south-west Germany" (Murray 1970:54). Until recently this culture was lumped together with the Pfyn culture group and considered to be very closely related to the Cortaillod culture. It is now regarded as somewhat younger than the Cortaillod and Pfyn groups from which it has been separated (Sauter 1976:55). Fossil poppy seeds were recovered from the Michelsburg settlements of Obermeilen on the eastern shore of Lake Zurich (Neuweiler 1946:129; Murray 1970:296) and Thayngen-Weiher (Guyan 1955; cited in Murray 1970:56, 296).

According to Sauter (1976:56), significant changes in Neolithic Switzerland occurred approximately 2500 B.C. when new groups moved into and settled among the established farming cultures of the region. This resulted in the gradual absorption of the Cortaillod culture into a new tradition, known today as the Horgen culture (ca. 2700–2300 B.C. uncorrected, Murray 1970:54). The Late Neolithic Horgen culture is not as luxuriant and diverse as the Cortaillod or Pfyn cultures which it superseded; indeed there is a clear degeneration in the pottery industry and architectural design (Sauter 1976:57; Murray 1970:54).

Although not much is known about the species cultivated by the Horgen culture group, fossil seeds of the opium poppy have been recovered from the following sites: Horgen (Neuweiler 1925; compare Bertsch and Bertsch 1949:198; Murray

1970:296), Mannedorf (Neuweiler 1925; compare with Bertsch and Bertsch 1949:198), and Utoquai on Lake Zurich (Neuweiler 1930:38).

In northern Italy, the late Neolithic Lagozza culture (ca. 2900–2700 B.C. uncorrected, Murray 1970:56) has also produced a few shreds of evidence for the prehistoric opium poppy. The settlements of the Lagozza culture were located on the flanks of the Alps above the central plain of the Po Valley which had been a relatively unattractive region for human habitation until the "extensive tracts of noxious marshes" were drained and flood problems were brought under control in the last century (Barfield 1971:31). At Lagozza, the type site of this culture, Sordelli (1880, cited in Neuweiler 1905:33; also see Bertsch and Bertsch 1949:199; Buschan 1895:246; Netolitsky 1931:60) found fossil opium poppy seeds among the plant remains of the lakeshore settlement. "*Papaver*, poppy, was cultivated; but since only seeds and no capsules were found, we cannot determine whether they come from *P. setigerum* or *somniferum;* probably the former" (Hoops 1905:291, trans. Inge Fischer 1976). The Lagozza culture was linked with the Swiss Alpine Foreland by way of the Saint Bernard Pass (Sauter 1976:53). According to Barfield (1971:16) there is evidence indicating that this pass was "already in use by the end of the Neolithic period." Perhaps the opium poppy was introduced into northern Italy from the Swiss Foreland during this period.

Table 14

The Archaeological Record of Neolithic Opium Poppy Fossils (from Schultze-Motel 1979:208–9)

Country and Site	Prehistoric Period	Author(s)
Spain		
Cueva de los Murciélagos near Albuñol, Prov. Granada	Late Neolithic	Neuweiler 1935
Italy		
Lagozza near Milano	Neolithic "pile dwellers"	Bertsch and Bertsch 1949
France		
Lac de Chalain	Neolithic "pile dwellers"	Neuweiler 1935
Switzerland		
Storen and Furren, Lake Greifen	Neolithic "pile dwellers"	Neuweiler 1925
Brise-Lames, Auvernier	Late Neolithic, Lake shore dwellers	Baudais-Lundstrom 1978
Moosseedorf, Canton Bern	Neolithic "pile dwellers"	Hartwich 1899
Thun	Neolithic "pile dwellers"	Bertsch and Bertsch 1949
Robenhausen by Lake Pfäffiker near Zürich	Neolithic "pile dwellers"	Heer 1865, Hartwich 1899
Utoquai Zürich	Late Neolithic "pile dwellers"	Neuweiler 1930
Horgen and Männedorf, Lake Zurich	Neolithic "pile dwellers"	Neuweiler 1925
Burgäschi near Solothurn	Middle Neolithic "pile dwellers"	Neuweiler 1905

Table 14 (continued)

The Archaeological Record of Neolithic Opium Poppy Fossils (from Schultze-Motel 1979:208–9)

Country and Site	Prehistoric Period	Author(s)
Seeberg, South Lake Burgäschi	Neolithic, young Cortaillod-Period, first half of the third millenium B.C.	Villaret-von Rochow 1967
Niederwil, Canton Thurgau	Neolithic "pile dwellers"	Hartwich 1899, Waterbolk and van Zeist 1967, van Zeist and Casparie 1974
Pfyn	Neolithic "pile dwellers"	Neuweiler 1946
Uerikon	Neolithic "pile dwellers"	Neuweiler 1946
Obermeilen	Neolithic to Early Bronze Age "pile dwellers"	Neuweiler 1946
Egolzwil and Schotz, Lake Wauwiler, Canton, Luzern	Neolithic "pile dwellers"	Neuweiler 1924
Oberkirch by Lake Sempacher	Neolithic "pile dwellers"	Neuweiler 1905
Lake Baldegger	Neolithic "pile dwellers"	Neuweiler 1905
Weier, Canton Schaffhausen	Neolithic "pile dwellers"	Fredskild 1978
Steckborn by Lake Unter/Lake Boden	Neolithic "pile dwellers"	Hartwich 1899
West Germany		
Sipplingen, Lake Boden	Late Neolithic "pile dwellers"	Bertsch 1932
Riedschachen settlement in Württemberg	Neolithic	Bertsch 1950
Reute near Waldsee, Württemberg	Neolithic "pile dwellers"	Bertsch and Bertsch 1949
Rhineland:		
Lamersdorf, Kr. Düren	Linear Pottery ("Bandkeramik")	Knörzer 1967
Oekoven, Kr. Grevenbroich	"Bandkeramik"	Knörzer 1971
Aldenhoven, Kr. Jülich	"Bandkeramik"	Knörzer 1971
Langweiler, Kr. Jülich (several findings)	"Bandkeramik"	Knörzer 1971, 1972, 1973, 1977
Garsdorf, Kr. Bergheim	"Bandkeramik"	Knörzer 1971, 1974
Poland		
Zeslawice near Kraków	Neolithic ("radial decorated pottery")	Giżbert 1960, Klichowska 1976
Czechoslovakia		
Vavrovice near the Opava River	"Epiatlantikum" (after 3,000 B.C.?)	Sikulová and Opravil 1974

Evidence from the Copper and Bronze Age Sites of West Central Europe

Separation of the Late Neolithic from the Early Bronze Age has often been more or less arbitrary. "In Switzerland, as elsewhere in Europe, the first metal appeared among groups which belong unquestionably to the Neolithic by their cultural context" (Sauter 1976:79). In the case of the Alpine Foreland, the earliest cultural association of metal objects with Neolithic farmers is probably a reflection of immigration or trade contact with external cultures because natural deposits of copper in Switzerland are quite rare. In any event, evidence of metallurgy first appears in west central Europe during the latter part of the third millennium B.C., initially in the form of copper implements (pure copper or copper alloyed with arsenic) and later in the harder metallic form of bronze (alloy of copper and tin) (Whitehouse and Whitehouse 1975:142).

Murray (1970:table 189) refers to the early age of metals in the region as the Swiss Copper Age and notes the recovery of opium poppy fossils from three Copper Age sites (Niederwil, St. Blaise, and Uerikon) where it is "known that emmer and club wheat, six-rowed barley, two varieties of millet, flax and possibly poppy were cultivated" (Murray 1970:94–95).

In southern Germany, the discovery of fossil opium poppy seeds has been reported from three Bronze Age sites including Wasserburg Buchau on the Federsee (Bertsch 1950), Mordorf Dullenried on the Federsee (Bertsch 1950), and the lake shore settlement of Langenrain on the Bodensee (Bertsch 1932).

At the Early Bronze Age Swiss Lake shore settlement of Arbon on the south side of Lake Constance, Neuweiler (1946:134) found opium poppy seeds along with fossils of twenty other species.

Neuweiler also reported fossil opium poppy seeds from other Bronze Age lake shore settlements in the Swiss Foreland including those at Alpenquai (Neuweiler 1919:632) and Wollishofen (Neuweiler 1919:645) on the shores of Lake Zurich, "Sumpf" (Late Bronze Age) near Zug (Neuweiler 1931:119), Saint Blaise on the side of Lake Neuenburger (Neuweiler, cited in Bertsch and Bertsch 1949:199), and Obermeilen on Lake Zurich (Neuweiler 1946:129).

In the latter part of the nineteenth century Uhlmann (1876:62–65) reported the discovery of opium poppy seed fossils among the plant remains found in the Bronze Age Swiss lake shore site at Morigen on Lake Bieler, and Buschan (1895:246) noted the discovery of opium poppy seed fossils at the Bronze Age lake shore settlement on the side of Lake Bourget near the Rhone River in the Savoy region of eastern France.

In 1956 a large amount of fossil plant remains was recovered from the Bronze Age lake shore settlement at Valeggio on the Mincio River in the northern Italian province of Verona. Among the fossils in the upper culture layer, Villaret-von Rochow (1958:106) identified two seeds of *Papaver argemone*, a weedy poppy species of south and middle Europe. This poppy species was also reported from the prehistoric lake shore site at Sipplingen in southwest Germany by Bertsch in 1932. Villaret-von Rochow also identified two additional, somewhat larger, fossil poppy seeds *(Papaver sp.)* from the uppermost part of the cultural layer; she suggested tentatively that these fossil seeds may be treated as belonging to the "lake-dweller poppy" of the "*Papaver somniferum P. setigerum* group." Unfortunately, these latter poppy seeds were subsequently misplaced (Villaret-von Rochow 1958:106).

In northern Austria, Werneck (1949:68; compare Neuweiler 1905:34, 116 and Willerding 1970:353) reported opium poppy seed fossils from the Hugelgraber (hill trench/grave) Early Bronze Age site at Mistelbach (Lausitzer Pottery, ca. 2000–1500

B.C. , Jürgen Schultze-Motel, personal communication, August 8, 1978). This site is located less than fifty miles from the Danube River but more than three hundred miles to the east from the nearest site in west central Europe where opium poppy fossils have been found.

In the Hungarian Bronze Age archaeological site at Dunapentele, Kosziderpadlas (b), Tempir (1964:72) reported the discovery of a small fragment of a seed, possibly of a poppy *("Papaver sp.")*. Along with this tiny shred of evidence, Tempir found other "weedy" fossils including five seeds of fat hen *(Chenopodium album)* which we already noted as showing up frequently with opium poppy fossils in other prehistoric European archaeological sites. In the following two sections I discuss the possible direct and indirect prehistoric associations between man and the opium poppy, focusing on their relationships in west central Europe during the Neolithic and Early Bronze Age periods.

The Opium Poppy in Prehistoric West Central Europe: A Cultivated Plant or Weedy Volunteer?

According to Villaret-von Rochow (1969:205), the

> general picture of Swiss Neolithic husbandry has repeatedly shown its mixed character, besides cereals, flax, opium poppy and other crops, considerable numbers of wild fruit have been collected such as hazel nuts *(Corylus avellana)*, raspberries *(Rubus idaeus)*, acorns *(Quercus spp.)* and sloes *(Prunus spinosa)*.

Others writing before Villaret-von Rochow have also referred to the opium poppy of prehistoric west central Europe as a crop or cultivated plant. Hartwich (1899) suggested that its cultivation is made plausible by the finds of charred poppy seed in the prehistoric lake shore settlements. Hoops (1905:297) argued that the prehistoric opium poppy in this region was both cultivated and important:

> The manner of utilization as well as the large number of seeds found, which is unrelated to the frequency of occurrence of the plant in the wild, prove with certainty that we are dealing with an important cultivated plant of the lake-dwellers. (trans. Inge Fischer 1976)

Much (1908:217) emphasized the importance of the relatively large quantity of fossil poppy seeds and the fact that they have been recovered from many individual settlements. He too argued that the opium poppy was cultivated by the prehistoric lake shore and bogside dwellers in this region. This position was once again stated by Schrader (1917–29:68). Andrews (1952:153) later phrased the same argument in the following manner.

> In western Europe cultivation of this species probably began as early as the Neolithic Age, for poppy seeds in large quantities have been found at various Swiss and north Italian pile-dwelling stations, and it is difficult to believe that such masses of seeds could have been secured by the casual collection of wild plants.

More recently, archaeologists working with fossils from the Neolithic Niederwil site in the Swiss Foreland have also suggested that the opium poppy was a cultivated and important species (van Zeist and Casparie 1974:426). Although the opium poppy probably was used, its cultivation has not been confirmed. In other words, the opium poppy may have been such an abundant volunteer in the fallow fields of these early farmers that it was unnecessary to sow the seeds. This could account for the relatively large quantities of fossil opium poppy seeds (as well as those of other potentially useful volunteer species such as the fat hen, *Chenopodium album*) which have been found in some prehistoric archaeological sites of west central Europe.

Critical proof of opium poppy cultivation would be the recovery of substantial numbers of ripe nondehiscent capsules (see Zohary 1969:59–60), this evidence has not yet been found. Moreover, it should also be pointed out that as far back as 1895, the exact status of the opium poppy (either as a cultivated or volunteer species) was already in doubt (Buschan 1895:246). Presently, many uncertainties surrounding the use of the opium poppy in the prehistoric period remain to be answered. The widespread and relatively abundant presence of fossil poppy seeds in archaeological sites of west central Europe is strong circumstantial evidence for its use among the Neolithic and Early Bronze Age cultures of the region. But just what were the uses to which this species was put? There is as yet no consensus.

Along with the extensive fossil plant remains recovered from the Robenhausen site (including a poppy seed "cake" and one unripe capsule), it should also be noted that Heer (1866:7) also reported the discovery of a layer of goat dung found buried beneath a six-foot-thick sediment deposit. Observations of the highly carbonized goat waste led Heer to interpret the mixture of dung and organic litter as a compost layer to be used on the cultivated fields of the ancient farming community:

> To one familiar with farming operations it is apparent that the knowledge of tilling the soil, and enriching it with the proper manures obtained from domestic animals, presupposes a long period of experimentation and an accumulated knowledge which could result only from an intimate acquaintance with crops and a prolonged and patient observation of agricultural cause and effect. (Ames 1937:17)

Perhaps the early farmers of west central Europe were well acquainted with the opium poppy and used it for many purposes including domestic animal fodder, composting material, medicine, ritual intoxication, and food. The following section focuses on perhaps the two most important potential applications of the opium poppy: drugs and food.

Applications of the Opium Poppy in Prehistoric West Central Europe

Drug Use

According to Lips (1947:173), both hemp (i.e., marijuana, *Cannabis sativa* L.) and opium were probably smoked by the "prehistoric Lake Dwellers, as extant implements indicate." This is a curious suggestion because there is no fossil evidence for the hemp plant among the prehistoric plant remains from west central Europe. Furthermore, there are no "implements" which have been identified as pipes or other apparatus from this region that could have been useful in the process of smoking either opium or hemp.

Nevertheless, even though there is no archaeological evidence indicating that a product of the opium poppy was smoked by the lake shore peoples, we need not abandon the idea that the opium poppy was used in some manner for its mind-altering recreational, ritualistic, or, most likely, medicinal value. Buschan (1895:248) suggested that the "lake-dwellers" may have extracted oil from the opium poppy seeds because its "narcotic properties" were considered a delicacy. Ten years later, Hoops (1905:297) reiterated this view, claiming that it was improbable that the "lake-dwellers" were unaware of the "narcotic qualities" of opium poppy seeds. Maurizio (1915:225) also concurred with this interpretation, asserting that in all probability the opium poppy was being cultivated by the "lake-dwellers" because of its slight narcotic action and not, as supposed for oil preparation.

On the other hand, Netolitzky (1931:61) argued that opium poppy seed and oil were not used as narcotics simply because these products do not have such effects. Netolitzky's opinion is supported by recent evidence: "Seeds of *Papaver somniferum* and *setigerum* are usually considered clean of the psychotropic alkaloids, i.e., the alkaloids are in a bound state" (James A. Duke, personal communication, December 2, 1976). Netolitzky did not completely reject the idea that the opium poppy may have had a medicinal role among the historic people of the region:

> Certainly the lakedwellers did not cultivate the poppy to gain opium; at the most they knew of the calming effect of the extract of unripe poppy heads, as does to this day the rural populace of areas where poppy is cultivated in large amounts . . . but it is not possible to conclude that the poppy was cultivated primarily because of its narcotic effect. Up until a few years ago, the entire poppy cultivation in Europe was devoted to the production of nourishing seeds, and the lakedwellers would not have done otherwise. (Netolitzky 1931:61, trans. Inge Fischer 1976)

As I noted previously, a recent interpretation of evidence from the Middle Neolithic Pfyn Culture site at Niederwil by van Zeist and Casparie (1974:426) suggests that these bogside dwellers probably used the opium poppy for medicinal purposes. However, I should also point out that this interpretation was based on circumstantial evidence (the presence of numerous opium poppy seed fossils in the cultural layers of the settlement site) and, therefore, must be regarded as speculative. In view of the lack of direct evidence, we cannot be sure that the drug plant was used for medicinal, recreational, or ritualistic purposes by the early farmers of west central Europe (for an interesting discussion of the problems involved in the evaluation of archaeobotanical evidence see Harlan and de Wet 1973:51–62).

Food Use

It has been widely assumed that the opium poppy provided edible products for the Neolithic and Early Bronze Age peoples of the Swiss Foreland region. However, the exact culinary use of this species by these peoples has been the subject of some controversy. Heer (1866:32; 1878:523), as I pointed out earlier in this study, did report the discovery of a whole opium poppy seed cake and suggested that the seeds were pressed for oil or may have been eaten, scattered over bread. In reference to this "cake" of poppy seeds, Hoops (1905:297) offered the following interpretation:

> A whole cake comprised of thousands of small, charred poppy seeds was found baked into a mass. It appears therefore that the seeds were enjoyed in the form of poppy cakes; however, it is probable that they were primarily used to yield oil. (Trans. Inge Fischer 1976; see also Buschan 1895:248)

Maurizio (1915:225) disagreed with an earlier comment by Braungart (1912:372) which referred to the use of opium poppy seeds by Neolithic peoples as flavoring for bread. Hoops (1905:298), Neuweiler (1924:316), Netolitzky (1912:228), and Andrews (1952:154) also maintained that opium poppy seeds were widely used in this region in prehistoric times, not for flavor, but as an important food resource.

Maurizio also suggested that poppy seed is made unclean by being mixed with earth and various harmless seeds. He pointed out that like other tiny seeds, when stored, the opium poppy "sweats" and thus must be thoroughly aired and moved around periodically. More importantly, Maurizio (1915:225–26) argued that dangerous contamination with another tiny common weed seed, Henbane (*Hyoscamus niger* L.), made the storage and use of opium poppy seed cakes (or oil?) quite precarious.

Netolitzky (1931:60) referred to the Slavs as the primary connoisseurs of various poppy food preparations. This statement was reinforced in my own experience when a fellow graduate student, Kurt Von Neida, gave me a recipe his Czechoslovakian grandmother had given him:

Poppy Seed Paste

(ingredients)

½ lb. poppyseed
½ cup sugar (or to the taste)
2 eggs
⅔ cup milk
¼ lb. butter (melted)
1 tablespoon of vanilla

(preparation)

> Grind poppyseeds and mix all ingredients. Cook over very low heat for 15–20 minutes, stirring constantly to prevent burning. Remove from stove and let cool and thicken. Spread thickened mixture on rolled, leavened, pastry dough, and then roll the dough covered with poppyseed paste into a loaf. Allow the loaf to rise for 30 minutes, then bake at 350 degrees for 40 minutes.

If we eliminate the vanilla and substitute honey for the sugar, we have ingredients

FIGURE 51. Poppy seed pastry.

that would have been available to the prehistoric lake shore and bogside dwellers of west central Europe. Perhaps this recipe is very much like one they knew and used over five thousand years ago!

Maurizio (1915:226) referred to ancient confections eaten in eastern Europe such as *Makagigi*, a very viscous mixture of poppy seed and honey which was still being consumed in parts of Poland and Russia. According to Netolitzky (1931:60), food use of the opium poppy in prehistoric Europe was restricted to the consumption of the nourishing seeds, probably in confections mixed with honey, e.g., *Naunt*, an old eastern European poppy seed food.

In reference to the possible extraction of oil from opium poppy seeds, Netolitzky referred to Walter Rytz who studied some of the charred poppy seed clump fossils and found no evidence of prior compression.

> The uncharred seed hulls have passed through the intestines, and it is improbable that they have previously been freed of oil, even partially, by a method of compression. (Netolitzky 1931:60, trans. Inge Fischer 1976)

Netolitzky argued that the charred poppy fossils found among the "lake-dwellers" are evidence of their storage habit rather than their use for oil. In fact, he did not regard the burnt seed clumps as either poppy cakes or compressed seed residue. Instead, he suggested that they represent raw food stashes accidentally destroyed by fire, but retaining the forms of the containers in which they were kept for later use.

In the Late Bronze Age site at Wasserburg Buchau on the Federsee in Württemberg, Germany, Karl and Franz Bertsch (1949:196) found a large vessel filled up with fossil opium poppy seeds. These seeds could have been stored for future consumption in a raw or baked state. Furthermore, masses of uncharred seed hulls whose surface sculpture remained in good condition were found in the open "culture layer" of a former lake at the Wasserburg Buchau site. These empty opium poppy seed hulls were found along with fossil strawberry, raspberry, and blackberry seeds. The Bertsch brothers concluded that (ripe?) opium poppy capsules were eaten whole like berries and that empty, undigested seed hulls were subsequently excreted along with other waste products into the lake or dump heaps near the settlement. They observed no evidence that the poppy seeds were compressed for oil extraction and therefore argued that the seeds were not used as an oil resource, but may represent the oldest kind of flavoring used in preparing "flat cakes."

Rytz (1949:84) suggested that the opium poppy was a major commodity of the prehistoric lake shore and bogside peoples of the Swiss Foreland. He came to this conclusion because large amounts of charred opium poppy seed were found:

> If I . . . classify them as staple plants, I am led by the consideration that the very state of being charred proves that these plant products could only be charred because they were kept in the lake-dwellings until they then suffered the fate of catastrophe by fire. (Rytz 1949:84, trans. Inge Fischer 1976)

Ascertaining the economic value of a prehistoric plant resource is by no means a simple matter. Many previous research attempts in this direction have been based on the assumption that those species represented by the largest amount of fossil specimens in the archaeobotanical sample naturally were the most important plants in the cultural community under investigation. However, since preservation of plant material is generally enhanced by carbonization, we should consider some pertinent variables here.

For example, some cereals such as einkorn or spelt require parching to release the grains from their seed hulls, and thus are more apt to be charred than other "free-threshing cereals such as bread-wheat" (Dennell 1978:9). On the other hand, opium poppy seeds, like legumes such as peas and lentils, can be consumed raw; therefore, the likelihood of their being carbonized is somewhat remote:

> They have become carbonized due to some accident in prehistory—being scorched in a parching oven, spilled by the hearth or singed in the conflagration of part of a building—and are thus only indicative of their own immediate circumstances, and not even of the entire harvest of that season (Renfrew 1973:21)

As previously noted, the relative abundance of fossil opium poppy evidence from the Swiss Foreland is largely a function of the excellent preservation conditions—in other words, anaerobic lake and bog environments in which charred and uncharred seeds have remained intact for thousands of years. If it is assumed that a majority of the prehistoric opium poppy seed harvests were consumed in a raw state, then we could hardly expect to find much evidence in the fossil record except under special circumstances such as those observed in the lake and bog environments of west and central Europe. In fact, except for the remarkable archaeological finds in the Neolithic and Bronze Age sites located in these environments, the fossil record for the opium poppy is essentially quite limited.

New retrieval techniques and the chance discovery of carbonized seeds and capsules may expand our record for the prehistoric existence and use of the opium poppy. However, it should be remembered that the discovery of fossil remains of useful plants does not necessarily indicate that they were actually being consumed. "Clearly, the presence of a potential plant food—whether defined on morphological grounds as wild or domestic—in an archaeological context need not imply that it was utilized" (Dennell 1976:245). Nevertheless, the cultural context of the fossil evidence

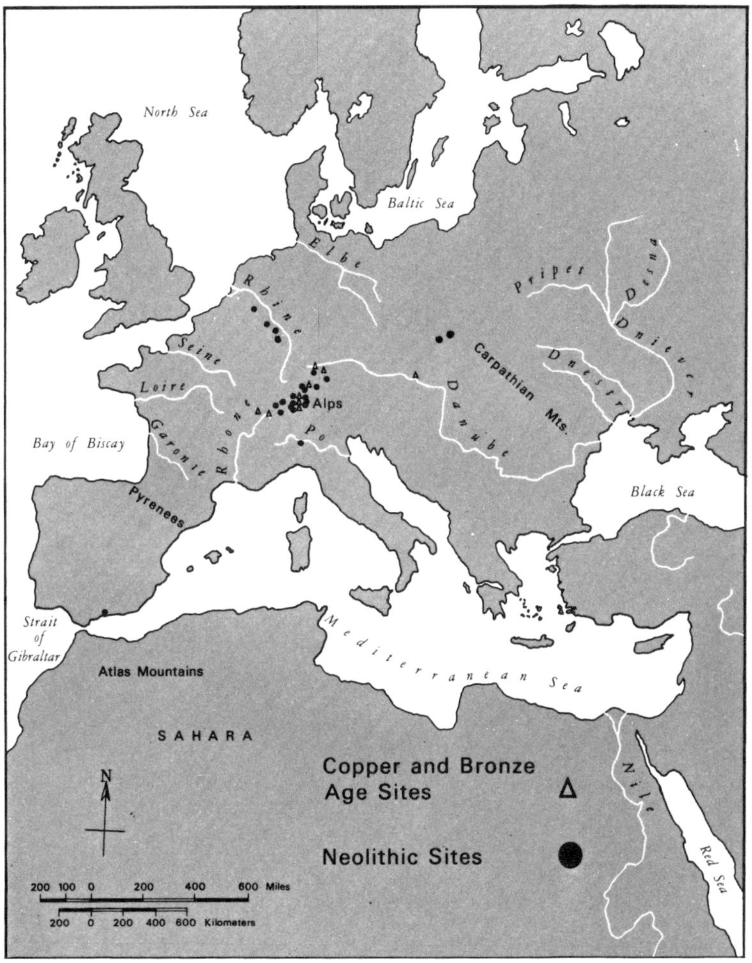

FIGURE 52. Distribution of opium poppy remains in human settlements prior to 2000 B.C. (Archaeological evidence from Neolithic, Copper, and Early Bronze Age sites).

may reveal some aspects of the plant's economic value and support certain assumptions regarding its actual use. For example, the discovery of masses of opium poppy seeds in vessels (storage jars?) "should represent plant food at a more advanced stage of preparation than that in a refuse dump or threshing area" (Dennell 1978:9). Therefore, if fossil evidence for the opium poppy is found in a specific context such as a "storage jar" or among "grave gifts," then the assumption suggesting that this species was utilized is certainly strengthened. Although fossil evidence of the opium poppy is in general very rare, especially in these kinds of obvious cultural associations, two extraordinary archaeological examples are described and evaluated in the following chapter which deals more specifically with the question of the opium poppy's origin in west central Europe.

6 • Possible Places of Cultural Origin for the Opium Poppy

In the previous chapter, I surveyed the evidence of the opium poppy reported from a large number of Middle Neolithic sites located. in west central Europe. In certain cases, the number of fossil poppy seeds recovered from settlement sites in this region has been relatively large. However, at least some of the evidence indicates that the species itself may not have been native to this part of Europe.

For example, the fossil record from the sites at Weier (see Fredskild 1978: 189–201) and Niederwil (see van Zeist and Casparie 1974: 415–28) support this hypothesis. Archaeobotanical interpretations of the resource use at both of these sites suggest that an early form of shifting agriculture was practiced and that this type of farming had a significant impact on the local vegetation and soil fertility. In fact, the neolithic settlement phase at both Weier and Niederwil appears to have been relatively short (about 75 years at Niederwil and about 100–200 years at Weier). The occurrence of the fossil opium poppy seed in the sedimentary deposits from these two Middle Neolithic bogside settlements is in both cases almost exclusively restricted to the "culture layers." The strong association of the fossil evidence with the short settlement phase at both Weier and Niederwil is an indication that the opium poppy was an exotic member of the prehistoric Swiss flora surrounding the habitation areas and suggests that the opium poppy was one of the crops or weeds brought into the Swiss Foreland approximately five thousand years ago by the early immigrant agriculturalists who colonized the region.

Archaeological evidence indicates that during the Neolithic period the Swiss Alpine Foreland was colonized by early farming people from two different routes (Sauter 1976:44). The southern sea route, which passed along the northern Mediterranean coasts, probably facilitated an important flow of goods and ideas up the Rhone river valley to west central Europe. The second, equally important, pathway was the eastern continental route, which connected the Alpine Foreland with the Balkans and the upper Danube Basin. This latter route skirted the northern margins of the great European alpine region, and was apparently used by the Danubian cultures to move into central Europe and establish the so-called Linear Pottery or Bandkeramik groups along the loess lowlands of the region. (Butzer 1971:571).

It is therefore suggested that the early farmers of the Swiss Foreland and adjacent regions emerged as a mixture of the indigenous Mesolithic people and several invading Neolithic groups which probably arrived in west central Europe by the abovementioned routes. For example, the Cortaillod culture, which extended from the Lake Neuchatel area east to Zurich and northward into the Black Forest region, is believed to have been a mixture of Neolithic traits that spread northward from southern France and Mesolithic (Azillian) traits that were indigenous to the western Alpine region (Hibbens 1958:151). On the other hand, the Pfyn culture occupied a region northeast of the Cortaillod groups, and is considered to be derived from an eastern European Neolithic tradition (e.g., Altheim and Mondsee) with which it shares a number of cultural similarities (Hibbens 1958:152–53):

> Pfyn and Altheim clearly belong to an eastern tradition, going ultimately back to the Balkan Copper Age, Cortaillod belongs to the western tradition. But what explains the common preference of Pfyn and Cortaillod for the same type of environment? Perhaps we may think of an earlier population, culturally much less differentiated, which had already adapted itself to the bog and lake-side environment. (Waterbolk and van Zeist 1966:577)

Lakesides were preferred sites for settlement because they offered plentiful water supplies and ready access to herds of game, fish, and wild plant foods. Therefore, indigenous Mesolithic peoples were probably already using these environments on a limited basis when their hunting and gathering culture merged with those of the immigrant farmers who apparently entered the region approximately 5,000 years ago. If one of the groups of colonizing farmers did bring the opium poppy into west central Europe, it is still necessary to determine its identity and whence it came. Hence, I review the derivation of the words for the opium poppy since names are frequently carried along as man migrates with his plants.

The Etymology of Words for the Opium Poppy

> There is little doubt, . . . that the opium poppy was a familiar plant of some significance at a very early epoch, for the names now current in Europe go back to common Indo-European roots. (Andrews 1952:152)

We know that the ancient Greeks were familiar with the plant because the Dorians referred to it as *Mākōn* (μάκων) and, in later times, the Classical Greeks knew it as *Mēkōn* (μήκων). The Indo European origin of these Greek words is indicated by their correspondence to the old Swedish *val-molghi* 'Mohn' (next to middle-high-German *mâhen*, *mân*, old-high-German *mâgo*, from which probably comes Lettish *magune*—and dark Lithuanian *aguna*—and Estonian *magun*, *magunas*, etc.) and the old Slavic *makŭ*, old Prussian *moke* (the last probably from Polish), so that a stepwise root *mâq-*, *meg-*: *mag-* (in Germanic with grammaric change) is present. *Val-* old Swedish *val-moghi*, is translated from old German *vala*—"narcosis," (Schrader 1917–29: vol. 2, p. 68, trans. Inge Fisher 1976)

Although it has been widely recognized that the early Greek word *Mēkōn* is cognate with several words of Indo-European origin, "the exact source" of the word is "unknown" (Andrews 1952: 152; Kritikos and Papadaki 1967b: 6). Alphonse de Candolle (1886: 399) referred to the "form *Mack*" by which peoples south of the Caucasus knew the opium poppy; according to Netolitzky (1931: 60) the Greek word *Mēkōn* is supposed to come from the South Caucasian word "Mak." The *Oxford Dictionary of English Etymology* states: ". . . The oldest Indo-European word for the plant (not Italic or Celtic) is represented by German *Mohn*, Greek *Mēkōn*, and Old Slavonic *Maku*" (Onions 1966: 697). Kostrzewski (1949: 48) pointed out that in Poland there are several place names associated with *Mak*, the "paleoslavic" word for the opium poppy, such as Mak, Makowiec, Makowisko, and Makow. Frisk suggested that the root of *Mākōn* is of Slavic or Germanic origin: "Thus in Russian it is termed *Mako* and in German *Mago*" (Kritikos and Papadaki 1967: 6).

According to Klein (1967: 1088), the modern word "opium" is of Latin derivation which in turn comes from the Greek word *opion*, meaning poppy juice; and *opion* is the diminutive of *opos*, meaning vegetable juice which in turn seems to be cognate with Old Slavic *soku*, Lithuanian *sakaĩ* (plural), Old Prussian *sackis*, and Lettish *svakas* ("resin").

Table 15

Names for the Opium Poppy from Various European Languages (from Veselovskaya 1976: 12)

Western Group	Eastern Group
Latin—papaver	Greek—mekōn; Dorian, makōn
Italian—papavero	German—Mohn; Old German—mage, maga, mahen, man
Portuguese—popoila	Rumanian—mac albastru
French—pavot	Hungarian—kekmak
English—poppy	Transylvanian—makos
Modern Greek—paparouna	Czech
	Polish —mak
	Serbian
	Ancient Slav—maki
	Russian—mak
	East Tatar—mak
	Cheremain and Mordovian—mako
	Latvian—maggom
	East Indian—maggunad, maggona
	Norway—mue, valmoe, vallmo, vallmoog
	Swedish—valmo; Old Swedish—valmyghi
	Danish—vallmue

According to Onions (1966:629), *opos* "has been referred to a base **ap*-water, represented in Indo-Iranian and Baltic (cf. old and modern French opium)." According to Skeat (1879–82:412), *opos* is "not connected with Old English sap; but rather with Sanskrit, *apas*, pl., waters."

The origins of the early "Italic" word for the poppy, *papaver*, are also obscure. According to Netolitzky, this word is of "pre-Indo-germanic" derivation (see Bertsch and Bertsch 1949:195). He suggested that the etymology of *papaver* may lead back to an Alpine source and may originally have been a "lake-dweller word" (Netolitzky 1931:61). Karl and Franz Bertsch (1949:195) suggested that *papaver* probably comes from France, which they believe is the native home of the "black sleepy poppy."

According to Partridge (1958:512), "poppy" comes from the Old English word *popig*, *popaeg*, or ultimately from Latin *papaver*, and is perhaps of Mediterranean origin since *peqer* in Egyptian means "poppyseed" (see also Skeat 1879–82:464). However, according to Klein (1967:1123), the Latin word *papaver*, "poppy," is a derivation from the "imitative base *pap-, 'to swell, blow', whence also *papula*, 'a swelling pimple'." Andrews (1952:152) suggested that *papaver* "is probably a participial formation, originally in the form *papa-wes, 'blown up, inflated' based on the root *pap-." Klein referred to *papula* as the etymological source of the Old Indian word *pippalī* ('berry, pepper corn'), the Lithuanian word *pāpas* ('nipple'), the Lettish word *pâpa*, *pāpis* ('swelling, pustule'), and the modern English word "pap" ('nipple'). The Old Indian, Lithuanian, Lettish, and English words are all of Indo-European origin; perhaps they all originally go back to an earlier common root whose geographical origins may be located in the European Alpine region or France, as suggested for the word *papaver* in the discussion above.

Reviewing this short survey of etymological information relevant to the opium poppy, it is apparent that the early Greek works *Mēkōn* (poppy) and *opos* (poppy juice) are cognate to and perhaps derived from Indo-European roots of an uncertain origin. The etymological source of *papaver* is more obscure. On one hand, its "pre-Indogermanic" derivation has been suggested along with a possible geographical origin in the European Alpine region; on the other hand, the imitative base (pap-) seems to have cognate affinities with words in some Indo-European languages.

If one assumes that the opium poppy has its earliest etymological roots in *papaver* and that this word is of southwestern European (France?) or Middle European (Alpine?) origin, then the Indo-European etymological derivations may have begun when the people of one or both of these regions were first contacted by the Indo-Europeans. It should be pointed out here that R. C. Thompson (1949:227) an Assyriologist, suggested the following possible origin for *papaver:*

> săm
> săm
> PA-PA-PA, *arat araru* 'the tops of the poppy', must presumably be the capsules. Can (PA) PA-PA be the origin of the word *papaver?*

Thompson's suggestion, which traces the etymological origin of *papaver* to an Assyrian word derived from the Babylonian or Sumerian language, has not been accepted (e.g., see Kritikos and Papadaki 1967:6 and Krikorian 1975:103).

Southeastern Europe or the Near East

Although Thompson's notion regarding the origin of the word *papaver* appears obsolete, on the basis of other etymological and scholarly interpretations, it has often been assumed that the opium poppy was native to some area in the Eastern Mediterranean or Western Asia (e.g., see de Candolle 1886:399; Veselovskaya 1976:168; Zeuner 1954:358–59; Collom 1957:614; Kritikos and Papadaki 1967:35; and Cohen 1969:76).

Ancient Mesopotamia

During this century, numerous scholars have referred to the very early knowledge and use of the opium poppy in the ancient Sumerian, Babylonian, and Assyrian civilizations of Mesopotamia. However, Krikorian (1975:95–114) challenged the traditional belief that the opium poppy was known and used in the ancient Near East, arguing that such conclusions have been based on secondary sources and faulty information. In the following sections, I first present some relevant background information, and then discuss the controversial evidence which has been used to support the idea that the opium poppy was present and utilized in Mesopotamia prior to the Christian Era.

One of the world's earliest civilizations developed in the fertile lowlands between the Tigris and Euphrates Rivers of Mesopotamia. This ancient civilization was probably developed by the Sumerians, whose technological innovations include the invention of the plow, irrigation, wheeled vehicles, sailing vessels, and writing.

As an advanced and powerful state, the Sumerian civilization developed a trade network and empire that reached to the frontiers of India and Egypt. Navigation up the Euphrates and Tigris rivers and camel caravans connected the major Mesopotamian markets with the important ports on the eastern coast of the Mediterranean Sea (Ezekiel 27:23 and Herodotus 2. 112). It has been suggested that both poppy seed and raw opium were among the products that were collected in baskets and transported by various means to sea ports and inland trade centers for thousands of years (Anslinger and Tompkins 1953:1; Malcolm 1971:49).

Ancient Sumerian clay tablets bearing the cuneiform script and dating back to approximately 3000 B.C. (Sonnedecker 1957:4) are among our most ancient written records. These tablets provide an important source of information and represent one of our very earliest bases of historical data. In fact, the Sumerian tablets reveal several of the earliest major advances in such fields as astronomy, mathematics, medicine, botany, zoology, and chemistry (Levey 1956:377). Nevertheless, although these written records (both Sumerian and subsequent Assyro-Babylonian) suggest that observation of empirical regularities and practical application of knowledge were remarkably advanced, they also indicate that theoretical explanations and formulation of general laws were still in their infancy. Observations in many fields were, in fact, often explained by supernatural forces which could be entreated with mystical ritual or controlled by means of magical tricks.

Given the opium poppy's psychoactive and therapeutic potential, perhaps its use played a role in the divination and magical rituals of ancient Mesopotamia. Several references to the medicinal use of this drug plant among the people of this and subsequent early Near Eastern civilizations can be found in the literature; the substantive nature and validity of some of these references are discussed below.

Medicine in the Ancient Near East

The field of ancient Near Eastern medicine combined both pharmaceutical and psychological procedures to diagnose causes of illness, prescribe cures, and restore health. When a person became ill, a priestly physician would diagnose the problem and prescribe a remedy. This was accomplished through a "magical" process of divination which would reveal the nature of the sickness, the course of its future, and the way to eliminate it.

Ultimately, however, the Babylonians and Assyrians believed that diseases were the result of invasions of the body by ever present evil spirits (an interesting analogy can be drawn between this superstitious belief and the modern concept of the ubiquitous presence of pathogenic bacteria). A major aspect of therapy in ancient Near Eastern medicine involved incantation and exorcism (Kilgour 1960: 19). Elimination of the demonic forces that produced the illness often required dream interpretation, magical pronouncements, imitative acts, use of charms such as amulets, and special drugs (Levey 1956: 377).

The importance of vegetable drugs is suggested by the mythological origins of medical magicians. They were believed to have been the descendants of the moon god, *Sin*, who supposedly supervised the cultivation of medicinal herbs capable of purging the demonic spirits of illness; *Sin* was the "lord of wisdom, dispeller of darkness and its evils, giver of dreams and oracles" (Budge 1926: 196).

The use of charms or amulets also appears to have been an important aspect of medical treatment in the ancient civilization of both Mesopotamia and Egypt (Budge 1926: 196). Imitative ornaments, pendants, gems, scrolls, and packages containing relics were worn for protection against sickness and magical dangers. Representations of amuletic pendants *(d/tudittu)* can be seen on Babylonian statuettes, figurine plaques, and cylinder seals: "That the 'd/tudittu' had some talismanic powers for warding off danger and evil is suggested by several passages [in the *Chicago Assyrian Dictionary*]" (Dales 1963: 32).

The rational side of the Sumerian and Assyro-Babylonian medical traditions included certain aspects of objective observation and practical treatment. An important part of this therapy often involved the use of drugs containing a variety of materials, many of which were of plant origin. In many cases, these drug potions probably played a key role in the recovery of a patient.

Many secondary sources refer to the opium poppy as one of the herbal drug sources used in ancient Near Eastern medicine. For example, Levey (1956: 380) includes the opium poppy with several other botanical sources such as the "lotus, olive, laurel, myrtle, and garlic" plants as well as some providers of powerful vegetable drugs from the deadly nightshade family such as hellebore, hyoscymus, and mandrake. Sonnedecker (1957: 6) listed the following representative plant drugs supposedly used in ancient Near Eastern medicine: "pine turpentine, styrax, galbanum, hellebore, myrrh, asafoetida, calamus, ricinus, mentha, poppy-opium, glycyrrhiza, mandragora, cannabis, crocus, thymus." In a provocative statement, Thorwald (1963) suggested that slaves and captive enemies in the ancient Near East sometimes served as medical "guinea pigs" in experiments designed to test the effectiveness of powerful drug concoctions such as opium, mandragora, henbane, and belladonna.

Evidence for the Opium Poppy in the Written Records

The spiritual center of the Sumerians was the Mesopotamian urban center of Nippur located on the Euphrates River less than one hundred miles south of modern Baghdad. More than seventy years ago, archaeologists from the University of Pennsylvania Museum discovered and excavated a very large and remarkable collection of clay tablets from Nippur. These thousands of tablets represent some of the earliest written records in the world. They contain information concerning Sumerian economics, law, politics, literature, natural science, and medicine. One of these tablets is regarded as the world's oldest recorded list of medical prescriptions (see figure 53).

This particular artifact (CBS 14221) is 150 mm long and 97 mm wide. The front side is so badly deteriorated that neither translation nor reconstruction is feasible. The reverse side, however, is well preserved and has yielded unique information regarding early Near Eastern medicine. "To judge from the script, it was written some time toward the end of the third millennium B.C. [ca. 2100 B.C.] by an anonymous Sumerian physician who . . . had decided to collect and record a number of what were probably his more valuable prescriptions" (S. N. Kramer 1954:77–78).

These prescriptions call for the use of animal and mineral materials such as sodium chloride, potassium nitrate, milk, snake-skin, turtle-shell, and parts of cassia, myrtle, thyme, willow, pea, fir fig, and date plants (Kramer 1954:78). According to some authors, reference to the opium poppy can also be found among the prescriptions recorded on this ancient tablet (e.g., Terry and Pellens 1970:54–55 and Kritikos and Papadaki 1967:37). If this interpretation is correct, then this reference represents the earliest written record of opium poppy use. However, in his analysis of the tablet, Civil (1960:59–64) made no mention of any ideogram which could be identified with the opium poppy.

Terry and Pellens (1970:54) referred to information given them by Professor Raymond Dougherty (1877–1933), a noted Near Eastern scholar, indicating that the Sumerians were acquainted with opium because they had a specific ideogram, *HUL·GIL*, "which signified the drug." Dougherty was asked about the meaning of this ideogram by Terry and Pellens and responded as follows:

> I should say that the basic meaning of the sign *HUL* is "joy," "rejoicing." *Gil* as single ideogram represented a number of plants, but its meaning in the ideogram for "opium" is difficult to determine with exactness. It may be suggested in a very tentative way that the Sumerians in their system of pictographic writing endeavored to depict the power of opium to produce a sense of delight or satisfaction. (Terry and Pellens 1970:55, quoting a personal communication from Dougherty)

FIGURE 53. This Sumerian clay tablet (ca. 2100 B.C.) is considered to be the world's oldest recorded list of medical prescriptions. It is 150 mm long and 97 mm wide and is now held in the University of Pennsylvania Museum. Some have suggested that the opium poppy is referred to on the tablet; others have rejected this interpretation.

Professor Dougherty suggested that although *HUL·GIL* could not be given a firm date of origin, he felt that it probably was developed in the fourth millennium B.C.—over 5,000 years ago (Terry and Pellens 1970: 56). Dougherty also suggested that the Assyrians incorporated many Sumerian ideograms into their own written record and referred to the abundant use of *HUL·GIL* in the Assyrian medical tablets found during a nineteenth century archaeological expedition in Mesopotamia. These later tablets were recovered from the royal Assyrian Library of Ashurbanipal (668–626 B.C.) at Nineveh.

The Assyrian tablets were eventually deposited in the British Museum where they were later studied for many years by R. Campbell Thompson. In the first half of this century, Thompson published several reports of his research and translations (Thompson 1904, 1907, 1924, 1934, 1949). He offered identifications for at least 250 vegetable drugs and 120 mineral drugs which were used medicinally in Assyria. These medicines and their uses were probably passed down from more ancient Sumerian and Bablyonian sources (Thompson 1949 : 2).

Thompson's translations of the Assyrian herbal suggested that the opium poppy was known and the various parts of the plant were used. According to Thompson (1949 : 225), ". . . ^{săm}*araru* and ^{săm}*ararŭ* are forms of poppy or opium." He argued that ^{săm}*araru* referred to the flower, ^{săm}UKUŠ—RIM referred to the capsules, and *irrŭ* referred to "an inspissated milky juice." He also suggested that the Assyrian medical tablets contain "various synonyms and bye-names for ^{săm}*irrŭ*, ^{săm}*arau*, about which, I think, as 'opium' and 'poppy' there can be little doubt" (Thompson 1949 : 226).

If Thompson's translations and interpretations of the Medical tablets from Ashurbanipal's Royal Library are correct, then we may assume that the Assyrians used the opium poppy for various medical purposes:

> Here, then, we have a drug of which probably more parts are used than any other simple plant in *MT:* simply, seed, ŠE-RU (shoot), root, fat, PA (tops), *saruru* (capsule?), "stone" (whatever this may be), powder, fruit. Its use is for eyes, feet, head and temples, stomach, swelling (common) externally; and stomach (frequently), urinary trouble, and in pregnancy internally; and as suppository in the anus. These coincide exactly with the use of the Poppy-drugs: apart from its ordinary use as an internal anodyne, especially in its use in genito-urinary diseases, irritability of the bladder or uterus, and in the passage of gall-stones (*BMP* No. 18), and externally "the decoction of poppy-capsules is a common anodyne and demulcent fomentation when applied hot to inflamed parts, bruises, sprains, and other painful affections." (Thompson 1949 : 228)

Over the years since publication of his etymological research, Thompson's interpretations have been widely accepted by many writers and often quoted as authoritative evidence in several secondary sources. For example, T. Williams (1947 : 88) relied on Thompson's translations and interpretations when he referred to the following use of an opium-based suppository in ancient Assyria:

> . . . if a man's belly unexpectedly is irritated . . . and while it hurts him he cries out . . . thou shalt reduce lion-skin, mix with lion fat (opium); let it dry again, mix with cedar oil, make into a suppository and put it on his anus. It may be mentioned that very similarly compounded opium suppositories are used to-day. This method of administration has the great advantage that absorption into the body takes place slowly.

More recently, Thompson's translations and interpretations, especially as they pertain to the opium poppy, have been critically re-examined by Krikorian (1975:95–113). First, Krikorian brought to light a misleading, but often used, reference to Thompson's *Assyrian Herbal* (1949:43) in which Thompson supposedly translated a section of an ancient medical tablet as follows: "Early in the morning old women, boys and girls collect juice by scraping it off the wounds (in the poppies) with a small iron scoope, and deposit the whole in an earthen pot, where it is worked by hand in the open sunshine until it becomes of considerable thickness." Several secondary sources have, in fact, used this passage to show that the harvesting system recorded by the Assyrians was very much like the collection methods in wide use today among peasant opium cultivators (e.g., Terry and Pellens 1970:54; Addens 1939:102; Sonnedecker 1962:277; and Wright 1968:62). Krikorian (1975:97–98), however, pointed out that in this particular passage, Thompson was not translating from an Assyrian medical tablet at all! Instead, Thompson (1924:43) was actually quoting from an anonymous, early nineteenth-century agricultural report ("Description of the Culture of the White Poppy and Preparations of Opium, as Practised in the Province of Bihar," *Asiatic Journal* [London] vol. 3, 1817:26–27) which was included almost verbatim about a half-century later in William Rhind's book, *A History of the Vegetable Kingdom, 1868:547).* Thompson had used the relatively modern passage to illuminate what he thought was a close relationship between it and the ancient one that he did, in fact, translate. According to Thompson (1924:43), the ancient passage on an Assyrian tablet does refer to the scraping of a drug plant juice by women and children; however, the plant which provides this juice has not been identified.

Krikorian also challenged the linguistic evidence that suggests an association of the opium poppy or opium drugs with HUL–GIL or any other ancient Near Eastern ideogram. He referred to a more modern dictionary of the Assyrian language, *The Assyrian Dictionary* (Chicago: The Oriental Institute, 1960) which, according to Krikorian (1975:101), renders Thompson's interpretation of HUL·GIL obsolete:

> . . . the cuneiform sign which Thompson read and transliterated as HUL is now transliterated as HUL when it means "to rejoice" (Sumerian *hul*, Akkadian *hadu*), but UKUS when (as in the present context) it means 'cucumber' (Sumerian *ukus*, Akkadian *qissu*). The sign is also fixed as a "determinative" or "semantic indicator" prefixed to a plant name to show it is a cucumberlike plant.

Krikorian (1975:101) also argued that there are inconsistencies in Thompson's translations involving GIL. He underscored his argument by quoting from a personal communication with an expert in the field, Erica Reiner of the Oriental Institute at the University of Chicago, who disputed the translation of HUL·GIL as "joy plant":

> Indeed, the first sign is UKUS 'cucumber'. In the lexical lists of the Babylonians, many varieties of this UKUS are listed and we really have no idea which variety is which. (personal communication of Reiner, quoted by Krikorian 1975:102–3)

This interpretation raises an interesting question: was UKUS-RIM or some other plant referred to as cucumber-like or a "variety" of a cucumber really the opium poppy? The drug plant does produce some "races" whose unripe capsules might be considered as resembling a cucumber. Thompson (1949:224) refuted the identification of HUL·GIL with the "stink cucumber," but did suggest that the poppy could be referred to as a "cucumber-like fruit." In any case, even with this question aside, Krikorian (1975:103) argued with emphasis that "... *there is no word in Sumerian or Akkadian or Assyrian that scholars are agreed upon that definitely identifies either poppy, opium poppy, or opium.*"

Evidence in the Art Work of the Ancient Near East

Ancient art work is another kind of evidence often referred to in support of the argument that the opium poppy was known and used among the Babylonians and Assyrians. However, as I point out, identifying ancient artistic representations of opium poppy plants or their parts has been controversial. An example of conflicting interpretations involves a series of ninth- to seventh-century B.C. bas-reliefs from Assyria which allegedly depict priestly physicians holding opium poppies. Figure 54 shows one of these bas-reliefs now on display at the Louvre Museum in Paris. Although a number of scholars have suggested that this relief and ones like it (such as those on display in the Musées Royaux d'Art et d'Histoire, Brussels, and the British Museum, London) do represent religious rites (e.g., see Kilgour 1960:19; Thorwald 1963; and Krikorian 1975:104), there has been a lack of consensus regarding the specific identification of the hanging objects held in the figure's hand. Krikorian (1975:104–6) has surveyed many determinations which have been made for these objects over the last one hundred and twenty-five years, emphasizing the wide variety of possibilities that exist.

First of all, he raised the question of whether or not the objects really represent plant material. Even if it is agreed that the objects do depict plant material such as "fruits," are identifications with the capsules of the opium poppy the most logical interpretation? Although most people believe that they represent pomegranate fruit (Krikorian 1975:105), some scholars think the hanging objects were modeled in the shape of opium poppy capsules (e.g., see Layard 1849:I, 47; Lenormant 1871:136; Zender 1928:379–82; Spleers 1938:122–36; and Weidner 1939:69, cited in Krikorian 1975:104–5). A recent and unequivocal opinion was published in a short article written by Kilgour (1960:19), who commented on the objects and their significance:

> This Assyrian scene portrays many of the features of medicine in the ninth and eight centuries B.C. A priest is engaged in exorcism before a sacred tree: it may be that the demons he is driving off are those of disease. Moreover, he holds in his left hand three mature poppy heads which, however, are not useful as a drug by themselves, and since the priest holds them upside down their seeds have dropped. Opium of course, is prepared from immature poppy fruits. Nevertheless, the plant the priest grips represents the source of a drug highly effective then and now.

If these objects do represent ripe opium poppy capsules and the figure is participating in a ritual sowing of the seeds, this would indicate that the drug plant had a special importance, perhaps as a fertility symbol.

In any case, Krikorian (1975:111) argued that an acceptable identification of botan-

FIGURE 54. Ninth-century bas-relief from Iraq, now held in the Louvre, Paris, France (from Kilgour 1960:20).

ical representations on any of the artifacts in question must involve careful consideration of the degree of detailed similarity between the art work and the suggested plant or its parts:

> To complicate matters, one should emphasize that these fruits(?), interpreted as pomegranates, show a direct attachment of pedicel to fruit with no circular flange (this point of attachment is known botanically as the *torus*). . . . The very real problem in the Assyrian bas-relief figures which nominally represent the poppy is that the "capsules" are directly attached to the "pedicel." There is a "torus," but it is not in the right place. In the opium poppy, and indeed all members of the Papaveraceae, this structure, which represents the point of attachment of the floral leaves, is *about ½ or 1 cm away from the base of the capsule* [see Krikorian and Ledbetter, 1975].
>
> In order to be convincing to botanists, the alleged poppy "stem" and its"capsules" represented in the reliefs would have to show a more or less realistic pedicel/ receptacle arrangement (i.e., *torus*). A further problem is that the representation of the 'stigmatic rays' is unrealistic.

Perhaps the hanging "fruits" represent neither poppy capsules nor pomegranates, but rather are depictions of pine cones which were used in the celebration of some Near Eastern cults (Isaac 1959: 182). It should be pointed out that artistic license, in the case of the hanging objects, may have reduced realism to such a degree that positive identification without sufficient corroborative evidence is virtually impossible. Krikorian (1975: 104) noted that the Assyrian bas-reliefs do "show figures holding something superficially suggestive of mature poppy capsules at first glance, but opinion has varied so much on their identity that the object(s) was even once described as a fly whisk."

Finally, in addition to his challenging review of earlier translations of the Assyrian medical tablets and the "highly speculative" interpretations of specific art work, Krikorian (1975: 113) referred to his personal communication with Hans Helbaek, a noted authority on man-plant relationships in Old World prehistory. In a letter Helbaek sent to Krikorian on April 23, 1973, he said that ". . . among all the gallons of grain from Persia, Iraq, Turkey, Lebanon, Syria, Jordan, and Egypt I have never established *P. somniferum*"; and that these fossils "reach in time from ca. 7500 B.C. to A.D. 1000."

In the conclusion to his article, Krikorian (1975: 113) suggested that we "reserve judgement on the question of the knowledge or cultivation of, or even the use of, the opium poppy and/or its products by the early peoples of the Near East."

Evidence from Anatolia

In his discussion of the excavations in the Early Bronze Age settlement at Thermi on the island of Lesbos (near the Ionian coast of Anatolia), Lamb (1936: 217) presented some of the archaeobotanical notes reported by "Mrs. Reid" who analyzed the few seeds that were recovered from the prehistoric site. Among the meager remains, Mrs. Reid found:

> "A mass of Poppy seeds. Probably *Papaver Rhoeas Linnaeus* [*sic*] certainly not the opium poppy." *From the Early Bronze Age deposit below House O, a deposit believed to belong to Town IV.* . . . "One or two poppy seeds." *These come from B, Town III.* "It is difficult to know whether the charred condition of most of the carbonaceous remains is due to the state of preservation or to intentional charring." (Lamb 1936:271)

In addition to this tantalizing but enigmatic evidence from Lesbos, a more challenging reference to possible direct evidence of ancient opium poppy use in prehistoric Asia Minor has been made. In a recent article describing a unique "opium pipe" discovered in Cyprus (dated ca. 1200 B.C. and discussed in more detail in a subsequent section), Karageorghis (1976:127) referred to poppy seeds "found in a footed bowl discovered in the ruins of the burnt palace at Beycesultan, Anatolia, in a context to be dated ca. 1900 B.C." According to the author, this information was supplied to him by the archaeologist R. D. Barnett. I wrote to Barnett about this discovery of prehistoric poppy seeds and he responded with the following information:

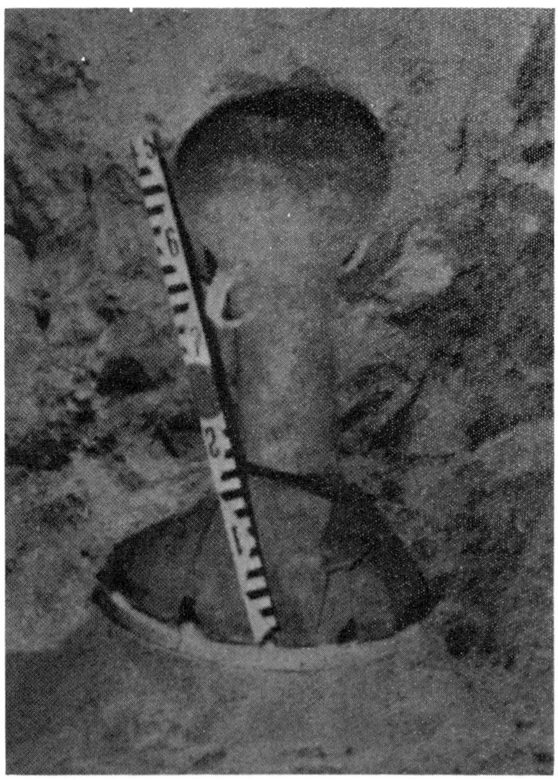

FIGURE 55. The vessel (bowl or drum) found in the ruins of the burnt palace at Beysultan, Anatolia, which reportedly contained opium poppy seeds (from Lloyd 1965, plate VIa; *Courtesy of Professor Seton Lloyd, London, England*).

> It is correct that a bowl-full of poppy seeds was found at Beycesultan in the excavations of the burnt palace while I was present in 1954, but by a most curious accident the description of them (but not of the bowl) was omitted by Professor Seton-Lloyd from his publication of the excavation. . . . The information, as far as I know, has not been published elsewhere, nor do I have photographs or other references, but you had best consult Professor Seton Lloyd. (Barnett, personal communication, September 22, 1976)

On Barnett's advice I wrote to Seton Lloyd who subsequently offered some more specific information regarding this intriguing discovery:

> The vessel to which Dr. Barnett referred is illustrated in my publication Beycesultan Vol. II. London 1965. Pl. VI(a). It was identified as a drum, but contained a quantity of seed which I myself and the peasant workmen easily identified as opium poppy. I sent samples of the seeds to the Kew Gardens laboratory in London for further identification, but received an unsatisfactory reply, suggesting that they must have been "intrusive," (which stratigraphical circumstances made impossible). For this reason the find was omitted from the publication. (Seton Lloyd, personal communication, January 14, 1977)

In response to my own queries regarding this and other fossil evidence for the opium poppy, Hans Helbaek related his personal skepticism:

> I am both an archaeologist and a botanist and as such extremely critical about archaeological conclusions. I have seen too much. I have worked several Cypriot finds and never found *Papaver*. Can you conscientiously quote an archaeologist on a botanical problem? . . . You may know that I published Beycesultan myself, but I never had the Bronze Age material sent. Now, I have investigated c. 6000 B.C. Catal Hüyük and Neolithic-Chalcolithic Hacilar in Turkey without coming across a single seed of *Papaver*, and these two finds are of exceptionally large volume, both as to cereals, pulses and weeds. Why should *Papaver* have been there at a certain time between Aceramic ages and the late Byzantine, but be completely unencountered during the periods between? It is up to your sense of logic to unravel this enigma, based on the opinions (I won't say observation) of archaeologists. (Helbaek, personal communication, November 19, 1976)

Although the "poppy seed" evidence from southwest Anatolia at Beycesultan may not be valid, as I noted before, many have assumed that the opium poppy was native to the eastern Mediterranean or western Asia. In fact, several scholars have suggested that the drug plant was spread from an eastern hearth to regions in northern, central, and western Europe. For example, as early as the nineteenth century, Alphonse de Candolle (1886:399) argued that the "variety" of the opium poppy cultivated in Europe during his time "came from Asia Minor, where the species has been cultivated, for at least three thousand years." Bazilevskaya (1928; compare with Veselovskaya 1976:166) also considered the eastern Mediterranean as the region where the origins of the "cultivated poppy" are to be found. Veselovskaya (1976:168) argued that cultivation of the opium poppy began

> in Greece and Asia Minor, and from there it spread to other territories of the Old World mostly as a cultivated plant. With the emigration of the Indo-European people from the coastal regions of the Mediterranean Sea to the north, poppy cultivation spread all over Europe.

Zeuner (1954:358.59) regarded the opium poppy as a domesticated plant of ancient origin, whose "natural home appears to be the Middle East and possibly Greece." Collom (1957:614) referred to Asia Minor as the native region of the opium poppy. Kritikos and Papadaki (1967:35) were more specific when they argued that the opium poppy's original home lies in Anatolia near Sinope on the south coast of the Black Sea. Meikle (compare with Merrillees 1968:154) suggested the possibility that the opium poppy *(Papaver somniferum)* may have developed as a hybrid of two other wild poppies, *P. setigerum* and *P. gracile*, in the Eastern Mediterranean area. Cohen (1969:76) argued that the original home of the opium poppy lies in Asia Minor. Hutchinson (1973:520) was less specific when he suggested that the opium poppy is native to the "eastern countries." Gimbutas (personal communication, February 27, 1978) has suggested that the opium poppy fossils found in the Middle Neolithic sites of the Swiss Alpine Foreland were derivatives of a species that was native to a region located southeast of west central Europe.

> Poppy seeds can of course be preserved only under special conditions. Therefore, it is difficult to speak about their origin. There is nothing of the kind in southeastern Europe as there is in the Swiss lakeshore sites. I doubt that the opium poppy will be discovered in neolithic debris of southeastern Europe. However, I would think that the opium poppy travelled from the southeast to central Europe, or from the southeast Mediterranean to central Europe, but not from the north to south.

Duke (personal communication, December 2, 1976) speculated that *Papaver somniferum* was originally selected out of a "middle eastern population." And de Wet (personal communication, June 1, 1976) assumed that ". . . the wild and weedy populations of the opium poppy are probably found somewhere in western Asia or possibly central Asia."

Two identical gold pins found in Anatolia and dated to the time of the Hittites (ca. 1400–1200 B.C.), have been recently identified as representations of dried opium poppy capsules (Merrillees 1979:168). A description and illustration of these intriguing artifacts is presented in chapter 8 (section entitled "Pinheads and Poppy Capsules").

If the opium poppy was brought into west central Europe from eastern Europe directly, it was probably introduced by farmers belonging to the Linear Pottery (i.e., Danubian) culture group. This group is regarded as one of Europe's earliest and most distinctive farming cultures. It may have originally developed out of an earlier European farming culture located in southeastern Europe referred to as the Starcevo culture (Gimbutas 1974; Fagan 1974:217). Murray (1970:111) refers to the Linear Pottery group as being "responsible for the transmission of agriculture across Europe from Hungary to Holland." More specifically, she credits this early farming culture with the introduction of emmer, einkorn, bread wheat, club wheat, barley, a species of millet, and flax into west central Europe.

Flax and the opium poppy were important plants in the Middle Neolithic cultures of west central Europe. Both species could have provided a useful oil as well as other valuable products. However, whereas the Neolithic flax of Europe is clearly a derivative of a Near Eastern species (Helbaek 1959:103–41; van Zeist and Bakker-Heeres 1975:215–19), the origins of the opium poppy are, as I previously pointed out, still unclear.

In any event, if the "seeds" in the vessel from Beycesultan were *Papaver somniferum*, it would indicate that the opium poppy was known and used in Asia Minor during the Bronze Age. However, even if this were the case, the drug plant may have been an exotic species which was only present in the region for a relatively short period of time before the palace at Beycesultan was burned. Although van Zeist (personal communication, 1980) has *no* knowledge of any opium poppy seeds recovered from Near Eastern archaeological sites, he has reported pollen core evidence indicating that the region surrounding Lake Beysehir in Turkey experienced a "deforestation phase" between ca. 1600 and 300 B.C. It is possible that the opium poppy could have been introduced to the highlands of Anatolia during the Bronze Age when the forests were first cut down. And, if the opium poppy was not present in southeastern Europe or the Near East until the Bronze Age (or even much later), perhaps its cultural origins can be found in a region to the north or northwest during the Neolithic period?

Northern or Northwest Europe

The relatively well-defined Linear Pottery (i.e., Danubian or Bandkeramik) culture group appears to have been firmly established in the middle region of the Danube River Basin during the Neolithic period, sometime between 5300 and 4600 B.C. (Milisauskas 1976:31; according to Gimbutas 1974:27 between 5500 and 5000 B.C.). From here, this early European farming culture spread rapidly out over a vast region to the north, east, and west along the major river valleys. Eventually settlements of Linear Pottery people reached Romania and the Ukraine in the east, eastern France, Belgium, and Holland in the west, Germany and Poland in the north, and the Alpine Foreland in the middle of this range (Gimbutas 1974:27; Rodden 1975:65).

Over the last few decades the archaeological sites of the Linear Pottery material have yielded an impressive legacy of pottery remains, stone artifacts, and distinctive settlement patterns. These reveal an apparently remarkable uniformity of material culture within the group. Although we still do not know very much about their economic, social, and ritual organization (Milisauskas 1976:32), we do know that the Linear Pottery folk inhabited "longhouses" and probably practiced a form of slash and burn agriculture in small plots near their settlements (Chard 1975:339). We also know that they established their communities on low-lying river and stream terraces, especially on or near loess or other easily tilled soils:

The settlements of the Danubian farmers are sharply restricted to loess sediments,

> which obviously provided greater soil fertility. At the same time, these often were areas with parkland vegetation and calcareous, chernozemic soils, or otherwise they had base-saturated forest soils under closely stocked, mixed oak woodland. It is no mere coincidence that primeval settlement, loess, calcareous or basic soils, dry lowland basins, and comparatively open, oak parklands or woodlands should provide a common denominator for the earliest agricultural lands of mid-latitude Europe. (Butzer 1971:577–78)

In the western region of its range, the Linear Pottery group eventually spread into and settled areas of west Germany, especially along the Rhine Valley (Murray 1970:35). By the beginning of the fifth millennium B.C., the Linear Pottery group had reached southern Holland (Fagan 1974:216). Although we do have fossil evidence indicating the presence of the opium poppy at two archaeological sites in the Netherlands (van Zeist 1974, 1975), it only dates back to the Roman period. Of more interest here are the reports of Neolithic fossil evidence of the opium poppy associated with Linear Pottery group sites on the loess-type soils of the Rhineland in west Germany.

During the past decade the German archaeologist K. H. Knörzer has published several reports dealing with his field research at a number of Linear Pottery group settlement sites in the lower Rhine Valley region of west Germany. Included in these reports have been several references to Neolithic poppy seed fossils found in the sites at Oekoven, Kreis (District) Grevenbroich; Aldenhoven, Kr. Jülich; Lamersdorf, Kr. Düren; Garsdorf, Kr. Bergheim; and Tangweiler, Kr. Jülich (Knörzer 1967, 1972, 1973, 1974, 1977).

Although Knörzer found only a few fossil poppy seeds in the first years of his work, he suggested that these Rhineland finds deserve special attention because they represent the earliest examples of the prehistoric "cultured" opium poppy in central Europe (Knörzer 1971a:34). Knörzer analyzed the poppy fossils, comparing them with recent samples of the opium poppy, "*Papaver somniferum*" (i.e., *P. somniferum* ssp. *somniferum*), the setaceous poppy, "*P. setigerum*" (i.e., *P. somniferum* ssp. *setigerum*), and other closely related poppies which are naturalized in this region. He concluded that the fossils showed the closest correspondence in seed size and surface structure to the setaceous poppy and should therefore be treated as prehistoric specimens of "*P. setigerum*" (Knörzer 1971a:37, see chapter 3). In 1974, Knörzer described two seeds of "*P. setigerum*" in 1977 he described four more seeds he also treated as "*P. setigerum*" (Jürgen Schultze-Motel, personal communication, August 19, 1978).

In any case, the fossils indicate that an opium- and oil-producing species was available in or near these early farming communities. In fact, because of their oil-rich seeds, Knörzer (1971a:38 suggested that this prehistoric poppy probably played a relatively important role in the diet of the Linear Pottery people inhabiting the Rhineland region. However, he also suggested that this species was probably introduced and not naturalized in the region. Knörzer argued that according to at least three sources (Oberdorfer 1970:402; Bertsch and Bertsch 1949:195; and Fournier 1946:372) the setaceous poppy is definitely a Mediterranean plant which is not found as a truly wild species in central or northern Europe and therefore could not have grown in the Neolithic fields of the Rhineland without the help of man. Furthermore,

Knörzer pointed out that clear archaeological proof of poppy cultivation is still lacking. He suggested that the recovery of a large mass of seeds such as found in the Alpine Foreland regions of northwest Switzerland and southwest Germany, or a find of seeds which are baked onto ceramic shards, would be a substantial indication of a domesticated species. In fact, Knörzer (1967 : 15) used this kind of evidence to confirm the domesticated nature of the oil-producing flax plant cultivated by Linear Pottery people in the Rhineland.

Although the opium poppy may have been nothing more than an introduced weedy plant growing in disturbed or abandoned habitats near human settlements, Knörzer (1971a: 38) referred to the fact that fossil seeds have been found at five out of the eleven Linear Pottery sites he investigated in the lower Rhine Valley. He suggested that this archaeological association could be used to support the idea that the prehistoric Rhineland poppy was cultivated. Knörzer did admit that his sample size for the prehistoric poppy was quite small in terms of the number of seeds recovered; but he argued that the scarcity of these finds can be explained by the fragility of the tiny seeds and relatively small chance of preservation by being charred or effectively covered over by a waterlogged soil.

If Knörzer's suggestion is valid and the prehistoric Rhineland poppy was a cultivated plant derived from a species native to the western Mediterranean region, then perhaps we should look to southwest Europe as the general area from where the opium poppy, or its possible root source, was introduced into west central Europe. In this case, the most logical route of its importation into central and northern Europe would have followed the Mediterranean coast to the mouth of the Rhone and then up the river valley to the pass leading into the Swiss Foreland.

According to Bertsch and Bertsch (1949: 195–96), the opium poppy is a "creation of Western Culture," which was brought from the western Mediterranean region through the valley of the Rhone, into the northern Swiss lake area, and up to the Bodensee and Federsee. They argued that the "list of finds presents ample testimony for this." If their interpretation is correct, the cultivation and use of the opium poppy may have been part of a cultural complex which spread from the southwest to the northeast of Europe during the fourth or third millennium B.C. (see Savory 1968: 83).

In chapter 3, I discussed many of the problems associated with the study of the opium poppy's natural origins. I pointed out that the various cultivated and semi-wild forms of this species, *Papaver somniferum* L., have not been located growing under truly wild conditions. I also noted that several botanists considered the cultivated forms of the opium poppy to be modifications of the "setaceous poppy" or "ancestral plant" which is either regarded as a subspecies of *P. somniferum* (i.e., *P. somniferum* ssp. *setigerum* (DC.) Corb.) or as a distant species (i.e., *P. setigerum* DC.). It is important to remember that the opium and setaceous poppies produce pain-killing, mind-altering substances. In fact, they are the only plants known to produce morphine. Furthermore, both produce edible, oil-bearing seeds. However, as I noted previously, some believe that the setaceous poppy, in contrast to the opium poppy, is a truly wild member of the Mediterranean flora in the western regions of the basin. In addition, the recent chromosomal study of the opium poppy taxa by Hammer and Fritsch (1977: 113–24) supported the suggestion that the opium poppy (i.e., *P. som-*

niferum ssp. *somniferum*) is a derivative of its very close relative, the setaceous poppy (i.e., *P. somniferum* ssp. *setigerum*), whose native range is considered to be located in the western Mediterranean region.

This set of assumptions regarding the natural origins of the opium poppy can be used, along with some limited, but extraordinary, fossil evidence from a cave near the coast of southern Spain, to build a case for the western Mediterranean origins of the opium poppy or its direct ancestor. In the following section, I describe the fossil capsules and seeds which were found in southern Spain and discuss their importance in relation to the cultural origins and prehistoric use of the opium poppy.

Southwestern Europe

Presently, we have no fossil evidence for the opium poppy in southwestern Europe prior to the establishment of agricultural communities in the region. In fact, the only confirmed evidence we do have from this region is restricted to a single cave site, Cueva de los Murciélagos, which was associated with a farming culture located near the southern coast of Spain. In the nineteenth century, Gongora (1868:36) described this cave site as a Neolithic burial sanctuary. This interpretation of the site's archaeological age was accepted without question for many decades and has only been challenged in recent years (Pellicer 1963:49). A brief survey of the Neolithic period in the western Mediterranean region will provide relevant background information and should help resolve some of the questions pertaining to the chronology of the opium poppy fossils found in the Cueva de los Murciélagos. This is followed by a discussion of the actual cave site and the prehistoric artifacts which it contained.

Early Farmers in the Southwest Mediterranean Region

Although the development of seed agriculture is generally considered to be an innovation that originated in western Asia and gradually spread northwestward into Europe, recent evidence from some Neolithic sites in the western Mediterranean region has yielded some relatively early dates. Unfortunately, the archaeological record in this region, especially in the far western areas, is both fragmentary and enigmatic. Therefore, even though it can be assumed that the pattern of livelihood in the Iberian peninsula slowly shifted from hunting and gathering to settled agriculture, the process "remains obscure because of the almost total lack of adequately recorded modern excavation of stratified open air settlements of the pure Neolithic phase, and the rarity of radiocarbon datings, which could be the basis of comparisons with the Near East" (Savory 1968:65). Presently we probably have less information from the Iberian peninsula than any other area in the Mediterranean during the Neolithic period (Whitehouse and Whitehouse 1975:123; Phillips 1975:46).

The lack of information regarding early farming cultures in the western Mediterranean region is not only a problem for Iberia; understanding of the development of agriculture in northwest Africa is also plagued by a paucity of evidence. In fact, only a very few Neolithic sites in north Africa have been documented. One of these sites is a farming settlement near Oran (Traventine Station) in Algeria, which has a confirmed date of 5810 B.C. ± 140 years (Camps 1975:190).

In France, the record is somewhat more detailed. Here, the earliest evidence of agriculture has also been found in the coastal regions of the Mediterranean. For example, we have early dates from Châteauneuf-les-Martiques (Bouches-du-Rhone, 5570 B.C. ± 240 years, uncorrected; see Bender and Phillips 1972:98) and two sites on the coast of the Bay of Marseilles at Cap Ragnon (6020 B.C. ± 150 years and 5700 B.C. ± 150 years) and Ile Rion (5650 B.C. .027 150 years and 5420 B.C. ± 160 years; see Phillips 1975:48). If these isotopic dates are valid, the Neolithic culture of southern France was long lasting, spanning a period of approximately 2,000 years (Whitehouse and Whitehouse 1975:140–42).

The first farming cultures of coastal Iberia, northwest Africa, and France are characterized by the Impressed (or Cardial) Ware pottery tradition, which is known almost exclusively from cave sites. In fact, all of the earliest agricultural communities in the Mediterranean Basin belong to this tradition (Whitehouse and Whitehouse 1975:123). Some have suggested that the Impressed Ware tradition arrived in the western Mediterranean region as a result of migrations originating to the east. "The affinities of the pottery and its distribution on islands and coasts never more than 50 miles inland indicated that these people were capable seafarers coming originally from somewhere in the eastern end of the Mediterranean" (Chard 1975:341). Others have suggested that this early Neolithic pottery tradition developed out of a synthesis of local and introduced cultures:

> . . . the Impressed Ware pottery of the Western Mediterranean region is in several areas so distinctive in both its form and decoration that it cannot be interpreted as purely a style brought in by 'immigrants'; and yet, this same pottery shows a degree of resemblance to that of the regions to the east that it is equally difficult to see the pottery as a native invention, independent of external stimulus. (Bailloud 1973:103)

Regardless of the geographical origins of the earliest Neolithic tradition in the Mediterranean region, there is no existing evidence for the prehistoric opium poppy at any of the Impressed Ware sites. Therefore, we have no direct evidence for the opium poppy during the early Neolithic period in the western Mediterranean region. The fossil capsules and seeds found at Cueva de los Murciélagos are associated with a *later* tradition of pottery decoration featuring "incised" markings:

> The recently published deep stratification from the cave of Carigüela de Pinar (Granada) makes it clear that this style [Incised Ware] succeeds the pure cardial decoration and must be dated in central Andalucia to the middle and late fourth millennium B.C. with some probable survival inland into the third millennium. (Savory 1968:75)

In addition to evidence from Cueva de los Murciélagos and Cueva de Carigüela de Piñar, incised ware has been found in other sites near the coast of southern Spain, such as Cueva de la Mujer, as well as at Furninha and Eira Pedrinha in Portugal

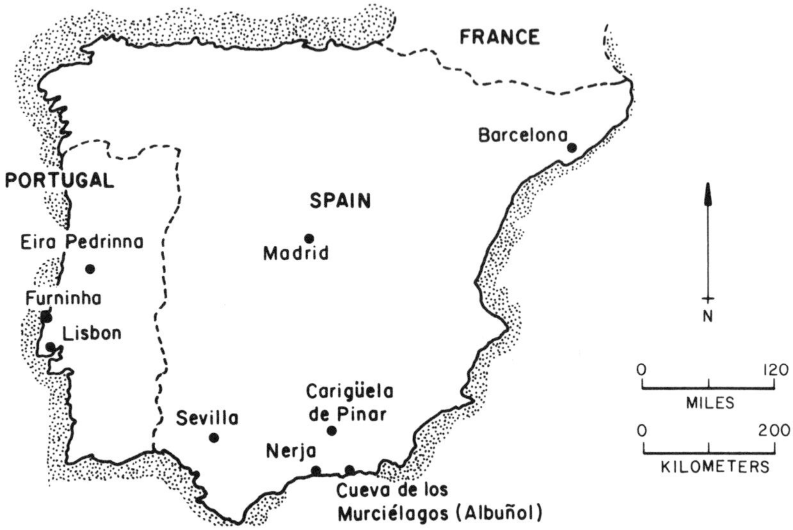

FIGURE 56. The distribution of the incised ware sites in the Iberian Peninsula (from Pellicer 1963:9 and Whitehouse and Whitehouse 1975:124).

(Whitehouse and Whitehouse 1975:123). Pellicer (1963:7) referred to the related group of sites located on the Andalusian coast which are characterized by the incised marking decorations on this pottery as the "Culture of the Costa del Sol" (see figure 56).

Fossil Opium Poppy Remains from Cueva de los Murciélagos

In 1868, Manuel Gongora published a survey of prehistoric Andalucia in which he described several weapons, utensils, and other important objects found in southern Spain. The very first section of this study (Gongora 1868:25–26) is devoted to Cueva de los Murciélagos. The following comments on the environment surrounding the cave site, the discussion of the historical aspects of its commerical use, the description of the prehistoric artifacts recovered from within the site, and the speculative interpretations of the extraordinary fossil opium poppy remains are essentially those of Gongora (1868:25–56). I have translated from the Spanish original and paraphrased information from those sections considered to be revealing and pertinent to this study.

A few kilometers to the east of Albuñol, a coastal town in the province of Granada, there is a deeply eroded and mostly barren maze of limestone ravines and gullies characterized by very steep cliffs and the presence of precipitated salts. The Cueva de los Murciélagos site is located beside one of the steep cliffs within this wasteland environment, about seventy meters down from the surface plateau and fifty meters above a dry ravine bed (see figure 57).

Cueva de los Murciélagos, as its name implies, was, at least at one time, the

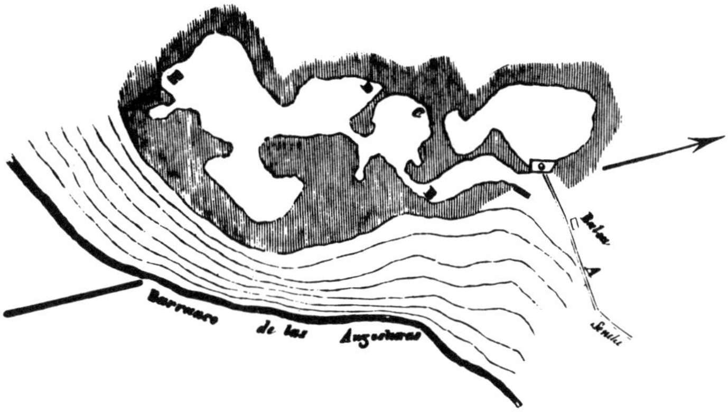

FIGURE 57 a. The plan of the Cueva de los Murciélagos located near Albuñol, Spain (from Gongora 1868: 50, figure 4). Scale = 1/1000.

FIGURE 57 b. Artist's sketch of Cueva de los Murciélagos (from Gongora 1868: 27, figure 2).

FIGURE 57 c. Artist's sketch of the dry, eroded environment surrounding Cueva de los Murciélagos, the site where 5,000-year-old opium poppy fossils were found among grave gifts (from Gongora 1868: 26, figure 1).

domicile of a bat population. The bats were apparently abundant and probably inhabited the cave for a relatively long time. In 1831, Juan Martin, the owner of a livestock fertilizer firm, found considerable deposits of bat guano inside the cave. Subsequently Martin and others explored some areas of the cave and located concentrations of dung which they extracted and sold. Eventually, in order to concentrate manure for commercial exploitation, the cave was used as an enclosure for domesticated animals. Extraction of bat guano and the impact of the enclosed animals probably had a significant effect on the artifacts and human remains which were left in the cave.

Around the middle of the nineteenth century, discovery of lead deposits in the cave spawned interest in a mining operation. In 1857, a company was formed and an organized effort to remove the lead was initiated. However, during their search for valuable mineral deposits within the cave, the miners made an unusual and provocative discovery: first they found three human skeletons buried together. Most notably, one of these, a male, was adorned by a coarse but pure gold crown.

As fortuitous as this original discovery may have been, it also had some very regrettable aspects. In their frantic attempts to find additional gold or other valuable grave gifts, the miners tore apart sections of the cave with overwhelming abandon and a great deal of violence. In their rash, excited state, they emptied prehistoric baskets, broke ancient jars, disarticulated many skeletons, and threw precious deposits over the precipice to the bottom of the gorge. Consequently the archaeological record for this remarkable "asylum of the dead" is woefully incomplete (Gongora 1868:36). Nevertheless, the relatively abundant human remains and associated artifacts recovered from the cave and described by Gongora are quite impressive. They provide us with several insights into the prehistoric culture of the people who deposited the dead bodies and grave gifts in the dark sanctuary of this Mediterranean cave.

In association with the communal burials of Cueva de los Murciélagos, the cave excavators found tools, weapons, utensils, and numerous woven objects made out of *esparto* grass (*Stipa tenacissima* L.). Many of the woven objects were handbags. Most of

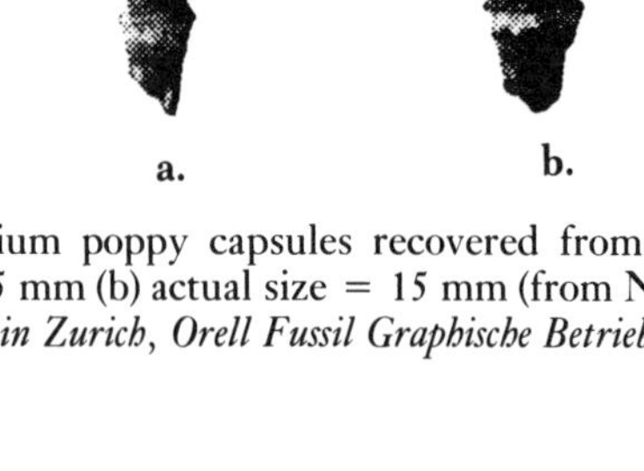

a. b.

FIGURE 58. Two of the opium poppy capsules recovered from Cueva de los Murciélagos: (a) actual size (diam.) = 13.5 mm (b) actual size = 15 mm (from Neuweiler 1935:109; *Courtesy of Naturforschende Gesellschaft in Zurich, Orell Fussil Graphische Betriebe AG, Zeitschriften-Abteilung, Zurich, Switzerland*).

these were relatively small, skillfully fabricated containers which had been placed at the side of each human burial. According to Gongora (1868: 35), the handbags contained a variety of funeral offerings including flowers, small plants, sea snails, fragments of attractive or transparent rocks, locks of human hair, many opium poppy capsules *(cabezas de adormideras)*, and a large quantity of opium poppy seeds *(semillas de adormideras)*. Gongora (1868: 55) suggested that all of the "poor, simple, poetic and eloquent" grave gifts in the handbags were left in the burial cave as "articles of remembrance and love" or for "tender memorial."

In his specific reference to the opium poppy remains, Gongora (1868: 56) argued that the number of capsules found in the cave site is quite large and, therefore, represents more than just a symbolic effort to conserve some viable seeds for an afterlife. He believed that the quantity of capsules indicates that they were placed in the burials because the opium poppy was "the symbol of a dreamfilled sleep and the image of the dead"!

In support of his argument, Gongora (1868: 56) referred to the great respect that the Romans had for the opium poppies of Spain which they knew under the name of "*papaver ibericum*" and from which they obtained a very powerful opium drug. Ac-

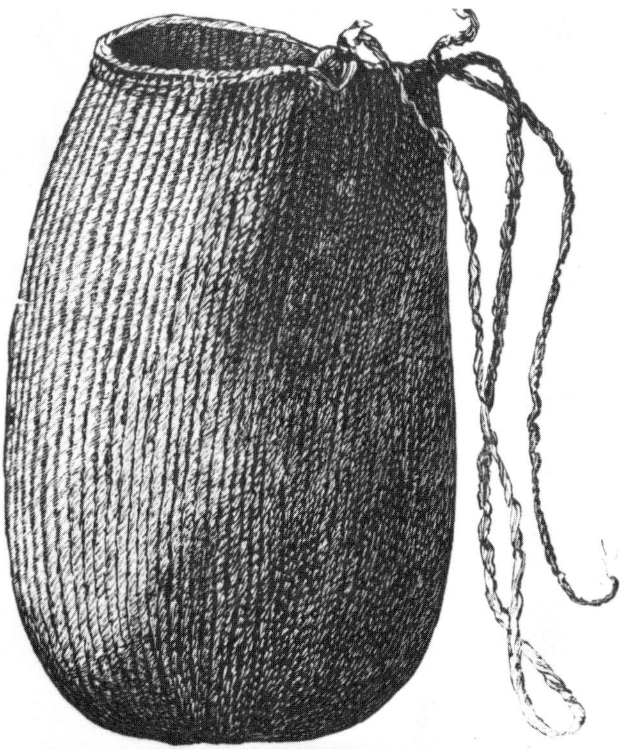

FIGURE 59. One of the handbags made out of esparto grass (*Stipa tenacissima* L.), in which were found many items including the seeds and capsules of the opium poppy (Gongora 1868: 55). Actual size = 35 cm.

cording to Gongora, Licinius, a Roman nobleman, saw such great power in the drug product of this plant that he used it "to put to rest with eternal dreams the perpetual and bothersome illness that had afflicted him."

Gongora's interpretation of the fossil opium poppy remains and their ritualistic role in the prehistoric burial cave near the southern Spanish coast strengthens my own general proposition that opium poppy's mind-altering qualities played an important role in the folk healing and spiritualism of some prehistoric peoples. I think the early people who utilized Cueva de los Murciélagos were such a people.

It should be pointed out here that the discovery of the opium poppy capsules in the context of the burials at Cueva de los Murciélagos is not the only archaeological evidence of psychoactive botanical objects being used as memorial grave gifts. For example, an Inca burial site at 17,712 feet on El Plomo peak in the Chilean Andes has yielded an extraordinary frozen mummy of a small male child whose funeral offerings included a feathered pouch full of coca leaves (McIntyre 1973:733). The coca plant (*Erythroxylon coca* Lam.) also has a lengthy history of drug use for medicinal, recreational, and religious purposes which dates back at least five, perhaps many more, thousands of years (Plowman 1979:103).

The Greek historian Herodotus (4.73–74) made a reference to the ritualistic use of hemp (*Cannabis sativa* L.) by Scythian nomads in the area northeast of Macedonia. He referred to the inhalation of hemp smoke as a means of cleansing after completion of a burial ceremony and described the peculiar tentlike apparatus which was employed in the ritual. Scythian tombs uncovered in the Altai Mountains of Central Asia have

FIGURE 60. Artifacts found in the Altai Mountain tombs of the Scythians (mound 2). "(a) The frame of an 18 in.-high tent in which the hemp smoke was collected. (b) A censer for burning hemp. (c) Pot containing hemp seeds" (from Godwin 1967:43, figure 1; *Courtesy of Sir Harry Godwin, Cambridge, England*).

yielded artifacts which confirm the authenticity of Herodotus's historical description:

> One particularly interesting apparatus was a kind of cone-shaped miniature tent, covered with a felt or leather rug, standing over a copper censer. Hemp seeds found on the spot suggest that this contrivance was a special enclosure that could be filled with narcotic smoke from burning seeds. (Artamonov 1965:108; see figure 60).

Although scholars have referred to the Scythian use of hemp as the inhaling of hemp *seed* smoke, it should be noted that the psychoactive potency of the seeds is much less than of flowers and leaves of mature, dried plants. Since the seeds are more resistant to combustion, they would remain after the rest of the plant was burned. "This would explain the preservation of seeds discovered by archaeologists" (Merlin 1972:60). In any event, the presence of the plant remains and the smoke-trapping device found in the context of an important burial, indicates that the Scythians did use hemp for mind-altering purposes (see Rudenko 1970:284–85). I believe that the coca leaves from the burial site in the Chilean Andes, the hemp remains from the frozen tomb in central Asia, and the opium poppy capsules from the necropolis in southern Spain may all be interpreted as grave gifts which symbolized the psychoactive powers of the respective plants and which also had important spiritual associations.

The Age of the Opium Poppy Fossils in Cueva de los Murciélagos

If we assume, as Gongora (1868:36) did, that the Cueva de los Murciélagos was used as a tomb during the Neolithic period, then can we also assume that the opium poppy was present in the western Mediterranean region during all of this initial agricultural period? The contents of the necropolis near Albuñol fail to confirm both of these assumptions. For example, the golden crown and other materials recovered from the grave indicate that this certainly was not an early Neolithic site. According to Clark (1952:108), the archaeological discoveries at Cueva de los Murciélagos furnish us with evidence indicating that the opium poppy was "being grown in Spain at a time when metal was coming into use." Pellicer (1963:49) specifically disagreed with Gongora and argued that the contents of the Cueva de los Murciélagos were not placed inside the natural tomb during the Neolithic period. According to Pellicer it is more likely that the grave goods were deposited in the cave during the Early Bronze Age. He suggested that his conclusion is supported by the evidence of metal use and the practice of communal burial, two traits which are associated with the Early Bronze Age of the Mediterranean coast of Spain.

In 1935, Neuweiler and Netolitzky examined four of the fossil opium poppy capsules found in Cueva de los Murciélagos. According to Neuweiler (1935:108), these fossils were sent to him by Professor Obermaier, a well-known Spanish prehistorian from Madrid. Obermaier told Neuweiler that the fossils from Cueva de los Murciélagos were approximately 4,500 years old (ca. 2500 B.C.). If we assume that Obermaier's date is roughly accurate in the uncorrected chronology, then a recalibration (according to published curves) would push the date of the contents of Cueva de los Murciélagos—including the fossil remains of the opium poppy—back to about 3100 B.C. This tentative conclusion is supported by a radiocarbon date from the closely related cultural deposit at Cueva de Nerja, which provided the first clear stratigraphic record for this region and the "Culture of the Costa del Sol" (Pellicer 1963):

> In Chapter 1 of the Nerja cave a pit 40 cm deep and a metre in diameter, perhaps originally lined with irregular limestone slabs, contained thirty litres of cereals and acorns. Just over 85% by weight were naked barley, and most of the remaining quantity wheat, probably Emmer (Hopf and Pellicer, 1970). Olives and acorns were also represented. A red-painted sherd was found on the floor of this pit, which has been dated by radiocarbon assay of the cereals to 3115 ± 40. (Phillips 1975:106)

Regardless of its absolute date of use, we may assume that the Cueva de los Murciélagos functioned as a communal funeral tomb during the latter part of the Neolithic or incipient stages of the Copper or Bronze Age. In a relevant note, Savory (1968:88) pointed out that the western Mediterranean region experienced "a climatic optimum in the moist Atlantic phase in the late fourth and the third millennium B.C." This indicates that the opium poppy probably could have been grown locally by the people who used Cueva de los Murciélagos as a communal burial place. If this was the case, how did the fossil opium poppy capsules and seeds arrive in the cave near Albuñol? Did these funeral gifts come from nearby, or were they exotic trade items from beyond the southern Spanish coast? If it was cultivated or wild in this area during the Late Neolithic or Early Bronze Age (which I think is supported by the large number of capsules and seeds found in the burial cave), then perhaps the opium poppy was a native plant in the western Mediterranean region.

The Origin of the Opium Poppy Fossils in Cueva de los Murciélagos

In his published report, Neuweiler (1935:111) argued that the poppy capsules from Cueva de los Murciélagos should be treated as *Papaver somniferum* var. *nigrum*. According to Fritel (1920:208), the fossil poppy capsule found in a tuff deposit at Aygalades near Marseilles in the lower Rhone Valley is also a specimen of *P. somniferum* var. *nigrum* (see figure 36).

This intriguing fossil capsule has not been associated with evidence of human activity and its exact age is unclear. Fritel (1920:208) argued that it belongs to the "Quaternary Period"; but Neuweiler (1935:112) called for a closer examination of the evidence before making a determination (see chapter 3). It should be pointed out here that the geographical origin of this fossil places it relatively near the Impressed Ware site of Châteauneuf-les-Martiques (5570 ± 240 B.C., uncorrected) and the "proto-Chasséen" site of Escanin 2, les Baux (3020 ± 250 B.C. uncorrected; see Bender and Phillips 1972:97–105). In other words, the fossil capsule site and the two Neolithic sites are all in the "Bouches-du-Rhone" region of southern France. If Fritel's chronological interpretation of the fossil capsule is incorrect, then the fossil may belong to the period when prehistoric farmers occupied the region. Neuweiler

(1935: 112) did note that the relatively recent age of the fossil is suggested by the "large-fruited" size of the capsule. In any case, too many questions must be answered before we can use this fossil capsule as reliable evidence of the opium poppy's age or origin in the western Mediterranean region.

Although the late Neolithic cultures of the western Mediterranean region were generally isolated from one another, there appear to have been a few strong, cross-cultural similarities. One of the most outstanding traits shared by a number of these culture groups was the use of collective burial tombs. According to Savory (1968: 90), the communal burials first appeared in Iberia in the caves of Andalusia and Murcia. Pellicer (1963: 49) argued that this would involve the "incised ware" sites of the "Culture of the Costa del Sol" to which Cueva de los Murcièlagos belongs (more than fifty skeletons were found in Cueva de los Murciélagos; see Gongora 1868: 15). This would also involve the famous, and roughly contemporaneous, Megalithic settlements of western Europe. One of these well-known monumental stone settlements, Los Millares (radiocarbon dated to about 3100 B.C.), is located on the Mediterranean coast of southern Spain only about fifty miles to the east of Albuñol. In fact, the practice of collective burials in natural caves, rock-cut tombs, or huge stone structures, (i.e., the Megalithic mausoleums) was a culture trait that became common in most parts of the Iberian Peninsula during the Early Copper Age (Savory 1968: 84); "Several thousand of them are found in Iberia, mostly in the South and West, dating from before 3000 B.C. to after 2000 B.C. in different regions" (Phillips 1975: 140).

This period of early metal use and collective interment in Iberia was also a period of relatively extensive trade development. According to Phillips (1975: 154), the rise of commerce was "based on sea communications." These communications probably played a role in the adoption of the communal burial trait and the exchange of certain goods such as some metals, some sea shells, obsidian, flint, beads, and amulets. The southern coast of Spain appears to have been one of the major areas involved in this maritime trade because of the abundant deposits of copper-ore. These deposits "were exploited early, leading to an established tradition of early metallurgy, trade networks, and a proto-urban situation by about 3000–2500 B.C. (Rodden 1975: 97).

According to Hibben (1958: 185), the site at Los Millares "became a great trade center" during this early period of metal use in southern Spain. Consequently, the people who used Cueva de los Murciélagos were probably affected by and may have even participated in this maritime trade. If so, the opium poppy may have been an item of trade or a species which was introduced into the area as a result of the trade development. If we assume that the opium poppies were placed in Cueva de los Murcièlagos on the eve of the Bronze Age and are approximately contemporaneous with the Megalithic cultures, then they may have been associated with the widespread pursuit of ancestor prestige, which became very important in the region during this period (see Phillips 1975: 152).

If the opium poppy is native to the Mediterranean region, but not specifically to the area around Cueva de los Murciélagos, then from where was it brought into the territory surrounding the cave site? Savory (1968: 77–78) suggested that the Incised Ware group (to which Cueva de los Murciélagos belongs) owes at least some important aspects of its cultural heritage to contacts with the Eastern and Central Mediterranean region during the middle of the fourth millennium. However, Savory (1968:82) also pointed out that during the "late neolithic a cultural tide began to flow from north-east Africa along the south Mediterranean coast into the [Iberian] peninsula, and that this was reinforced during the early Copper Age." A third opinion suggests that the Iberian Copper Age and the associated cultural developments evolved locally along the coast of Iberia (Whitehouse and Whitehouse 1975:128).

Although we must remember that a relative lack of scholarly interest in archaeobotanical data (see chapter 1) may help explain why we have fossil opium poppy evidence from only one site in the western Mediterranean region, the fact remains that this is the only confirmed evidence of the plant's existence in the region during the prehistoric period. As is the case for the other regions I have considered in this chapter, the paucity of direct evidence hinders the acceptance of speculations regarding the cultural origins of the opium poppy. Nevertheless, if it is assumed that the opium poppy was native to the western Mediterranean region and that its earliest use developed there, then perhaps the cultural diffusion of the drug plant was part of a larger cultural complex which spread from the southwest to the northeast of Europe during the Late Neolithic period. According to Savory (1968:83) ". . . the development of a . . . range of plain pottery forms about this time in the Chassey, Lagozza, and Cortaillod cultures north of the Pyrenees is most probably the result of influences from south-western Europe connected with the spread of megalithic architecture."

However, it can be argued that the cultural contact between Iberia and west central Europe, which Savory referred to, may have actually been responsible for the introduction of the opium poppy into the western Mediterranean region. In other words, the drug plant may have been brought down from the Swiss Foreland during the Late Neolithic or Early Bronze Age as a result of increasing communication and the exchange of trade items. Perhaps the opium poppy was first used in west central Europe. We certainly have ample evidence of its early association with human groups in this region.

West Central Europe

Our archaeological review of the Middle Neolithic sites in the Swiss Foreland indicates that the opium poppy was well known in the region at an early date.

> Personally I have established *Papaver somniferum* as a very common plant (weed?) from earliest times [?] in Switzerland. One single seed in Iron Age Britain. That is all—and I have seen a lot. Not in Scandinavia, not in Italy, not in Turkey etc. Negative enough for me to be positive proof of absence. (Hans Helbaek, personal communication, November 19, 1976)

Native hunters and gatherers in west central Europe may have first learned how to farm from a culture group to the west, south, or east, and then incorporated an indigenous (or inadvertently introduced) opium poppy population into their complex of cultivated plants. Is it possible that native Mesolithic or Upper Paleolithic peoples found this species in restricted habitats such as open landslide areas, morainal deposits, or talus slopes where the plant could have evolved an adaptation to such a naturally exposed environment? The opium poppy is a heliophyte, needing an open habitat, which would have been prevalent in the Post-Glacial period.

On the other hand, the opium poppy may have been introduced to this region from elsewhere, but evolved new characteristics after its arrival in the Swiss Alpine Foreland. As in the case of wheats in Neolithic Switzerland, Murray (1970: 56) stated that there is no wild form of the naked wheats which could have been derived from emmer:

> The Cortaillod culture is one of the earliest known European cultures to center its cultivation upon naked wheats, and it is suggested here that this chromosome aberration may have taken place in the central alpine zone of Europe as a result of environmental conditions. It is possible that this took place in the highest cultivation zones, which would be exposed to more ultraviolet and cosmic radiation, and when the advantages of a glumeless wheat were realised, the grain was generally traded for sowing at lower altitudes.

The opium poppy may have evolved under circumstances similar to those suggested for the naked wheats by Murray; in other words, it could have developed as a result of a series of mutations out of an introduced population of the setaceous poppy (i.e., *P. somniferum* subsp. *setigerum* or *P. setigerum;* see chapter 3).

In any case, it has been pointed out earlier (see chapter 1) that the concentration of fossil poppy seeds could be explained by local ecology rather than the actual prehistoric distribution of the species. The relatively abundant fossil remains recorded in west central Europe (especially in the Swiss Foreland) may be the result of the excellent preservation conditions in this region, as well as a general interest in archaeobotanical remains, and does not prove that the opium poppy was native there.

Czechoslovakia and Poland

It should be noted here that two additional Neolithic sites in Europe have yielded fossil seeds of the opium poppy (see figure 52). Recently, Sikulová and Opravil (1974: 124–25) reported the discovery of fossil opium poppy seeds among the prehistoric animal and plant remains found in the ancient settlement at Vavrovice near the Opava River and the Polish border in central Czechoslovakia: "According to the short German summary it seems that the material comes from the 'Epiatlantikum'" (Schultze-Motel, personal communication, August 19, 1978). This chronological association suggests that the fossils date from some time after 3000 B.C., or the end of the Atlantic Period of prehistoric Europe.

Furthermore, Gizbert (1960: 349–54; see also Klichowska 1976: 27–67) reported the discovery of twelve fossil opium poppy seeds at Zeslawice, near Krakow in south central Poland. This Polish find is associated with the Neolithic culture of "radial decorated pottery" (Schultze-Motel, personal communication, August 19, 1978) and is located within one hundred miles of the Neolithic Czechoslovakian site at Vavrovice. Moreover, both of these sites are near the important prehistoric culture route of the Danube River.

The absolute dating of these unusual fossil opium poppy finds and their chronological comparison with the Neolithic fossils from west central Europe and the Rhineland may give us some significant insight into the cultural origins and spread of the plant in Europe. It may have moved into what is today Czechoslovakia and southern Poland from the south via the Danube and Morava River Valleys, thus skirting the White and (much larger) Carpathian Mountains; or perhaps it moved in the opposite direction and was cultivated or known as a weed in northeastern Europe at an earlier period than in the important settlements in the Swiss Foreland or Rhineland.

The prehistoric data base increases and the distribution of fossil finds becomes more extensive, but the cultural origins and prehistoric spread of the opium poppy remain essentially unclear. Additional evidence is needed to produce a more complete record of the distribution and age of the opium poppy in Europe and western Asia. In fact, we have to have an increased application of the sophisticated techniques now coming into widespread use among the students of archaeobotany before we can expect to make a definitive determination of the opium poppy's origins.

In this chapter, I have reviewed the data now available. Although there are a number of possible interpretations regarding both the natural and cultural origins of the opium poppy, I hope that this survey will stimulate interest and encourage further research.

7 • Origins of the Opium Poppy in the Eastern Mediterranean Region during the Late Bronze Age

For more than a century archaeologists and historians have utilized a convenient dating system based on the spreading use and production of metals. This scheme has enabled scholars to bring some sort of chronological order to the study of prehistory. Under this temporal system, archaeological sites are defined primarily in terms of the absence (i.e., the Stone Age) or presence of certain kinds of metal objects (i.e., the Copper, Bronze, and Iron Ages).

The age of metals begins with the use of copper (i.e., the Copper, or Chalcolithic, Age). Later, craftsmen learned how to make *bronze* by combining copper with small amounts of arsenic (in an approximately 9:1 ratio). This new metallic alloy was considerably harder and easier to cast, yet still retained the fine ductility and malleability of pure copper (Muhly 1973:169).

During part of the *Early Bronze Age* there was a relatively extensive use of arsenic and copper to produce bronze. However, working with arsenic was quite dangerous and thus its use in the production of bronze was limited. Consequently, tin, an uncommon but safer metal, gradually displaced arsenic in the formulation of bronze. By about 1600 B.C., the widespread use of copper alloyed with tin marked the beginning of the *Late Bronze Age*.

In this study the Late Bronze Age is to be understood as that period in which the predominant composition of bronze was a combination of copper and tin *and* during which extensive trade in metals such as tin and copper and other valuable luxury goods such as amber, obsidian, faience, and perhaps the opium poppy (or its products) became relatively abundant and important (i.e., beginning about 1600 B.C.).

The Late Bronze Age was definitely a period of many important and related developments. For example, this period corresponds to the Wessex Bronze Age in southern England during which several famous megalithic structures, such as Stonehenge, were built. This is also the time when (1) local bronze cultures developed into full bloom in Scandinavia and the Baltic region; (2) important trade contacts between Greece and other parts of Europe were established; (3) the great Alpine mountain copper mines of Europe were initially utilized on a relatively large scale; (4) the magnificent Minoan civilization disappeared; (5) the heroic and first truly Greek civilization of the Mycenaeans emerged as a formidable power; and (6) the foreign reign of the Hyksos rulers was overthrown in Egypt and the New Kingdom regained native control in the Nile Valley. The Late Bronze Age was a very complex and revolutionary period in the eastern Mediterranean region because of increasing cultural contact and international trade (Chard 1975 : 246).

Earlier in this study, I discussed the possibility of the opium poppy being native to the Near East or western Asia and presented the "evidence" suggesting that its earliest use occurred in this region. As I noted in chapter 6, many scholars have referred to the Near East as the general area in which the species evolved and was first used by man. However, I also pointed out that these assumptions have been elegantly challenged in a recent article by Krikorian (1975). I am inclined to agree with Krikorian regarding the earliest use of the opium poppy in the Near East region.

Nevertheless, during my own research I have recognized a familiar chronological pattern for the data which I consider to be indirect evidence of the opium poppy's presence in some parts of the eastern Mediterranean region, including Greece, Crete, Cyprus, Egypt, and possibly Turkey. I believe that the opium poppy was not known in the eastern Mediterranean region until the Late Bronze Age (i.e., about 1600 B.C.), when it was introduced from either central Europe or the western Mediterranean region. Furthermore, I suggest that the transmission of the opium poppy into the eastern Mediterranean region was stimulated, consciously or unintentionally, by the important European and Mediterranean trade developments involving the exchange of precious materials and new ideas. The Late Bronze Age evidence that I will present includes symbolic representations of the plant's capsules in the form of pendants, pins, beads, statues, juglets, and painted objects.

In order to understand the context in which the opium poppy spread into the eastern Mediterranean region during the Late Bronze Age, I first describe the basic commodities of the Late Bronze Age trade and the great increase in cultural exchange that occurred during this period. Then, I suggest possible reasons why the opium poppy was introduced into the eastern Mediterranean region during this period, either intentionally in the form of a valuable drug and food product, or unwittingly as a weedy component of the exotic flora.

The Tin Trade

Certainly one of the major trade commodities of the Late Bronze Age was tin, the uncommon ingredient needed to produce true bronze. Although access to alluvial tin (the only form of tin available to the prehistoric smith) was hampered by the limitations of prehistoric technology, only relatively small amounts of tin were required to forge true bronze because just one part of tin was necessary for every nine parts of copper: "It was possible, therefore, to transport small amounts of tin over considerable distances, thereby bringing tin from a distant source to a nearby source of copper" (Muhly 1973: 169). Although the sources, routes, and markets of the tin trade will be referred to in the context of the following discussion, it is important to point out here that movement of this vital commodity may have been the stimulating factor responsible for the movement and trade of other items as well.

The Amber Trade

Amber has been held in high esteem for ages because of its attractive and malleable nature. During the Bronze and Iron Ages it was a valuable raw material and item of exchange. According to Malinowski (1974: 195), " . . . the story of its commercial and artistic uses shed important light on the nature of trade in ancient Europe."

By far the major European source of amber has been the fossil resin commonly found in accessible deposits on the southeastern shores of the Baltic region. This northern European amber was a very special substance in the ancient world. It was known as "northern gold," a precious material which supposedly contained unique occult energy:

> It was not just the amber that we moderns use for jewellery; for our forefathers it was a "substance of the sun," *elektron*, as the Greeks called it—an electric shining substance, endowed with mystical powers. As precious as gold, it was sought by both the barbaric north and the civilized south. For the Baltic area it indeed became a magical "metal"—*metallum sudaticum*, or "exuded metal," as Tacitus called it—a key to prosperity and a basic link with the rising central European culture and Mycenaean Greece. (Gimbutas 1963: 55)

The fact that amber was believed to be imbued with special supernatural powers and thus worn in the form of beads is relevant to this study, for I suggest in subsequent discussion that amuletic pendant beads in the symbolic form of opium poppy capsules were also worn by true believers who desired either the health-sustaining and -restoring effects or the more mystical benefits of the drug plant.

Before about 1600 B.C., very little amber moved out of the main source areas in the Baltic region. However, once the trade began, it developed into a relatively important cross-cultural activity involving peoples in western, central, and eastern Europe. Figure 61 depicts the principal amber trade routes that developed during the Late Bronze Age between 1600 and 1200 B.C.

> The beginnings of a transcontinental amber trade and of the Baltic Bronze Age occurred simultaneously. Amber was demanded by the rising central European Únětician culture and by Middle and Late Helladic Greece. When the Úněticians

FIGURE 61. The major trade routes for amber during the Bronze Age (1600–1200 B.C.). The source areas were located along the Baltic Seacoast (from Gimbutas 1963:58; *Courtesy of Holt, Rinehart and Winston, New York, NY*).

> established their commercial relations with the Mycenaeans sometime before or around 1600 B.C., the amber trade rapidly reached an amazing volume. (Gimbutas 1963:56)

In Greece, amber beads are first found during the Mycenaean period: "Amber artifacts have been discovered in the Peleponnesus and on Crete in Mycenaean burials dating from the thirteenth to the seventeenth century B.C.," (Malinowski 1974:195).

In fact, amber beads have been recovered from the earliest group of "shaft graves" at Mycenae; for example, Mylonas (1966:104) described the amber bead necklace recovered from Shaft Grave Omicron at Mycenae. This necklace has been dated from the "transitional period" (middle Helladic III/late Helladic II), or about 1600 B.C. (Muhly 1973:432, n.185). In addition, many amber beads were found in the shaft graves that Schliemann excavated and which date to the period 1580–1510 B.C.; furthermore, large quantities of other such amber beads were found in several Mycenaean *tholos* tombs (Gimbutas 1963:59).

The Baltic origin of the amber found in ancient Greek sites has long been demonstrated:

> The Baltic amber has a high succinic acid content, from 3 to 8 per cent, which distinguishes it from the amber from other sources. As early as 1885 amber beads from the Mycenaean shaft-graves were chemically analysed and proved to be of

> Baltic origin. Since then many analyses of subsequently discovered beads in Greece and Italy have shown the same results. (Gimbutas 1963:59)

Evidence of the long-distance trade in such commodities as amber and metal can also be found in the Baltic region itself where some unusual artifacts have been recovered:

> Thus, for instance, a statuette representing a Hittite deity of the air or lightning god was discovered in Sernai near Klaipeda in Lithuania. It must have come from Syria or Anatolia via Greece in the Thirteenth century B.C., since almost the same kind of figurines have been found in Mycenae and Tiryns. (Gimbutas 1963:59)

The Faience Trade

Besides amber, tin, and gold, another precious material that became important in the Late Bronze Age trade was the brilliantly decorated, glazed faience pottery, "especially a particular type of segmented blue faience bead" (Muhly 1973:275). During the Late Bronze Age such faience beads were in use from Egypt to England as well as other areas of northern and central Europe; according to McKerrell (1972:300) the British faience beads probably are of "Eastern," most likely of Egyptian (18th Dynasty), derivation: "A date around 1450 B.C. for their arrival in Britain fits well with the avilable evidence."

The distribution of the blue faience beads in the Late Bronze Age suggests that they moved along two trade routes connecting various cultures from the eastern Mediterranean region to the British Isles. One route crossed through central Europe and the Únětician culture area via the Garonne River to the Atlantic Coast where it probably then turned north by sea to Brittany (Fox 1964:82).

The Role of Central and Western European Peoples in the Late Bronze Age Trade

The contacts between the northern European and eastern Mediterranean regions during the Late Bronze Age were to a certain extent facilitated, perhaps precipitated, by the emergence of the earliest central European Bronze Age culture which is named after the site of Únětice (formerly Aunjetitz) in western Czechoslovakia. Previously considered to be introduced from the Near East, the origin of bronze metallurgy in central Europe is now considered to be indigenous (Whitehouse and Whitehouse 1975:162; Gimbutas 1965; J. M. Renfrew 1973).

Muhly (1973:274–75) refers to the Únětice Bronze Age culture as the crucial force in the development of prehistoric Europe in the second millennium B.C. and quotes Gimbutas in this regard:

> The rise of metallurgy produced a surplus of means of subsistence and it was the basic factor for changes in economy. The Uneticians controlled the central Euro-

> pean copper sources and created a unique metal culture. Located in the center of Europe they developed a vital trade in bronze, amber and gold across the European continent, connecting the Baltic area with the British Isles and Mycenaean Greece. (Gimbutas 1965 : 250)

According to Muhly (1973 : 275), the British Isles may have supplied both the gold and Cornish tin involved in this trade, while Ireland furnished gold, and the Baltic Region provided the amber.

On the western margins of Europe, the great Wessex Bronze Age culture of southern England (most famous for the construction of Stonehenge and other related megalithic sites) seems to have had trade contact (if only indirectly) with peoples of faraway lands:

> A trade route between Britain and Saxo-Thuringia has been reconstructed, going down the Rhine, across the North Sea, and up the Thames. The trade perhaps involved tin and amber, the tin of Cornwall being exchanged for the amber of the Baltic. (Muhly 1973 : 273)

Trade between the Mycenaean and Wessex cultures supposedly moved along the Rhine, reaching Greece by way of the Brenner Pass and the Adriatic Sea (de Navarro 1950 : 101 and Strong 1966 : 7). According to Muhly (1973 : 276), Cornish tin may have come south to the Mycenaean realm along this same major amber trade route, but that exchange (of whatever commodities) was handled indirectly in the overland section by middlemen: "The objects involved do not seem to be direct imports; they look more like local imitations of exotic foreign work."

It is important to point out that metallurgical trade in Greece did not really begin on any significant scale until the rise of the Mycenaeans in the Late Bronze Age (Muhly 1973 : 188–89). Although we do not have much evidence for local metal production in Greece during this period, both the amber trade and the establishment of contact between Greece and central Europe began during the early part of the Late Bronze Age. Furthermore, the substantial copper resources of the Alps were also first exploited during this period: "Is it possible that these developments are all related and that Alpine copper was a factor in the development of the Late Bronze Age metal industry in Greece?" (Muhly 1973 : 188).

Muhly argued that there is a lack of significant evidence indicating much real central European influence on the "metal work" of the Mycenaeans, but he did suggest that metal resource trade played an important role in their foreign contacts during the Late Bronze Age: "The contacts of Greece with [other parts of] Europe are, at least partly the result of a quest for metals, the tin of Cornwall, perhaps the gold and even the copper of Ireland, and at least the copper of Sardinia if not also the copper of the Alps" (Muhly 1973 : 190).

Late Bronze Age Trade between the Western and Eastern Mediterranean Regions

Significant trade between the eastern and western Mediterranean regions also begins during the Late Bronze Age. It now appears that it was during this period (i.e., from about 1600 to 1200 B.C.) that the Mycenaeans expanded their trade westward as far as southern Italy and Sicily, where goods of eastern and western origin were exchanged (Whitehouse and Whitehouse 1975:99; see also Brea 1966 and Taylour 1958). This East-West trade in the Mediterranean may have taken place on an island in the central Mediterranean. For example, according to Muhly (1973:280), Lipari Island (north of Sicily) "served as a free port for the western trade, the Mycenaeans bringing their industrial products to Lipari, there to exchange them for the raw materials, the metals, and the amber of the west."

It should also be noted that even though Italy was not a source area for metals during the Bronze Age, in the latter part of this period northern Italy became an important center of metal work. Rock drawings in the Camonica Valley testify to this, and lie in a geographical position along a major amber route leading southward to

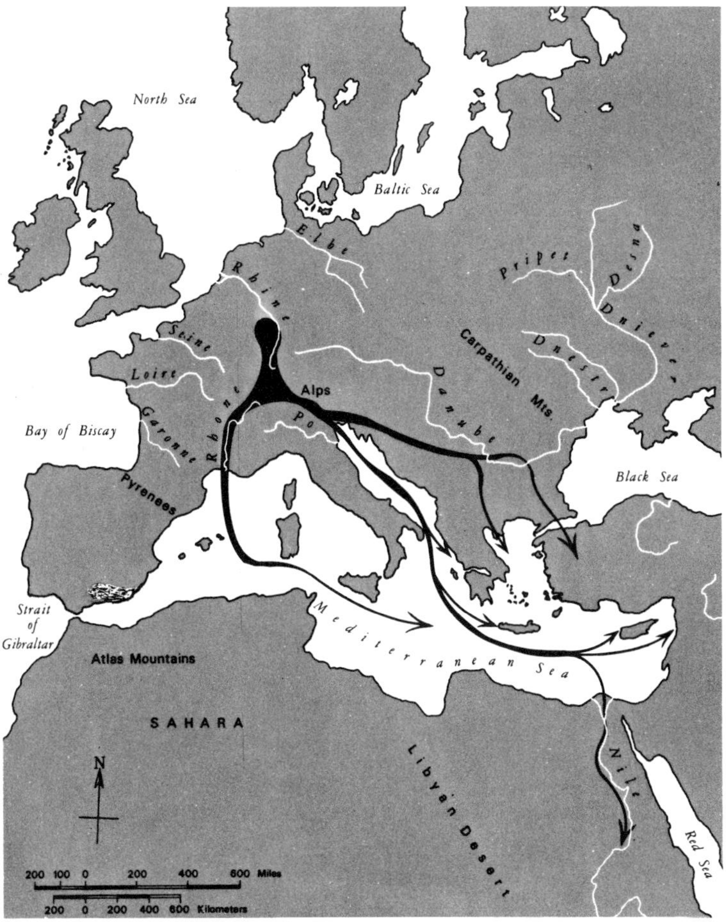

FIGURE 62. Hypothetical paths of diffusion from west central Europe to the eastern Mediterranean region along Late Bronze Age trade routes.

the Mediterranean: "The strong Indo-European element coming down from the north, brought with it the amber of northern Europe as well as alpine copper and Cornish tin" (Muhly 1973:282).

In conclusion to this cursory survey of the important and far reaching trade developments that characterized the Late Bronze Age, I turn once again to Muhly (1973:335) who summarized the major aspects of trade during this crucial period in Europe and the Mediterranean regions:

> Greece and the western Mediterranean only begin to use tin-bronze around 1600 B.C. This date coincides with the commencement of the amber trade in the Aegean and the first extensive exploitation of the tin deposits of Cornwall. It is the introduction of Cornish tin which makes possible the development of a bronze industry in western Europe and the Aegean. The new trade routes brought Cornish tin and Baltic amber into the Aegean and resulted in the development of a European civilization in the Aegean, the civilization of Mycenaean Greece. Commercial relations with the east were still maintained and Cyprus was the major source of copper used in the Aegean, but the establishment of contact with Europe introduced a new factor which was to dominate the history of Greece till the eighth century B.C. (Muhly 1973:335)

Trade in the Late Bronze Age, therefore, involved widely separated cultures which were exchanging highly esteemed goods such as amber, faience, and tin. I believe that the opium poppy, its organic products or amuletic models of its capsules also played a role in the Late Bronze Age trade. In fact, I think that the relatively significant increase in cultural contact (i.e., stimulus diffusion, trade, or migration) during the Late Bronze Age was responsible for the spread of the opium poppy from either central Europe or western Europe to the eastern Mediterranean region (see figure 62).

Late Bronze Age Transmission of the Opium Poppy into the Eastern Mediterranean Region: A Possible Trade Item, Valuable Crop, or Unintentional Weed

Before I discuss the possible reasons for the introduction of the opium poppy into the eastern Mediterranean region during the Late Bronze Age, I think it important to consider the pertinent, but skeptical, comments of a scholar who has taken a keen interest in the field of European prehistory:

> We are wont to make trade, spreading of customs, migrations of peoples, or of artisans, accountable for the diffusion of implements, of techniques, materials, and forms. Besides these agents there is another of less rational order: those who write the history of prehistoric times should sometimes think of John 3:8. "The wind bloweth where it listeth, and thou hearest the sound thereof, but canst not tell whence it cometh, and whither it goeth." It should also be remembered that the diffusion of plants offers similar problems. In other words, we shall not always, rarely indeed, be able to answer the question where the cradle of given form lay, through what agents and on what routes it spread within this belt, stretching from

the Near East to Scandinavia, in which from at least the Middle Bronze Age down to Hallstatt there was interchange in the sphere of arts and crafts. (Jacobsthal 1956:183–84)

Although we may not be able to locate specific "cradles," "agents," and "routes" of the opium poppy, we may be able to enhance our understanding of an herb which seems to have played a role in the ancient civilizations of the Mediterranean. My own hypothesis concerning the early spread of the opium poppy into the eastern Mediterranean region during the Late Bronze Age (for medicinal, if not for religious, purposes) is admittedly dependent upon a large degree of speculation. However, I will show that it is supported by a considerable amount of circumstantial evidence. In this regard, I believe that the position of Cornford (1934:220) is especially pertinent:

> Many literary critics seem to think that an hypothesis about obscure and remote questions of history can be refuted by a simple demand for the production of more evidence than in fact exists. The demand is as easy to make as it is impossible to satisfy. But the true test of an hypothesis, if it cannot be shown to conflict with known truths, is the number of facts that it correlates and explains.

There can be little doubt that the opium poppy reached the eastern Mediterranean region by the Classical period (i.e., about 2,500 years ago). This assumption is supported by the strength and weight of references to the drug plant in the ancient Greek literature (e.g., Kritikos and Papadaki 1963 and 1967). In fact, I believe that the opium poppy was known and used by the Greeks at least a thousand years before the Classical period, well beyond the chronological limits of the written records. In the following section, I consider how and why the opium poppy might have arrived in the eastern Mediterranean region near the end of the prehistoric period, during the Late Bronze Age.

According to Anderson (1967:14), when "man moves about the earth, consciously and unconsciously, he takes his own landscape with him." If we assume that the opium poppy was part of the "landscape" which was introduced into the eastern Mediterranean region during the Late Bronze Age, then the question is whether it was a "conscious" or "unconscious" introduction.

On one hand, it is possible that the introduction of the species was an *unconscious* affair, involving the carriage of the diminutive poppy seeds as a contaminant of the preferred grains. This type of unwitting dispersal has historic parallels. Even in more recent times the opium poppy itself seems to have been spread under such conditions into various parts of England (Harry Godwin, personal communication, May 21, 1976, and Herbert Edlin, personal communication, August 2, 1976). I suggest that what I wrote some years ago regarding the hemp plant (*Cannabis sativa* L.) is also appropriate to many other economic annuals including the opium poppy:

> Traveling about the world, man has been and still is a dominant dispersal agent. His herding of grazing animals and his disturbance of natural communities has vastly influenced the diffusion and establishment of many plants. Some, the so-called weeds, thrive in the wake of man's disturbance. *Cannabis sativa* is just such a plant. The means and routes of its dispersal are closely related to cultural history. (Merlin 1972:32)

On the other hand, if we assume that the opium poppy was *consciously* introduced into the eastern Mediterranean region, for what purposes was its use originally intended? As I pointed out earlier, the opium poppy is a multipurpose plant. Therefore, its introduction into the eastern Mediterranean region may have been stimulated by any one of a number of uses: for example, its seed value as a vegetal fat source or flavoring; its medicinal value as an effective pain-suppressant sedative; its use as a hedonistic euphoriant; or its use in magico-ritualistic ceremony.

Isaac (1959: 179–86) has already undertaken research related to the association of trade and the dispersal of a cultivated plant with religious importance. He suggested that the early spread of the citron (*Citrus medica*) occurred "in the course of trade" (Isaac 1959: 181). More specifically, he argued that the citron was introduced into the "Roman Mediterranean" by the Jews because it was an essential object involved in their ritual Feast of Booths ceremony: "The history of the citron is to me a striking illustration of the part, far too little recognized, played by religion in transforming the landscape" (Isaac 1959: 179).

The prehistoric relationship between religion and another more well-known and important cultivated plant, the grape vine and its product, wine, has also been discussed recently by Stanislawski (1975: 427–44). He has described how cult associations with wine consumption played a significant role in the spread of viticulture and the resultant changes in the economic geography of the prehistoric Mediterranean world: "The long prehistoric existence and association of the Great Mother cult and Dionysus have important bearing on the acceptance of the Dionysus cult and wine, and on expanded commercial production" (Stanislawski 1975: 427). According to Stanislawski (1975: 428), the special mind-altering properties of fermented grape juice were a major aspect contributing to its widespread acceptance and religious involvement:

> In its early phases the worship of Dionysus constituted a subcult, and its association with the vine might be compared to that of the olive with Athena and those of various plants and animals with other gods and goddesses; but in the case of Dionysus and wine another factor was of primary importance—that of the euphoria induced by alcohol and the feeling that the worshipper was being transported into the presence of the deity. Its votaries became so utterly involved that it was a complete religion to them. In that fact lay the distinction between the Dionysus subcult and others: the association of the plant and the god was more than a pro forma sponsorship.

Religious and ritualistic associations with euphoria or other mind-altering effects, induced by consumption of a plant product such as wine, may have also been part of the relationship that existed between the opium poppy and some culture groups in the eastern Mediterranean region during the Late Bronze Age. However, if such a relationship did occur, I think it was probably much less overt and observable than that which involved the use of fermented grape juice. In any case, I believe that the use of the opium poppy in religious ritual could have been the principal factor responsible for its cultural diffusion during the prehistoric period.

Whatever the reason(s) for its introduction, it is important to note that although the drug products of the opium poppy (i.e., the capsules and alkaloid-rich sap) are relatively small in size, they are quite potent in relation to quantity. This is significant because the means of transportation did not allow bulky items to enter into trade in the Late Bronze Age. In consideration of size: value relationships, we know that gold, tin, amber, and faience were items that met this prerequisite. The drug products of the opium poppy would have also met this size: value qualification, allowing their involvement in the commercial trade of this period.

> Overland trade could only be a trade in small and relatively costly objects and materials. The international art style of the mid-second millennium B.C. developed through the circulation of such objects and the great silk and spice caravans of later times are excellent examples of the type of goods carried overland (Muhly 1973: 279)

I should also point out that the spice trade was an important activity in the Mediterranean region during the Late Bronze Age; this trade involved the transport of such spices as anise and coriander (Brothwell 1969: 158 and Stubbings 1973: 67). Importation of opium poppy seeds for flavoring could have been part of the Late Bronze Age trade.

In summary, I suggest that the opium poppy reached Greece and other parts of the eastern Mediterranean region during the Late Bronze Age as an exotic species and probably became associated with indigenous cults. In the following chapter, I first describe some of the major aspects of the Late Bronze Age in the eastern Mediterranean region, including the cultural foundations upon which the Mycenaeans of ancient Greece built their civilization and rose to power during this period. This will serve as an introduction to my detailed discussion of the possible relationships which developed between man and the opium poppy in this region during the latter part of the prehistoric period.

8 • The Opium Poppy in the Greek Realm during the Late Bronze Age

Neolithic and Early Bronze Age Greece

The emergence of "effective village farming" in the regions of coastal Greece, coastal Italy, and Crete occurred approximately 9,000 years ago. According to Gimbutas (1975 : 201), this early Neolithic tradition in the Mediterranean region was part of a larger prehistoric cultural unit, the autonomous, matriarchal civilization of "Old Europe": "The area covered is southeast Europe, the Balkan peninsula (Greece, Bulgaria, Rumania, Jugoslavia, and the part of the Ukraine adjacent to Rumania), northward to the Carpathians and to southern Poland and Moravia, and it includes the Aegean Islands and the islands between Italy and Greece, Sicily, and also Malta." Gimbutas suggested that Old Europe played a larger role in the development of Western Culture than any of the contemporaneous, but separate, early civilizations of Anatolia, Mesopotamia, and Egypt. She also argued that the Old European civilization was basically an indigenous development: "There is no archaeological proof that immigrants accompanied by domesticated plants and stocks came from the Near East" (Gimbutas 1975 : 201).

Although we may accept Gimbutas's concept of the Old European civilization and assume that it was a native development, the factors that stimulated the internal development of this early Neolithic tradition are far from clear. In the Mediterranean parts of the region, access to the sea and maritime contact probably played a major role in the incipient stages of the Old European civilization. There can be little doubt that seaborne cultural exchange had a significant impact in many areas of the eastern Mediterranean region during the latter part of the prehistoric period. On the other hand, even though the effects of maritime trade may have been important in the development of Neolithic Greece, the extent to which they affected, perhaps were responsible for, the origins of the Bronze Age culture in Greece is also enigmatic and contested (e.g., Colin Renfrew 1972 and Muhly 1973).

The Early Bronze Age in Greece and the Aegean region lasted from about 3200 to about 2000 B.C. It was an early period of metal use and one in which village economy gradually gave way to incipient urbanism in northwestern Anatolia, the Cyclades, Crete, and southern and east central Greece (Whitehouse and Whitehouse 1975:99). Two schools of thought have formed around opposing explanations for the source of this Bronze Age culture in the Greek realm. Some believe it to have been an Indo-European culture imported from Anatolia (e.g., Alexiou et al. 1968); others argue that it was an indigenous development (e.g., Colin Renfrew 1972). Still another perspective suggests that this culture represents a synthesis of native and foreign elements in Greece (e.g., Chadwick 1976:1–3).

According to Samuels (1966:3), whatever the origins of the Neolithic and Early Bronze Age traditions in Greece may have been, by the end of the third millennium B.C. (i.e., the Middle Bronze Age), the Greek mainland was occupied by the "forebears" of the earliest truly Greek people. However, while these patriarchal "proto-Greeks" were establishing a revolutionary culture on the mainland during the latter part of the Middle Bronze Age, the influential, non–Indo-European, matriarchal civilization of the Minoans was still thriving on the island of Crete.

Although the Minoans are not considered to have been Greeks, they did make a considerable contribution to the Greek culture and certainly affected the latter period of prehistory in the eastern Mediterranean region. According to Gimbutas (1974:236), the Minoans were the last representatives of the Old European civilization in the Mediterranean. She argued that the "mythical imagery and religious practices" of the Minoans were extensions of the Old European culture, which was a "worthy parent of the Minoan culture."

The Minoan Civilization: Florescence in the Middle Bronze Age

Arthur Evans's famous archaeological discoveries on the island of Crete during the early part of this century brought to light the existence of the maritime Minoan civilization. Histories of the Greek realm usually begin with a description and analysis of this highly developed Cretan society which was contemporaneous with the early Egyptian dynasties. The Minoan civilization is usually divided into three general periods: the Early Minoan (ca. 2600 to ca. 2000 B.C.); the Middle Minoan (ca. 2000 to ca. 1600 B.C.); and the Late Minoan (ca. 1600 to ca. 1200 B.C.) (e.g., Evans 1964; Simpson 1965:5; Whitehouse and Whitehouse 1975:96–97; and Alexiou et al. 1968).

A revised chronological interpretation of the pre-pottery layers at the major Minoan site of Knossos indicates that Crete was originally colonized by migrating farmers about 9,000 years ago (Gimbutas 1965:201). By the Early Bronze Age (i.e., the beginning of the third millennium B.C.), maritime trade in the eastern Mediterranean region had reached a relatively advanced level of development (Ward 1963:41

and Colin Renfrew 1972). The role of the Minoans in this seaborne activity seems to have been substantial; even by the Early Minoan period, Cretan metallurgy had already become an important industry and had been affected by Minoan contacts with the Near East.

The growth of the settlement at Knossos is representative of Minoan development; it evolved from a small village of but 2,500 square meters in the "aceramic" Early Neolithic period to a palatial urban center of approximately fifty hectares in the height of the Minoan civilization—ca. 1900 B.C. in the Middle Bronze Age (Whitehouse and Whitehouse 1975:96). By this time Crete had become the central base of a major seaborne civilization. This was the golden age of the Minoans, "Pax Minoica"; palace towns had been built at Knossos, Phaistos, Mallia, Gournia, and Zakro; and widespread maritime contacts had been established with several areas of the eastern Mediterranean region, as indicated by the discovery of Minoan pottery on Melos in the Aegean, at Lerna in the Peloponnese, and on the coasts of Egypt and Syria (Alexiou et al. 1968:3).

During this period of florescence (i.e., ca. 2000 to 1600 B.C.), the Minoans probably received imported copper from Cyprus and may have even established colonies abroad (Alexiou et al. 1968:3 and Muhly 1973:190), such as the trading base at Ugarit on the Levantine coast (Culican 1966:47). According to Stanislawski (1975:433), the palace-center at Zakro on the east coast of Crete was constructed in order to facilitate the Minoan commerce: "That Zakro was chosen for trade is obvious from its usefulness as a harbor in a surrounding area of almost no productive importance."

Although the maritime commerce of the Minoans in the eastern Mediterranean region seems to be well accepted, evidence (i.e., "stylistic parallels") indicating that the Cretan civilization also had significant intercourse with areas to the west is now disputed. For example, suggestions made by Branigan (1968:53) regarding Minoan importation of copper and tin from western Mediterranean sources in the Early Bronze Age have been challenged by Muhly (1975:191–92, 370–71, n.143, 143a), who argued that Branigan "seriously exaggerated the importance of Minoan contacts with the West."

In any case, the strong maritime orientation of the Minoan civilization in the eastern Mediterranean region did form the basis for a great expansion of overseas contact and cultural exchange, which flowered after the decline of the Cretan hegemony in the beginning of the Late Bronze Age. *I believe that this maritime orientation facilitated, perhaps precipitated, the spread of the opium poppy into some parts of the Eastern Mediterranean region during the Late Bronze Age.*

Most importantly, the Late Bronze Age was a period in which expanded land and sea trade brought the Mediterranean world in closer contact with the rest of Europe and parts of western Asia and northern Africa. However, the beginning of the Late Bronze Age and the associated expansion of seaborne trade did not occur until the Minoan Empire was essentially overrun by and absorbed into the earliest Greek civilization, which evolved on the mainland and was controlled by a people known as the Mycenaeans.

Although there is evidence indicating the presence and use of the opium poppy on the island of Crete during the latter part of the prehistoric period (discussed in detail in later sections of this chapter), it is important to note that all of this evidence belongs to the Late Bronze Age (i.e., after 1600 B.C.) and is probably associated with the culture of the Mycenaean interlopers who came to control Crete and parts of the Aegean, either by military overthrow or a less violent takeover.

Another important point to keep in mind is that the Late Bronze Age was characterized by a major revolution in the social prehistory of the Old European civilization. According to Gimbutas (1975 : 201), it was the matriarchal nature of the social organization which most predominantly differentiated the Old European civilization from the superseding, patriarchal (Indo-European) civilizations that usurped political control and introduced many new cultural traits into the former realm of Old Europe during the latter part of the Bronze Age.

Nevertheless, Gimbutas did point out that the superimposition of the patriarchal culture apparently did not obliterate all of the religious attitudes and symbolic expressions of the indigenous peoples in the Old European civilization. She argued that in some key respects, the cultural continuum persisted: "The earliest European civilization was savagely destroyed by the patriarchal element and it never recovered, but its legacy lingered in the substratum which nourished further European cultural developments—the Old European creations were not lost; transformed, they enormously enriched the European psyche" (Gimbutas 1974 : 238).

Therefore, although much of the Old European culture did disappear during the latter part of the Bronze Age, some important elements of the original civilization were combined with those of the new patriarchal society, especially in the realm of mythology and sacred symbolism. For example the fundamental religious figure—the Great Earth Mother or Vegetation Goddess—was not lost, but remained as a significant part of the subsequent, patriarchal Greek culture. Late in this chapter, I will discuss how *the opium poppy may have become a part of the Old European Mother Goddess cult and intimately involved with her worship in the Greek realm during the Late Bronze Age.*

Greece in the Late Bronze Age: The Rise and Fall of the Mycenaeans

The Mycenaeans were the first Greeks and theirs was considered a Heroic Age. That their intrepid, aristocratic heroes were at or near the root of the Greek ethnic group is reflected in their genealogies; most of these human idols "could trace their line up through three generations, then comes a god" (Kitto 1951 : 18). And from their earliest beginnings, there appears to have been a strong maritime tradition. In fact, knowledge and experience with the sea has played an especially important part in Greek culture history, leaving a significant imprint on their character and their spirit: "The Greeks were sailors from the dawn of their history, and, because they were bred to ships, they were saved from sinking into the narrow, parochial round which would otherwise have been the lot of dwellers in small city-states" (Bowra 1957 : 24–25). However, although their epic myths and ancient maritime tradition kept the legend of the earliest Greeks alive, the prehistory of these people was virtually unknown until the great discoveries of Troy and Mycenae during the latter part of the nineteenth century. The earliest Greeks did have a written language with Indo-European affinities (Linear "B"), but it seems to have been lost before the first millennium B.C. The oral histories and epic sagas of these people were passed on, but over the centuries they became embellished by folklore and transformed into myths that were adorned with an "aura of the supernatural" (Stubbings 1973:7).

In the Homeric tradition, the earliest Greeks were referred to as Achaeans, Danaans, and Argives. Today, the name of these early Greeks and their late Bronze Age civilization is derived from Mycenae, the first archaeological site on the Greek mainland which was associated with their culture. This famous site was discovered over one hundred years ago by the wealthy retired German merchant Heinrich Schliemann. Through his extraordinary archaeological efforts to demonstrate that Homer's *Iliad* was not fantasy, he helped "legend become history" once again (Taylour 1964:15). And yet, for all their intriguing connection with established archaeological fact, the legends and epic stories of the Mycenaeans have only limited value from the historian's perspective.

It is an outstanding feature of the Mycenaean culture that we know so little about so many aspects of it. We have found, and effectively translated, some of their written records. These are certainly helpful, but they do not tell us about Mycenaean origins, political history, and religion. We also have some material remains of their artistic efforts and other archaeological evidence, including remnants of their monumental, "cyclopean" architecture, fragments of their ceramic industry, and indications of their extensive maritime contact and probable colonization in the eastern Mediterranean. However, in spite of this diverse information, the bulk of Mycenaean history has yet to be understood (compare Stubbings 1973:85; Samuel 1966:2–3; and Chadwick 1976:15–34).

The Mycenaean era of power and influence in the Greek realm has been classified as the Late Helladic (LH) period and divided into five subperiods: LH I, ca.1600–ca.1450; LH II, ca.1450–ca.1400; LH IIIA, ca.1400–ca.1325; LH IIIB, ca.1325–ca.1210; and LH IIIC, ca.1210–ca. 1100 B.C. (Simpson 1965:5).

During the Late Helladic I period, important contact and trade with those areas of Europe beyond Greece was initiated by the Mycenaeans. I believe that it was during this period that the opium poppy and its use were introduced into the Greek realm. In the Late Helladic II period, the Minoan Empire based on Crete was subjugated by the Mycenaeans, whose influence in the Aegean became widespread. It was during this period that the opium poppy and its use were probably spread into the eastern Mediterranean region beyond the Greek mainland. In the Late Helladic IIIA and IIIB periods, Mycenaean trade and colonization in the eastern Mediterranean flourished. This was the golden age of the palace-centered early Greek civilization: "At this time . . . there is an obvious access of wealth, and the Mycenaean economy 'took off,' as modern economists say" (Chadwick 1976:6). During this later period the opium poppy was probably cultivated in some parts of the Greek realm and distribution of its drug products (even if only within a select religious group) was relatively well established.

For reasons that scholars presently disagree upon, the Mycenaean civilization appears to have fallen apart sometime in the late twelfth century B.C., during the Late

Helladic IIIC period, when the Mycenaean trade network disintegrated, their palaces and towns were deserted, and their aristocracy disappeared (Whitehouse and Whitehouse 1975:99). According to Kitto (1951:18), the Mycenaean "genealogies peter out two generations after the Trojan War, which would bring us to the traditional date of the Dorian invasion, about 1100, at which time (as excavation has shown) all existing cities on the mainland were destroyed." By the end of the twelfth century B.C., I believe that the opium poppy had become part of the semiwild vegetation (i.e., associated with environments in or near cultivated areas) in various parts of the Greek realm.

Before reviewing the indirect opium poppy evidence derived from Mycenaean archaeological sites, I will first consider some of the prehistoric man-land relationships in Greece during the Late Bronze Age. This should help develop an understanding of the environmental context in which the opium poppy became established in the Mycenaean Realm.

Environmental Change in Greece during the Late Bronze Age: Woodland Exploitation, Commercial Agriculture, and Introduced Weeds

Greece and most of the terrestrial environments of the Aegean region have arid climates and relatively porous geological formations. These conditions are not as suitable for the preservation of organic remains as the waterlogged environments in the Swiss Foreland, and thus limit the potential for finding the direct evidence of plant fossils, especially those diminutive types such as opium poppy seeds (see chapter 1). The relative lack of suitable sites notwithstanding, Greig and Turner (1974:177–94) analyzed the fossil pollen in core samples taken from Lake Philippi in Macedonia and from Lake Kopais in Boeotia.

Their interpretation of the data from the site in northern Greece at Lake Philippi shows that the first significant indications of anthropogenic vegetation change in the surrounding area involved woodland exploitation and occurred during the period from about 1900 B.C. to about 1360 B.C.: "This time span can be equated with that of the Middle Bronze Age and Mycenaean [i.e., Late Bronze Age] periods of central and southern Greece" (Greig and Turner 1974:186). During the following period from about 1350 B.C. to about 1000 B.C., there was an "increase in weedy pollen types." The data from Lake Kopais in southern Greece tend to support this chronological association with human disturbance.

Therefore, it seems that by the time of the Late Bronze Age, exploitative land use was beginning to have a significant impact on the native Greek ecosystem. According to Greig and Turner (1974:192), poor environmental management in the drier southern areas during this period apparently precipitated permanent effects. This largely destructive exploitation of forested lands in Greece during the Mycenaean or Late Bronze Age probably represents the expansion of commercial agriculture, involving perennial cultigens such as the grape vine and olive tree. According to Stanislawski (1975:434), the territorial expansion of viticulture, even into marginal terrain, had serious effects on the society, straining both the economic and social order: "Common lands that had been available to ordinary people were being appropriated by large landowners for the cultivation of specialized commercial crops, especially the vine, which was spread even to rocky slopes."

In addition to the effects of expanded viniculture the Mycenaeans also appear to have significantly altered their forests through increased production of olives. Expanded planting and cultivation of olive trees actually coincided with, and indeed probably was, a major factor affecting the significant woodland clearing of mainland Greece during the Middle and Late Bronze Age.

According to Colin Renfrew (1972:280), the Greek Bronze Age cultures were based primarily upon the production of three staple crops: wheat, olives, and grapes—the "Mediterranean Triad." Greig and Turner (1974:192) argued that this hypothesis is supported by their analysis of fossil pollen samples from Lake Philippi and Lake Kopais. They suggested that this Bronze Age agriculture (composed of the three staples) may have evolved in response to the radical, man-induced change in the prehistoric ecosystem, which produced "dry and unstable soils following forest clearance."

I believe that the removal of forest vegetation and the cultivation of annual grain crops may have resulted in the coincidental or conscious introduction of the opium poppy into the Greek realm during the Late Bronze Age. As I mentioned earlier, the tiny seeds of the opium poppy are known to mix easily with the staple grains and thus become a weedy pest in some areas. Perhaps the relatively large-scale establishment of grain fields, vineyards, and olive orchards provided the open environments in which the opium poppy became part of the semi-wild Greek flora.

An interesting point made by Greig and Turner concerns the rise and fall of olive cultivation and its relative importance in ancient Greece. During the Peloponnesian Wars, the Spartan army attacked the Athenian olive groves: ". . . and this may not have been the first time that olives had been a target of hostilities" (Greig and Turner 1974:193). It is important to remember that olives were most probably a major oil source for the Mycenaeans as they were among the later Greeks. In this case, what substitute could have replaced the oil extracted from olives? If importation of olive oil from an external source was precluded during the nonproductive replanting period, then perhaps a locally produced vegetable source was required and utilized. Even if the importation of organic oil resources was not curtailed, it may not have been olive oil that was imported. The pollen diagrams examined by Greig and Turner indicate that olive cultivation in Macedonia was intermittent, with two well-defined periods: "The chronology of these events is fairly clear at Philippi, the first phase being of a date equivalent to the Middle Bronze Age/Mycenaean period and the second equivalent to Late Dark Age/Early Classical times" (Greig and Turner 1974:192).

It should also be noted that, even though the pollen record is supportive, there is, at present, only a small amount of archaeological evidence for the olive in Bronze Age Greece; even the large ceramic ware supposedly used as separating vats and storage jars for olive oil is suspect (Greig and Turner 1974:192). In any case, vegetal oil could have been produced for consumption and use in Bronze Age Greece from olives, linseed, or perhaps, the opium poppyseed. Even if the opium poppy was not specifically cultivated for its oily seeds, as an adventitious species growing around settlements or in grain fields, its seed capsules could have been collected casually in times of need. The ancient use of poppy seeds as a famine food in the Greek realm is in fact verified in the historical record. According to Thucydides (book 4, ch. 26), in 425 B.C., during the Peloponnesian War, Athenians blockaded an army of Lacedaemonian soldiers on the island of Sphakteria, which lies south of Pylos, close to the west coast of Messenia. To the dismay of the Athenians, the Lacedaemonians were able to hold out on the island. "The secret of this protracted resistance was a proclamation issued by the Lacedaemonians offering large fixed prices, and freedom if he were a Helot, to any one who would convey to the island meal, wine, cheese or any other provision suitable for a besieged place" (Godolphin 1942:750). Although some brave persons successfully used boats to break the blockade, others "dived and swam by way of the harbour, drawing after them by a cord skins containing pounded linseed and poppy-seeds mixed with honey" (compare with Godolphin 1942:750). This shows us that ". . . a Greek rustic population in classical times was conscious of the value of poppy-seed as a sustaining and concentrated food in an emergency situation" (Seeberg 1969:11). I believe that earlier Greek populations, even during the Late Bronze Age, had access to and used the seeds of the opium poppy during periods of agricultural abundance as well as in times of need.

The Written Record of the Mycenaeans

> The most important single fact that archaeologists and historians learned through the decipherment of Linear B is that the Mycenaeans were Greeks, Greek in language and in religion. (Stubbings 1973:84)

In this discussion of the Mycenaean evidence it seems reasonable to begin with a consideration of the possibility of finding reference to the opium poppy or its products in the written records of the earliest Greeks. The non-Greek Minoan civilization that preceded the Mycenaeans did have their own written script, "Linear A"; but as of now it remains undeciphered. Nevertheless, the script is apparently not of Indo-European origin and thus is not related to the modern Greek language.

In contrast, the Mycenaean script, "Linear B," has been deciphered effectively, offering considerable insight into some aspects of early Greek culture.

For example, even though the Linear B records found on clay tablets do deal primarily with commercial matters, additional information concerning identification of Mycenaean deities and other social and religious concepts have been meticulously gleaned from the recovered inscripts (Chadwick 1976). Furthermore, the Mycenaean language does have Indo-European affinities and indicates "that Mycenaeans, unlike the Minoans, spoke Greek" (Whitehouse and Whitehouse 1975:99).

Recently I wrote to John Chadwick, who is well known for his study and interpretations of the Linear B Script and Mycenaean history. In light of the suggested Indo-European origins of the classical Greek word *mēkōn* and Dorian *mākōn*, which refer to the opium poppy, I asked Professor Chadwick about possible reference to the opium poppy in the Mycenaean written record. He responded:

> The Greek word (classical *mēkōn*) does not appear on the Mycenaean tablets, nor has any ideographic sign been so identified. Before the decipherment in 1952, this value was suggested for an ideogram, on the basis of the form in Linear A; but it has now been satisfactorily identified as wheat, and if it had been poppy, the Mycenaeans would have been addicts on a gigantic scale. (John Chadwick, personal communication, June 21, 1976)

After reading Chadwick's recent publication on the Mycenaeans (1976), I wrote again asking him to comment on the possibility that the Linear B ideogram identified with wheat (), by nature of its shape, might be a symbolic representation of the opium poppy. Again he responded:

> The ideogram in Linear B  definitely has the value wheat; its Linear A form  was what suggested the association with poppy-heads. But it is quite certain that this too is a major grain crop. That of course does not prove that the opium poppy was unknown to Minoans and Mycenaeans; as you rightly note there is archaeological evidence which suggests it was known. (John Chadwick, personal communication, August 22, 1977)

I also asked Professor Chadwick if one of the spices identified in the Linear B script could refer to the opium poppy rather than the presently suggested identifications, especially in regard to one ideogram transliterated as *po-ni-ki-jo*. Chadwick also responded to this question:

> So far, none of the spices we have listed on Linear B documents has been identified as poppy-seeds, but as you say it might lurk under one of the unidentified names. You will find a very sensible discussion of *po-ni-ki-jo* by C. M. Murray and P. Warren in *Kadmos* 15 (1976), 40–60. They conclude that it may be the red florets of safflower, which were much in demand as a dye. I am not wholly convinced by this, because its context suggests that it is something used to flavour food, though one cannot be sure. They argue, correctly I think that since it is measured by weight it must be a substance which could not be accurately measured by volume. So far, as we know, volume measure was normally employed for any seeds, e.g. coriander is always so treated. I think this is against poppy-seed. The quantities are usually in the range 2–10 kg. but there is one entry of 34 kg. This would surely be large for poppy-seed. (John Chadwick, personal communication, August 22, 1977)

If we assume that the opium poppy served as an oil source (as well as a food and drug source) and may have been imported for seed supplies, then perhaps the weights referred to in the Mycenaean records are not unreasonable. In any event, it must be admitted that presently we have no confirmed evidence for the opium poppy in the available written records of the earliest Greek people.

Poppies or Pomegranates?

In a number of the following sections, I refer to prehistoric artifacts such as beads, pendants, amulets, vases, painted figures, and pinheads. Although I believe that these artifacts represented the psychoactively potent opium poppy capsule, a number of them have been identified with one or more other plant species. The most commonly suggested alternative has been either the calyx or fruit of the pomegranate, *Punica granatum* L.

As the sole member of the monotypic family *Punicaceae*, the pomegranate is a large perennial shrub or small tree which may reach a height of approximately four to six meters. It produces fruits up to 12 cm in diameter that are generally much larger than those of the herbaceous annual opium poppy. In addition to its use as a food source, the pomegranate has provided a number of medicinal products derived from its roots, fruit, seeds, flowers, leaves, and bark.

The pomegranate and the opium poppy both had associations with the concept of

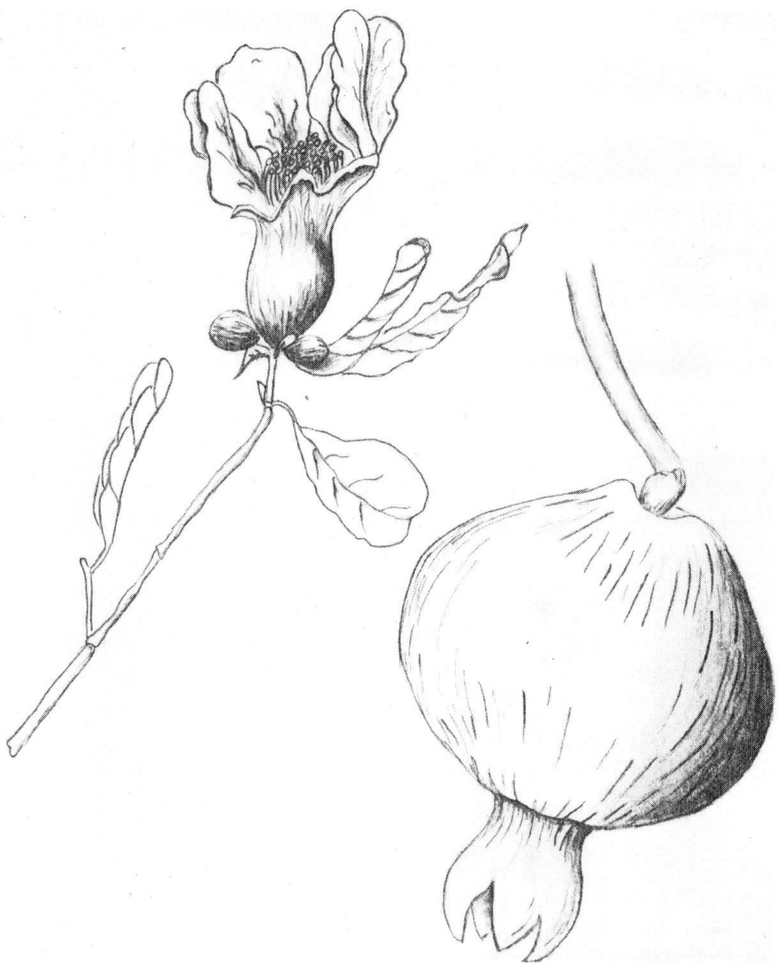

FIGURE 63. The leaves, flowers, and fruit of the pomegranate *(Courtesy of Denise Peltier, Los Angeles, California)*.

FIGURE 64. Ancient Greek depiction of the pomegranate fruit: a clay model of a pomegranate fruit, ca. seventh century B.C. (Ashmolean Collection, Oxford, England #1934.312; *Courtesy of Michael Vickers, Ashmolean Museum, Oxford, England*).

FIGURE 65. Ancient Greek depiction of the pomergranate fruit (or is it an opium poppy capsule?) *(Courtesy of Mrs. Suzanne Labiner, Beverly Hills, California)*.

immortality in the ancient Mediterranean world (compare Seeberg 1969, Kritikos and Papadaki 1967, and Jacobsthal 1956). In addition, both species produce a multitude of seeds and, therefore, have also had a long-lasting association with fertility. For example, according to Cook (1940: 814), the pomegranate is included "in a Florentine spell against sterility in women." Unlike the annual opium poppy, the pomegranate shrub has also been linked with fecundity because it has a relatively long span of reproductive potential and longevity: "It is very long lived, a specimen in Paris being known to have fruited for over 200 years" (Ansley 1935: 1420).

The antiquity of man's association with the pomegranate plant is indicated by the fact that there are words for it in several ancient languages. According to Hoffner (1974: 119) it was identified as *nu-ur-ma* in Sumerian, *1/nurmu* in Akkadian, *1rmt* in Ugarit, *nuranti* in Hurrian, *1rmt* in Egyptian, and *rimmon* in Hebrew. In the Bible, the pomegranate is the plant of immortality in the Garden of Eden. Thompson (1949: 314–16) presented information indicating that the pomegranate had a number of medicinal applications in the ancient Near Eastern region. Hoffner (1974: 120) argued that the Hittites (second millennium B.C.) listed pomegranate among their fresh and dried fruits (perhaps under the name *nurati-*); he suggested that during the period of the Hittite civilization, pomegranates were probably imported into Anatolia from the Levant and may have also been growing along the southern coast of Asia Minor. Hoffner (1974: 120) also noted that the Hittite written record indicates that pomegranate "juice (?)" was combined with "other substances" in order "to produce something which congeals (*n-at mahhan igaitta*)." This reference to a congealing compound is intriguing. Perhaps one of the "other substances" was the coagulating sap of the opium poppy. Hoffner did suggest that the compound may have been a medicine.

In any case, although I believe that both the pomegranate and the opium poppy could be found growing in the eastern Mediterranean region during the Late Bronze Age, the promegranate may have arrived in this region prior to, perhaps much earlier than, the introduction of the opium poppy. However, regardless of the sequence of their arrival, I do believe that Late Bronze Age craftsmen in the eastern Mediterranean region used both the fruits of the pomegranate and the capsules of the opium poppy as models for ornaments. In relation to this argument, it is important to remember that many ornaments (amulets) have been worn with the belief that they were imbued with the occult properties of the actual objects which they represented: "Much early jewelry originated as forms of the amulets, and much early decorative art employed magical designs or motifs" (Ansley 1935: 63). This premise can be applied to prehistoric amuletic representations of both the pomegranate and the opium poppy (Jacobsthal 1956: 188).

Unfortunately, a major problem which I have had to deal with in this study involves the identification of the botanical prototypes for the various artifacts that were used as evidence. Although there are a number of key morphological features which could be used to differentiate the species taxonomically, many of these characters are not useful because the designs of the artifacts have often been stylistic generalizations. In most cases, the key botanical characters which could have been used were not even represented on the artifacts. The problem of identification has been further complicated by the fact that the fruits of both the pomegranate and the opium poppy manifest a substantial degree of intraspecific morphological diversity.

Nevertheless, for many individual artifacts, there is sufficient evidence to make a strong case for a specific identification. For example, some artifacts can be associated with pomegranate fruits. This interpretation is strongly suggested by the characteristic remnants of the calyx protruding from an area near the apex, "without interfering—at least not much—with its contour" (Seeberg 1969:8). In contrast, small roundish artifacts with a top that is flat, raised, and has indented (or simply rounded) margins probably represent the opium poppy capsule and its stigmatic disk. Examples of what I believe to be Late Bronze Age pendant beads copied in the form of opium poppy capsules can be seen in figure 97.

In the subsequent discussion I will examine several of the conflicting identifications involving modeled artifacts which were inspired by the shape of poppy capsule or pomegranate fruit. However, it should be kept in mind that many, perhaps most, of the identifications can be accepted only on a tentative basis. In his extensive and well illustrated study of Greek pins, Jacobsthal (1956:37) refers to the difficult problem of identifying the plant source of inspiration for specific artifacts:

> We have to deal here not with likenesses of definite species but with fruit ornament. There is gradual transition from abstract forms to fruits: only pomegranate, poppies, and apples can be verified—and even these are not always with certainty; otherwise these pin-heads recall fruits vaguely, and botanists whose advice I sought had difficulty in seeing through the ornamental disguise and identifying the models.

Although one might be tempted to reject all artifact associations with opium poppy capsules as unrealistic and try to reidentify them with the fruits or calyx of the pomegranate (or some other object), I believe that it would be a mistake to do so because there is a good deal of corroborative evidence which supports the idea that the opium poppy was present in the eastern Mediterranean region during the Late Bronze Age, and probably did inspire some artistic or amuletic creations.

Mycenaean Signet Rings and Seal Beads

By 1876 Schliemann had followed his great discovery of Troy with the equally important location and subsequent excavation of Mycenae. Considered to be the principal palace-center of the Mycenaean civilization, the ruins at Mycenae have yielded an extraordinary wealth of golden artifacts. One has only to visit the National Archaeological Museum in Athens to become aware of the vast amount of gold and bronze metal work that the Mycenaean craftsmen produced.

Among the more impressive Late Bronze Age objects uncovered at the acropolis of Mycenae (under the direction of Schliemann's engineer Drosinos) is a well-known "gold seal-ring" engraved with intaglio work (Schliemann 1878:354; see figure 66).

This "great gold ring" is one of the largest of its kind and has been widely acclaimed by scholars and artisans: "The design on its bezel, cut in intaglio, is remarkable both

FIGURE 66. Greatly enlarged representation of the impression from the gold seal-ring recovered from the archaeological site at Mycenae. Note the objects held up by the seated "goddess."

for its technical perfection and its subject" (Mylonas 1966: 149). Although Schliemann did not discover the ring himself, he was the first to publish a description of the scene depicted on the ring. According to him, it portrays a woman seated under a luxuriant tree holding "in her uplifted hand [right hand in the impression of the seal-ring] three *poppies*, which she appears to offer to a tall woman, splendidly dressed, who is standing before her and extends her right hand [left hand in the impression] toward the flowers" (Schliemann 1878: 356, emphasis added). Most scholars who have associated the objects in the hand of the seated woman with the opium poppy have referred to them as capsules, not flowers. For example, Mylonas (1966: 149) suggested that they represent "poppy heads on long stalks." In addition to Schliemann and Mylonas, several other scholars including A. Evans (1928: 194, vol. 1; 1930: 458, vol. 3), Marinatos (1948: 82–83), Nilsson (1950: 347), and Kritikos and Papadaki (1967: 27) have also referred to these same objects as opium poppy capsules. Although a majority of scholars have concurred with this interpretation, at least one disagreed: "Botanical observation makes me virtually certain that they are not poppy-heads but the seed-capsules of the water-lily or lotus" (Butterworth 1966: 175). Emboden (1979: 12–13) has pointed out that *Nymphaea caerulea*, a well known waterlily of ancient Egypt (depicted on numerous sacred artifacts), produces flowers containing narcotic properties: "I was able to trace the first Old World use back to the Antilles as early as 1822, at which time *Nymphaea* was regarded as an effective substitute for opium. Many later reports confirmed that use."

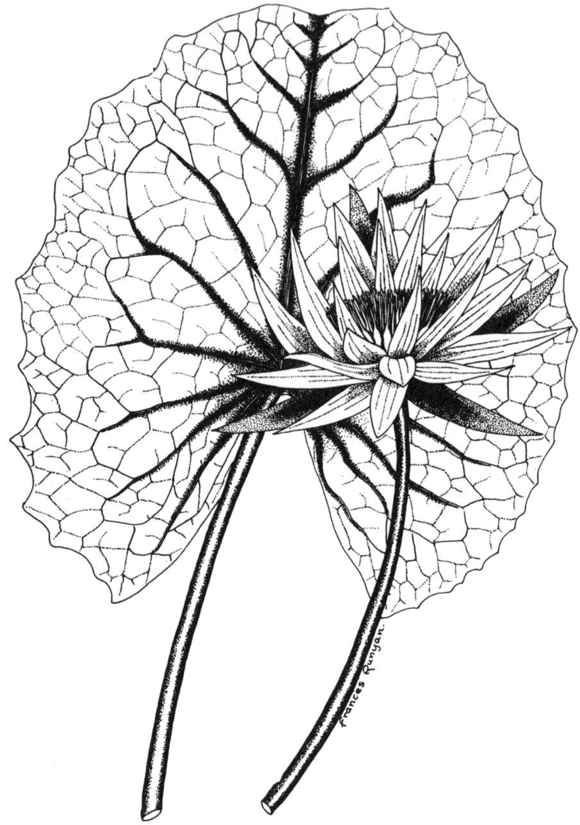

FIGURE 67. Line drawing of the water lily *Nymphaea ampla* (drawn by Frances Runyan). Aporphine (easily converted to apomorphine) has been isolated from *N. ampla* (see Emboden 1974 and 1979:50, figure 1; *Courtesy of Professor William Emboden, Department of Biology, California State University, Northridge*).

Arthur Evans (1930:458, vol. 3) refers to another "gold signet-ring" from the Late Bronze Age Mycenaean town of Thisbe near the Gulf of Corinth on which is depicted a female figure who holds out "the poppy capsules, so appropriate to an Earth Goddess" (A. Evans, *Ring of Nestor* 1925:11).

Evans went on to describe a related Mycenaean gold artifact from Thisbe which also associates the "Earth Goddess" with opium poppy capsules:

> This particular attribute, indeed, appears in her hand on the very important bead-seal from a Thisbe tomb in which she rises from the Earth like Gaia. She is helped in this case by an attendant—in a courtly fashion quite in keeping with the Minoan atmosphere. Sprouting shoots of vegetation, like nursling palms, appear on either side. This, surely, is the same Goddess who sits beneath the fruit-tree on the gold signet from Mycenae, holding the same poppy capsules. (Evans 1930:458, vol. 3).

Although the Thisbe rock tombs were not investigated by Evans himself, he did make

FIGURE 68. Line drawing representing intaglio of signet-ring from Thisbe, Greece. The artwork on the ring depicts a seated goddess being offered opium poppy capsules. The ring was probably worn by a woman during the Late Bronze Age; scale 3/1 (from Evans 1925:11, figure 11).

FIGURE 69. Intaglio on gold bead-seal from Thisbe, Greece. The artwork on the bead-seal shows a Late Bronze Age "Goddess helped to rise from the Earth by young male attendant." Note the three objects in her left hand; ". . . the poppy-heads that she holds in her hand, besides being emblems of fertility, may also in their soporific quality bear a certain allusion to her winter sleep" (Evans 1925:15–16). Scale 3/1 (from Evans 1930:Vol. III, 458).

firsthand observations of several of the relics recovered from them; and in spite of the fact that no scientific account of the associations of the individual specimens can be made because of the looting of the tombs by treasure-seeking peasants, Evans (1930: vol. 3, 457) believed that the intaglios "all were unquestionably genuine." Indeed, he assured us that all of the Thisbe tomb artifacts can be dated to the period of the Late Bronze Age and even suggested that the association between the Late Bronze Age Earth Goddess and poppy capsules was a forerunner of a similar relationship manifested many hundreds of years later during the Classical Greek period: "These are, in fact, the same poppy-heads that we see over a thousand years later in the hands of [the Earth Goddess] Hellenic successor Persephone, at the moment of her ascension from the earth" (Evans 1925: 16). It should be noted that Evans (1930: vol. 3, 457) emphasized the importance of the female deity represented in the artistic motifs of the Late Bronze Age: "Clearly, the Goddess was supreme, whether we are to regard her as substantially one being of varied aspects, celestial, terrestrial, or infernal, or whether we have to deal with separate, or partly differentiated divine entities."

The Great Earth Mother Goddess of Old Europe

We know from the mythological tradition that the goddess Persephone was the daughter of Demeter, the Classical Greek Mother Goddess. We also know that Demeter was the daughter of Rhea, the pre-Hellenic, Cretan Earth Goddess (compare Fairbanks 1910: 209 and Seyffert 1956: 177). According to legend, Rhea, the "Mother of the Gods," is also known as Gaia (Gaea or Gē), the prehistoric, older form of the Earth Mother Goddess (compare Mackenzie 1917: 174 and Seyffert 1956: 542). Gaia in turn is a manifestation of the prehistoric Old European Vegetation Goddess who "emerges with the inception of agriculture" and symbolized the divine aspect "of seed and the sown field"; she is at one with the Great Earth Mother Goddess:

> The task of sustaining life was the dominating motif in the mythical imagery of Old Europe, hence regeneration was one of the foremost manifestations. Naturally, the goddess who was responsible for the transformation from death to life became the central figure in the pantheon of gods. (Gimbutas 1974: 236–37)

This primal goddess of fertility and regeneration represents a tradition that goes back at least 9,000 years in the southeast European part of the Mediterranean region.

Kritikos and Papadaki (1963, 1967a, and 1967b) presented evidence from the ancient eastern Mediterranean region which associates the opium poppy with a number of female deity figures including those responsible for fertility and regeneration. They suggested that ". . . the symbolic meaning of the poppy, or more precisely of its capsules, varies each time according to the nature of the divinity portrayed"; they also listed a number of these suggested symbolic associations between goddess figures and the opium poppy:

(1) That of life or wealth by reason of the oil taken from its multitude of seeds;

(2) That of healing, because of its narcotic, hypnotic, and general curative properties;
(3) That of euphoria, voluptuous ecstasy, and increased fertility;
(4) That of sleep and death, which can be induced by the plant. (Kritikos and Papadaki 1967:9)

I would like to stress this point: evidence for the above mentioned relationships does not precede the Late Bronze Age in the Greek realm. Previously I suggested that the opium poppy probably arrived in Greece sometime around the middle of the second millennium B.C., either as a consciously imported crop plant or as an unsuspected weed which subsequently became part of the adventive flora in and around the cultivated fields or settlement areas. After its arrival in the Greek realm, from whatever direction by whatever means and for whatever motives, I believe that it became an "accretion" to a number of indigenous cults and identified with various deities (e.g., Earth Goddesses) symbolizing fertility, health, life, death, and mystical consciousness.

The Shaman Cult in Mycenaean Greece

Although the opium poppy's association with the Great Earth Mother Goddess continued down through the epic, archaic, and classical periods of ancient Greece, it did not begin until the Mycenaean civilization emerged and the shamanistic influences of the patriarchal Indo-European culture had entered into the Greek realm. When these influences arrived in Greece, in the Late Bronze Age, some were synthesized with the indigenous traits of the matriarchal culture dominated by the Earth Mother Goddess. Thus I believe that the opium poppy may have been introduced into the Greek realm by shaman seers who utilized this species (as well as other hallucinogenic plants) for ecstatic, divinatory (and/or medicinal) purposes. This could explain why there is no confirmed evidence for the opium poppy in the available written records of the Mycenaeans. It may have been a calculated omission. Perhaps the opium poppy's use as a cult drug was secretive (at least throughout the beginning of the Late Bronze Age). If so, knowledge of the magic plant source may have remained hidden within the confines of an elite religious order.

Such an idea appears to contradict my suggestion that Late Bronze Age ornaments (amulets) were modeled after the shape of the opium poppy capsule. Nevertheless, I think it is worth considering the possibility that these ornaments were only available to an elite or royal class in the early Greek realm. If this assumption is correct, then the movement of the opium poppy in the context of internal Greek trade during the Late Bronze Age may not have been very important. Instead, the drug plant may have been introduced as part of the complex of psychoactive plants utilized by the shamans in their roles as the magical healers of the divinatory personages.

In any case, even if the opium poppy was not a secret part of the shaman cult in Greece during the Late Bronze Age, I do think the drug plant played a role in its practices, and therefore deserves attention. Dodds (1951:136), however, has eloquently described the difficulties which one must encounter when trying to understand the nature of this ancient Greek cult:

> Discussion of the origin and history of this pattern, and its influence on ancient culture, could easily occupy an entire course of lectures or fill a volume by itself alone. All that I can do here is to consider briefly some aspects of it which crucially affected the Greek interpretation of nonrational factors in human experience. But in attempting even this, I shall have to traverse ground which has been churned to deep and slippery mud by the heavy feet of contending scholars; ground, also, where those in a hurry are liable to trip over the partially decayed remains of dead theories that have not yet been decently interred. We shall be wise, then, to move slowly, and to pick our steps rather carefully among the litter.

In spite of the myriad limitations which Dodds and others have pointed out regarding our understanding of the shaman cult in ancient Greece, again I believe that corroborative evidence does support my assumptions. For example, Dodds, himself (1951: 140–42) has suggested that in the course of trade and colonization, the Greeks were significantly affected by the shamanistic culture they encountered in "Scythia, and probably also in Thrace." According to Dodds, this culture introduced some important new elements into Greece, including a "revolutionary conception of the relation between body and soul" (compare with Meuli 1935).

In contrast to the earlier "collective" tradition, the religious experience associated with the shamanistic elements introduced from the north involved an individual "experiment in psychic excursion" (the Divine Dream) and a belief in rebirth (Dodds 1951: 143):

> . . .passages from Pindar and Xenophon . . . suggest that one source of the puritan antithesis might be the observation that "psychic" and bodily activity vary inversely: the *psyche* is most active when the body is asleep or, as Aristotle added, when it lies at the point of death. This is what I mean by calling it an "occult" self. Now a belief of this kind is an essential element of the shamanistic culture which still exists in Siberia, and has left traces of its past existence over a very wide area, extending in a huge arc from Scandinavia across the Eurasian land-mass as far as Indonesia; the vast extent of its diffusion is evidence of its high antiquity [see my discussion in chaper 4].

Meuli (1935) argued that the shaman cult had a significant effect on the Greeks during the epic period (i.e., Mycenaean Age); but Dodds (1951: 164; note 47) rejected this assumption as a "hazardous" speculation and suggested that the influence of this cult did not enter the Greek realm until the seventh century B.C., at the end of the Archaic Age. More recently, Butterworth (1966: 133) suggested that there is good reason to believe that the influences of shamanism *were* felt in Greece during the Mycenaean period:

> It is worth observing in the first place that, if we may take myth for evidence, there had been some contact in Mycenaean times between Greece and the lands lying round the Black Sea. Greek myth, as we shall see, quite unmistakably shows that

shamans were known to the ancestors of Mycenaean rulers. Indeed, it seems that in shamanism is to be found the origin of at least three of the most important clan-cults contending for existence in mainland Greece during the fourteenth and thirteenth centuries.

If Meuli and Butterworth are correct and shamanistic traits did have an impact on Mycenaean culture, then perhaps one of these traits could have been the drug use of the opium poppy in hallucinogenic ritual or sleep-dream interpretation. Dodds (1951:140) pointed out that in the ancient cult, the shaman or magical seer is able to contact the "spirit world" by passing "into a state of mental dissociation"; and from such a psychic excursion or divine dream ". . . he derives the skill in divination, religious poetry, and *magical medicine* which makes him socially important" (emphasis added). In Dodds's words, the shaman thus "becomes the repository of a supernormal wisdom." I think the opium poppy was a source of the shaman's "magical medicine."

Drawing upon the work of earlier scholars (i.e., Meulie 1935, Dodds 1951, and Kirchner 1952), Butterworth (1966:153) argued for "the existence of a shamanist context common to both Mycenaean Greece and to twentieth-century Siberia." Furthermore, he suggested that the religious conceptions of the central and northern Asian shamanistic culture had great influence "upon the spiritual life of Europe and the Mediterranean region in prehistoric times" (Kirchner 1952:245; compare with Butterworth 1966:153). As for the means by which these conceptions arrived in the Mediterranean region, Butterworth referred to a view attributed to J. Wiesner and quoted by Kirchner (1952:245–48) which suggests that

> . . . a wave of shamanist horsemen ("diese reiterlichshamanistische Welle"), impelled by the pressure of peoples in the steppes further east, reached the Mediterranean principally by sea from the region of the Euxine and Asia Minor (though some may have passed by land north of the Black Sea and through the Danube countries). The wave will have reached central Europe and left its marks at Hallstadt and Villanova about the turn of the second millennium B.C. There is, however, evidence of shamanist practice in late Neolithic times, that is, a thousand years earlier. (Butterworth 1966:153)

Dodds (1951:146–47) summarizes this spreading shaministic culture as

> a tentative line of spiritual descent which starts in Scythia, crosses the Hellespont into Asiatic Greece, is perhaps combined with some remnants of Minoan tradition surviving in Crete, emigrates to the Far West with Pythagoras, and has its last outstanding representative in the Sicilian Empedocles. These men diffused the belief in a detachable soul or self, which by *suitable techniques* can be withdrawn from the body even during life, a self which is older than the body and will outlast it [emphasis added].

In reference to the social intercourse between Greece and "lands lying round the Black Sea" during Mycenaean (i.e., Late Bronze Age) times, we should recall the very old epic legend of Jason and the Argonauts. According to tradition, the "extremely ancient" legend of the Argonauts occurred a generation before the Trojan War and the famous voyage of Odysseus; ". . . even Homer speaks of it as universally known" (Seyffert 1956:63).

In search of the Golden Fleece, Jason and his intrepid Argonauts ventured far from their native homes in Greece. Their ultimate destination was "Colchis, lying at the farthest end of the Black Sea" (Seyffert 1956:61). In a garden at Hekate near Colchis, the ancient Greek travelers sighted a variety of powerful and dangerous drug plants:

> And there were overhanging grasses with roots sunken low, and asphodel, and famed and lovely maiden hair, yew and camomile and the *black poppy*. Mighty plants, and white hellebore and monkshood too, and many other baneful plants that grow upon the ground. (John Gottlob Schneider, *Orpheus's Argonautica* [1803] 11.914–15, cited by Kritikos and Papadaki 1967a:18; emphasis added)

The "black poppy" in the passage above referred to the opium poppy which produces dark seeds, previously identified as *Papaver somniferum* var *nigrum*.

The presence of the opium poppy and its use in prehistoric Anatolia is suggested by objects thought to be opium poppy seeds found in the "footed-bowl" or "drum" excavated at Beycesultan (in southwest Asia Minor) and dated to about 1900 B.C., which have already been discussed (see chapter 6). I have also already presented the negative evidence suggested by the extensive research of Helbaek in the region and the recent study of the ancient Near Eastern evidence by Krikorian (1975). Although we should remember that some areas of Anatolia have long been, and continue to be, important regions of opium poppy cultivation, we do not know if this agricultural tradition goes back in time as far as the Bronze Age. If the opium poppy was brought into Greece, the Aegean, and the eastern Mediterranean region during the Late Bronze Age, as I suggest, then perhaps the drug plant was introduced into these areas *directly* from Anatolia. If this was the case, then perhaps the opium poppy was carried down from central or eastern Europe to Asia Minor, from where it was subsequently transported into the Greek realm.

In the following sections, I examine the possible associations between the opium poppy and some specific Greek cults and deities. In all cases, the role of the opium poppy as a provider of mind-altering drugs is suggested: "The mystery cults of the Hellenistic world sought to continue and to intensify the archaic faculties of man, to keep human imagination directed to the worlds of hallucination and of dreams" (MacDermot 1971:32). I believe that, at least during the early period of their culture, the illusory perceptions and animated dreams induced by opium intoxication played a role in the Greek perception of the spirit world. It may have even influenced their concept of life after death.

Hermes and Mekone

Butterworth (1966:154) argued that Hermes, a god of various aspects including agriculture and fertility, represented shamanistic power and that his pronged caduceus was primarily an instrument used in the conveyance of souls. In his estimation: ". . . there can be little doubt that the soporific quality of the staff was a symbol of the shaman's trance, and its capacity to restore a man to a state of youth and beauty is to be associated with the same shamanist context: . . . in a word, rebirth." Hermes could have been one manifestation of the introduced shamanistic belief in transcendental consciousness and the importance of sleep, dreams, and spiritual regeneration:

> He is, besides, the god of sleep and of dreams; with one touch of his staff he can close or open the eyes of mortals; hence the custom, before going to sleep, of offering him the last libation. As he is the guide of the living on their way, so is he also the conductor of the souls of the dead in the nether-world (*Psychopompos*), and he is as much loved by the gods of those regions as he is by those above. (Seyffert 1956:287)

Hermes was the "intermediary between the upper and lower world" and was therefore the object of sacrifice at funerals: "His worship early spread throughout the whole of Greece" (Seyffert 1956:287).

In special reference to the power of Hermes' caduceus, Butterworth (1966:154) discusses the source of this power: "If we remember that Hermes was 'the Kyllenian' and that the former name of Kyllene was Mekone ('Poppy Town'), we may infer that the sleep sent by Hermes' wand was sometimes induced by opium. Its true nature, however, is of another kind."

Although Butterworth seems to be rejecting the role of opium in Hermes' mystical sleep-producing activities, it should be pointed out that the Greek god of sleep, Hypnos "was always represented crowned with poppies or lying surrounded with them" (Nash 1927:21). In addition, Morpheus, the Greek god of dreams, was often associated with opium poppies and "commonly represented as a fat, winged child bearing a vase in one hand and a bunch of poppies in the other" (Williams 1947:88). Morpheus's association with Hermes is evident if it is remembered that Hermes was one of the principal gods to "have authority over dream gods" (Seyffert 1956:201).

The association between Hermes and the ancient Mycenaean settlement at Mekone is especially interesting. Indeed, some additional, intriguing aspects of this prehistoric "Poppy Town" near Corinth should be noted. According to Skalet (1928:41), the physical geography of the surrounding region, the archaeological evidence, and references in the Homeric literature indicate that Mekone (modern Sikyon) was "a Mycenaean habitation of considerable importance." In Hesiod (*Theogony*, 536), the ancient town of Mekone is referred to as the place where ". . . Prometheus brought down fire from heaven, the first sacrifice was made and the sacrificial customs of the Greeks were instituted" (Skalet 1928:41; also see Kerényi 1963:42).

According to McMurty (1889:268), Mekone, or Poppy Town, was "an appelation which undoubtedly originated from the abundant growth of wild poppies, which still, at the present day, are scattered over the plateau upon which the old city was built." Callimachus (*Hecate*, fragment 195) and the *Great Etymological Dictionary* (of Greece) "assign the origin of the name to the fact that here Demeter first found the poppy" (Skalet 1928:41 and Kritikos and Papadaki 1967a:18). According to Kerényi (1963:42), the geographical location of "this Place of Poppies (from *mekon*, 'poppy') was thought to be situated in the vicinity of the Peloponnesian city of Sikyon near

Corinth, and mythologically in the realm of the poppy goddesses, Demeter and Persephone."

The early association between the opium poppy and goddesses that represent fertility at Mekone is especially noteworthy. According to Pausanias, the Greek traveler and geographer from Lydia who explored Greece in the second half of the second century A.D., the sanctuary of Aphrodite in Sikyon (Mekone) contained a seated gold and ivory image of the honored goddess holding an apple in one hand and a "poppy" in the other (Pausanias 2. 10.4ff). Although Aphrodite is best known as the Greek goddess of love, she is also "the goddess of gardens and groves, of Spring and its bounties, especially tender plants and flowers, as the rose and myrtle; hence, as the fruitful and beautiful, . . ." (Seyffert 1956: 39). Aphrodite can thus be seen as related to the Earth Mother or fertility goddesses. According to Kritikos and Papadaki (1967b: 10), the symbolic meaning of the opium poppy capsule in the hands of Aphrodite, the goddess of pleasure and fertility, has a special significance: "In this instance the symbolic meaning is based primarily on the pleasure and, by extension, the fertility derived from the poppy."

Pausanias (2. 10.2) also referred to a temple of Asclepius in Sikyon which had statues of "Sleep and Dreams." Asclepius was the Greek god of medicine. According to Castiglioni (1941: 121), the Asclepius Myth resulted from the long-term synthesis

FIGURE 70. Ancient Greek cameo showing "Nyx distributing poppy capsules" (Emboden 1979: 27, figure 22). The opium poppy had sacred associations with the ancient Greek gods Hypnos, god of sleep, his son, Morpheus, god of dreams, Thanatos, god of death, and Nyx, goddess of night. *(Courtesy of Professor William Emboden, Department of Biology, California State University, Northridge).*

of several underworld deities. I think it is relevant that one of the ancient geographical centers of the Asclepius cult was located "at Titano near Sikyon, where the Thessalian serpents live" (Castiglioni 1941:122), and where, as I have noted, by virtue of its ancient name the opium poppy appears to have been well known during Mycenaean times.

It is important to recognize that the essential aspect of all Asclepian therapy (and the preceding medical tradition?) focused on the dream or hypnotic state. The power of suggestion, which is still used on semiconscious or hypnotized patients was a fundamental basis of Asclepian treatment (Castiglioni 1941:125). I believe that sometime, perhaps very early, in the development of Greek culture, the opium poppy was used by Asclepian priests (or their predecessors) to induce dream and hypnotic states, as well as therapeutic deep sleep and pain relief. It was not coincidental that it their mythology, the Greeks dedicated the opium poppy to *Morpheus*, the god of dreams, *Hypnos*, the god of sleep, *Nyx*, the god of night, and *Thanatos*, the god of death!

Helen of Sparta and the Plant Source of Nepenthes

Helen of Sparta, the famous daughter of Leda and Zeus, and Artemis, the protectress of young females, have both been identified with similar attributes such as the "she-cult" (Butterworth 1966:174). According to Plutarch (*Theseus* 30, 31), their connection was indeed a close one, for as a child, the divinely beautiful Helen was abducted from Artemis's sanctuary at Sparta by her father Zeus. During the excavation of the sanctuary of Artemis Orthia at Sparta, Dawkins (1929) found a number of objects which Kritikos and Papadaki (1967a:30) believe to be symbolic representations of opium poppy capsules.

These objects include a lead figurine "earring" with hanging "poppy capsules" and several "bone pendants . . . which were obviously worn around the neck" and which "also seem to represent poppy capsules" (Kritikos and Papadaki 1967a:30).

The lead figurine earring is one of several "discs with or without pendant buds below"; although these disks have been dated to the latter part of the eighth century B.C., they were "all found within the area of the early sanctuary between the early temple and the archaic altar, and at the same level as Proto Corinthian pottery" (Wace, in Dawkins 1929:251). According to Wace (Dawkins 1929:253), some of these disks have decorations similar to gold jewelry found in a Shaft Grave at Mycenae; therefore, their connections with the Mycenaean period, if only on an inspirational level, may be inferred.

In 1908 and 1909, on the hills just south of Sparta, British archaeologists excavated what they thought was "the Menelaion" or "the shrine of Helen and Menelaus mentioned by Herodotus and Pausanias" (Wace et al. 1909:108). The votive offerings found in their excavations, along with the ancient references, suggested to these archaeologists that Helen was regarded as "an old mother or nature goddess"; among the artifacts found in this site were lead pins which, according to the archaeologists, resembled the gold ones recovered from the Sanctuary of Artemis Orthia (Wace et al. 1909:133). In fact, they have interpreted these pins, which were all "pierced for suspension," as "pendants . . . in the shape of pomegranate buds or poppy heads"; dates for these pins have been set by the archaeologists between the latter part of the ninth century B.C. to the last quarter of the seventh century B.C. (Wace et. al. 1909:133).

According to Homer, after the fall of Troy, Helen returned to Sparta with her husband Menelaus. Among the areas Helen visited during her extended return trip was Egypt (*Odyssey* 4.219ff.). Homer tells us that while waiting for a shift in the winds on the Egyptian coast, Helen met Polydamna, an Egyptian woman famous for her knowledge of drugs in a country generally renowned for its therapeutic products.

The symbolism of Polydamna's name is intriguing, for *damna* is a Greek verb "that means to tame, subdue, gain mastery over and is frequently used in connection with drugs"; Polydamna, the "tamer of many," is used as a proper name only in Homer's *Odyssey* (4.228; compare with Caldwell 1970:56–59). It is from Polydamna that Helen receives some strong herbal drug potions that she takes back with her to Sparta.

Some years later, during a party at Menelaus's home, friends began talking about the Trojan War and the long past disappearance of Odysseus. Sympathy for the lost Greek hero was still strong; tears were shed and general depression threatened the party. For distractive purposes some sugggested that the gathering eat dinner and restrain their sorrow; but this placed Helen, the hostess and not an especially notable cook, in a difficult position (Caldwell 1970:56–59):

> But the admirable Helen had a happy thought. She lost no time, but put something [*nepenthes*] into the wine they were drinking, a drug potent against pain and quarrels and charged with forgetfulness of all trouble; whoever drank this mingled in the bowl, not one tear would he let fall the whole day long, not if mother and father should die, not if they should slay a brother or a dear son before his face and he should see it with his own eyes. That was one of the wonderful drugs which the noble Queen possessed, which was given her by Polydamna the daughter of Thon, an Egyptian. For in that land the fruitful earth bears drugs in plenty, some good and some dangerous; and there every man is a physician and acquainted with such lore beyond all mankind, for they come of the stock of Paieon the Healer. (*Odyssey* 4.219–32, trans. Rouse 1937:43–44)

Paieon (Paeon) was the Egyptian physician of the gods. Indeed, he was himself a god of health and healing charms from whom the physicians of Egypt claimed descent. According to Butterworth (1966:177), even though much of the Greek medicinal repertoire was probably indigenous to Greece, many features of Egyptian cults seem to have been introduced from Egypt and incorporated into Greek medicine and ritual. "It is even possible that some of the 'events' of legend may themselves be no more than the memories of general trends, crystallised into the narrative form of personal exploits" (Stubbings 1973:85). In other words, Helen's acquisition of the "wonderful drugs" from the Egyptian herbalist(?) may be representative of a "general trend" which involved the importation of the medicines from Egypt to Greece during the late prehistoric period.

Although Homer did not specifically indicate what the nature of the *nepenthes* drug was, many scholars have referred to it as a preparation containing opium (e.g.,

Sprengel 1817:38, Lewin 1964:41, Wallnöffer and Rottauscher 1959:76, Kritikos and Papadaki 1967b:9, Emboden 1979a:23, and Ruck ("probably") in Wasson et al. 1978:91). If we assume that the opium poppy did provide at least one of the psychoactive ingredients, then we might ask the following question: Why did Helen obtain the *nepenthes* drug from a female Egyptian sage when it was suggested earlier in this study that the opium poppy was probably introduced to Greece from the north and therefore should have already been present in the Greek realm before Helen's time?

This interpretation could be explained in at least two ways: if the drug was a mixture containing opium, it may have been a novel combination previously unknown to Helen. Perhaps, in Greece, its special drug use had been previously restricted to the shaman's domain and only used in special divinatory or funeral rites. In this case, it may have spread to other regions outside the Greek realm, such as Egypt, where I believe it became known it the Late Bronze Age (i.e., the era of the New Kingdom) and possibly fell out of the special purview of the shaman-seer class and into a more widespread use as a magical medicine. It could then have been reintroduced to Greece as a more popular item of pharmacological importance, a position it attained here at least by the classical period, if not well before. Helen, herself, may have been merely a mythological manifestation of a real aristocratic lady, who in legend became deified. If so, her association with the Trojan War would place her chronologically toward the end of the Mycenaean period. Later, during the Roman period, Galen, the famous Greek physician, referred to opium drugs as the gift of Thoth, the Egyptian god of the moon and of wisdom and learning (Kritikos 1960:61).

Of course, there is an alternative view that *nepenthes* did not contain opium at all. Some ancient writers actually believed that *nepenthes* was not a drug at all. For example, Plutarch, Athenaios, and Philostratos suggested that it "allegorically represents Helen's charm and fascination"; furthermore, Theophrastus, the father of botanical science, argued that *nepenthes* was a figment of Homer's imagination (Kritikos and Papadaki 1967b:8). Other scholars (e.g., Snijder 1936:134–40) have suggested that the active ingredient came from a solanaceous species such as henbane, *Hyoscyamus niger* L. (see figure 71), which we know was used along with other psychoactive solanaceous plants such as mandrake, *Mandragora officinarum* L., and deadly nightshade, *Atropa belladonna* L., for magico-religious, mind-altering purposes in Europe during the Middle Ages (Harner 1973:125–47). These solanaceous species definitely have psychoactive potential, but I do not think any one of them was the sole ingredient of *nepenthes*. In fact, Dioscorides (4. 68.3), the noted Greek student of pharmacology, referred to henbane as a better soporific than opium, but suggested that *nepenthes* was most likely a mixture of opium and henbane because magic potions were usually concoctions. According to Catkins (1871:25), Plato knew of an "electuary compound" that contained opium and henbane juice.

Probably the strongest case for identifying *nepenthes* with the opium poppy has been presented by Kritikos and Papadaki (1967b:8–9). Dismissing suggestions that view *nepenthes* as a poetic invention or that opium drugs are too bitter to be consumed easily, these authors argued that it could only be opium, and if its identity was not known to such ancient botanical specialists as Theophrastus, this was probably "be-

FIGURE 71. Henbane, *Hyoscyamus niger* L., a psychoactive member of the solanaceous family, Solanaceae. Henbane and the opium poppy share a common ecological niche as weedy herbs with small seeds and powerful, but dangerous, hallucinogenic-medicinal properties. Their cultural spread and use seem to have been intertwined, so to speak—perhaps unwittingly in the process of cultural migrations, or perhaps consciously among a shaman cult movement (from Rhind 1868).

cause its use remained a secret of the sanctuaries and revealed to only a few" (Kritikos and Papadaki 1967b:9). They pointed out that when combined with wine, as it was in the passage from Homer, the drug, if it was opium, would not be bitter. According to Carl Ruck (in Wasson et al. 1978:42),

> . . . ancient wine, like the wine of most early peoples, did not contain alcohol as its sole inebriant but was ordinarily a variable infusion of herbal toxins in a vinous liquid. Unguents, spices, and herbs, all with recognized psychotropic properties, could be added to the wine at the ceremony of its dilution with water. A description of such a ceremony occurs in Homer's *Odyssey*, where Helen prepares a special wine by adding the euphoric *nepenthes* to the wine that she serves her husband and his guest. The fact is that the Greeks had devised a spectrum of ingredients for their drinks, each with its own properties.

It is worth noting that opium and alcohol mixed together (now known as tincture of opium) was a popular medicinal in western Europe during the eighteenth and nineteenth centuries, when it was known as *laudanum*.

I believe that the opium poppy probably can be associated with *nepenthes*; however, I would tend to agree with Dioscorides' idea that the active ingredient of the opium poppy may not have been the sole ingredient of the potent drug referred to in the *Odyssey*. This position is supported by the fact that opium was often combined with a number of other ingredients to make an assortment of medicinal potions during the Classical Greek and Roman period. For example, Claudius Galen (A.D. ca. 138–201), the immensely influential doctor from Asia Minor, who was the personal physician of Roman emperors and author of numerous books, recorded the recipe for *theriaca*, a well-known, ancient opium-based drug:

> *ELECTUARIUM THERIACAE MAGNUM*
>
> Root of Florentine iris, licorice, 12 ounces each; of Arabian costus, Pontic rhubarb, cinquefoil, 6 ounces each; of Ligusticum meum rhubarb, gentian, 4 ounces each; of birthwort, 2 ounces; herb of scordium, 12 ounces; of lemon grass, horehound, dittany of Crete, calamint, 6 ounces each; of pennyroyal, ground pine, germander, 4 ounces each; leaves of lourus cassia, 4 ounces; flowers of red roses, 12 ounces; of lavender, 6 ounces; of St. John's wort, 4 ounces; of lesser centaury, 2 ounces; saffron, 2 ounces; fruit of amyris epobalsamum, 4 ounces; cinnamon, 12 ounces; cassia lignea and spidenar, 6 ounces each; Celtic nar, 4 ounces; rape seeds agaric, 12 ounces each; seeds of Macedonian parsley, 6 ounces; of anise, fennel, cress, seseli, thlaspi, amomum, sandwort, 4 ounces each; of carrot, 2 ounces; *opium*, 24 ounces; epobalsamum, 12 ounces; myrrh, olobanum, turpentine, 6 ounces each; storax gum arabic, sagapenum, 4 ounces each; asphaltum, opoponax, galbanum, 2 ounces each; juice of acacia and of hypocist, 4 ounces each; castor, 2 ounces; lemnian bole, calcined vitriol, 4 ounces each; trochiscs of squill, 48 ounces; of vipers, of sweet flag, 24 ounces each . . . Triturate the balsams, resins and gums in a sufficient quantity of wine to form a thin paste, and incorporate the whole with 960 ounces of honey. (Macht 1915:479, quoting Galen 14. 107, emphasis added; see also Nicander, *Theriaca* 946)

Some of these supplementary ingredients may have had mind-altering potential, but they also could have been included for flavor, fragrance, sweetness, and other therapeutic qualities absent from opium. Perhaps some were included for superstitious reasons, or simply to mask the special psychoactive substances such as can be found in the opium poppy or henbane.

Butterworth (1966:175–76) apparently believed that Helen's *nepenthes* contained opium, but he did not think that opium use was restricted to the cult of Helen. He suggested that indulgence in opium drugs "is rather to be seen as an index of the importance generally attached to dreams." According to Butterworth, Helen and the other Greek deities share a special connection with another world beyond everyday reality:

> a world in which supernatural powers can communicate with human beings in dreams, or in which, through the medium of the drug, men can attain to an analogy of bliss. In the case of Helen we hear only of the use of opium (as we may assume

the drug to have been) for inducing this latter condition.

However, although Butterworth feels that the cult of Helen was "centered upon, or at least sought experience of the Transcendental," he rejected the idea that opium indulgence was the basis of the cult: "To believe that a drug-induced condition lay at the root of the cult of Helen would be a.fundamental error." Nevertheless, Butterworth (1966: 175, note 14) went on to suggest that ". . . the use of opium may be regarded as an accretion to the cult, or a degenerate extension of it to a wider public." I believe that this connection with a "psychic excursion," or divine dreams, may once again be seen as a possible attribute of the introduced shamanistic influences.

Although we can at present only speculate as to the identity of the shaman's magical plants in Mycenaean Greece, in my opinion the opium poppy certainly remains as a logical possiblity, along with henbane or other solanaceous species that were available in the region during the Late Bronze Age. It is, of course, possible that the Indo-European shamans did not bring the opium poppy into Greece; instead, they may have first encountered the drug plant in this region and then absorbed it into their sacred body of knowledge on account of its psychoactive powers. In any case, whether the use of opium-based drugs was an indigenous trait or an introduced, "accretionary" aspect of the shaman cult, several lines of evidence suggest that it was a part of Greek culture, probably by the time of the Late Bronze Age. Furthermore, I believe that one of the opium poppy's more important associations with ancient Greek culture involved its connection with the Earth Mother cult of Demeter.

The Cult of the Goddess Demeter

One of the many remarkable developments of ancient Greek culture was its pantheon of gods. It stands out as a striking example of insight into the ironies and intricacies of human nature. In a spirit of high idealism, the Greeks viewed their religious conceptions as models for human behavior. Both good and bad, their gods represented the union of religion and morality. Their deities were endowed with supernatural powers; but they were also divinities just as susceptible to corruption as humans. Therefore, although the Greek gods symbolized the capricious aspects of the human condition and a mortal desire for protection from inexplicable dangers, it was with a minimum of dogma and much joy that the Classical Greeks decorated their cities with shrines and filled their calendar with days held sacred to their identifiable gods (Fairbanks 1910: 14). The examples of the Greek gods thus provided magnificent lessons of worldly virtue in the form of an awe-inspiring mythology which has long survived the passing of their own culture.

And yet, in spite of all the rational, lofty development in fields of rhetoric, philosophy, and humanistic thought, some of the long-lasting superstitious traditions shared by commoners and nobles persisted, especially their devotion to the mysterious agricultural rites of occult initiation. An excellent example of this can be found in the cult of Demeter, the ancient Earth Mother Goddess, which remained important in Greece for about 2,000 years as a powerful (but contradictory) feature of Greek society. In fact, the cult's overt presence was not eliminated until the pressures of an intolerant Christian ideology finally brought on its demise in the fourth century A.D. (Ruck, in Wasson et al. 1978:76; also see Mylonas 1961:285).

In light of the extraordinarily rational character of the ancient Greek civilization, one may ask the following question: Why did the belief in occult ritual endure, and why with such importance? For one thing, while aesthetics and intellectual discourse advanced at a relatively fast pace in ancient Greece, the traditional patterns of domestic agriculture still supported the society, as they had for thousands of years during the prehistoric period. However, since adequate success in farming was not always realized, the fear of an insufficient harvest persisted as a real anxiety among the general populace. This perennial apprehension probably played an important part in stimulating faith in an immortal existence: "The universal testimony of Greek literature teaches us that the most important result of the mysteries was this clear hope of a real life after death" (Fairbanks 1910:136). As a consequence, mystic ritual symbolizing the related concepts of farming, fertility, and immortality continued to be a significant activity; in fact, cult rites celebrating the synchronization of cultivation practices and seasonal cycles were observed throughout the ancient Greek and Roman periods.

Jane Harrison (1922:272) suggested that during the ancient period it was the Greek women who did most of the work in the cultivated fields, while the men hunted, fought, and indulged in politics. If this was the case, then it would have been natural for the Greeks to deify their successful harvest in the form of a goddess (in a fashion similar to the preceding agricultural societies of "Old Europe"; see Gimbutas 1974). This female Greek deity was, of course, Demeter, the goddess of marriage, good health, fertility, and the agricultural basis of civilization. Her cult encouraged and reinforced the mystical belief in immortality which was manifested in the cyclical concept of birth, death, and rebirth.

I believe that there is ample evidence indicating that the ecology of the ancient Greek grain fields, the cult of the goddess Demeter, and the opium poppy were connected in a significant cultural relationship. According to Brown (1938:8), the ancient Greeks "found the poppy growing in their grain fields and dedicated the plant to Ceres (Demeter)." For both the ancient Greeks and Romans, it was this Earth Mother Goddess who reigned over the cultivation of cereal grains and growth of opium poppies, plants which were her special botanical attributes (cf. Seyffert 1956:178, Mylonas 1961:217, and Gow 1965:169). Kritikos and Papadaki (1967a:34) referred to "ears of wheat and heads of poppies" as "the two principal symbols of the worship of Demeter."

Artistic representations of Demeter were sometimes crowned in garlands of poppy capsules interwoven with barley or wheat (Nash 1927:219); other artifacts depicting the Earth Mother Goddess show her holding sheaves of grains and poppy capsules. For example, excavations in the ancient marketplace or popular assembly area of Athens, the *Agora*, uncovered a seated statue of Demeter holding an ear of wheat and a poppy capsule; the statue "dates from the end of the fifth or beginning of the fourth century B.C." (Kritikos and Papadaki 1967a:33; see figure 72). Additional ancient Greek statues of Demeter holding grains and poppies can be seen in museums located

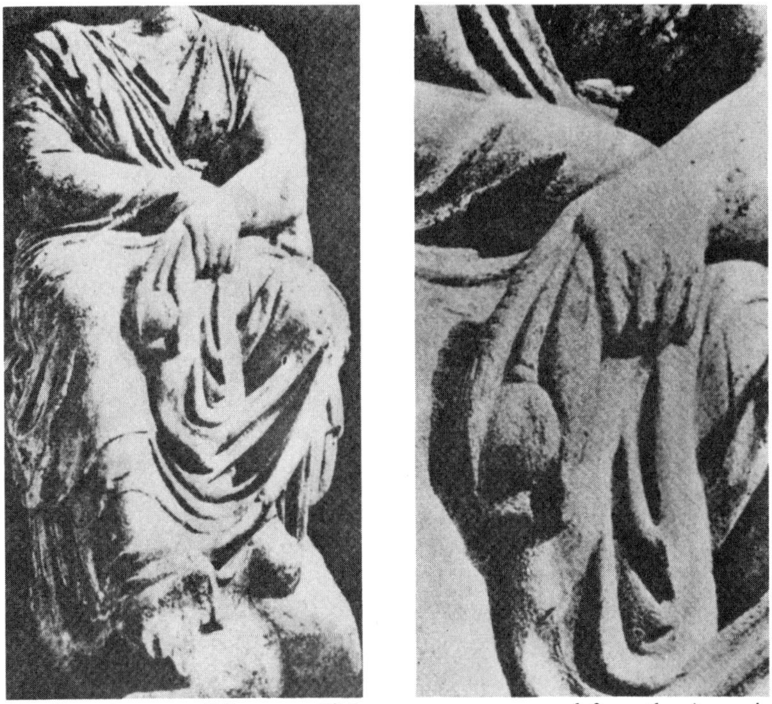

FIGURE 72. Seated statue of Demeter. This statue was recovered from the Agora in Athens, Greece. Probably carved in the fifth century B.C., it shows the Earth Mother Goddess holding an ear of wheat and a poppy capsule (from Kritokos and Papadaki 1967a:33; *Courtesy of Archaiologike Ephermis, Hellenike Archaiologike Hetairia, Athens, Greece*).

in the Vatican City and Athens (Kritikos and Papadaki 1967a:33). Furthermore, there are a number of other ancient Greek and Roman statues which shown women (goddesses?) bearing ears of grain and poppy capsules. One example, which has been tentatively dated back to the first quarter of the first century A.D. (C. K. Williams II, Director, Corinth Excavations, American School of Classical Studies at Athens, personal communication, April 29, 1976), shows a female figure on one side of a rectangular base. This relief image depicts (what I believe is) an Earth Mother Goddess "holding stems with poppy pods and heads of grain" in her right hand (Williams and Fischer 1975:23–24).

The association between Demeter and the opium poppy can also be found in ancient Greek verse. For example, Theocritus, the pastoral poet (310–245 B.C.), tells us in his *Idylls to the Temple of Demeter Alois* (Idyll 7:155) that the first-fruits of the harvest were brought to "Demeter of the Threshing-floor" on the island of Cos and that "she smiles on us with sheaves and poppies in either hand" (Gow 1965).

The agricultural cult of Demeter, and the "mysteries" surrounding it, were among the most important features of ancient Greek ritual. In the *Homeric Hymn to Demeter*, this great cult and its association with life, death, and rebirth are presented in allegorical terms. The central theme involves the abduction of Demeter's daughter, Persephone, by Hades, the God of the Underworld or the Dead. After Demeter realized that her daughter had been seized, she set out in an attempt to find the maiden. According to at least one interpretation, following a long, unsuccessful quest for Persephone,

> . . . the gods caused poppies to spring beneath (Demeter's) feet. She knelt to look at them and their breath caused her to fall asleep. Thus she obtained rest for her weary limbs. (Brown 1938:8; also see Seeberg 1969:11 and Kritikos and Papadaki 1967b:10).

Coates (1968:196) provided another interpretation of this association between the Earth Mother Goddess and the opium poppy:

> According to classical mythology, [the poppy] was created by Somnus [i.e., Hypnos], God of Sleep, to ease Ceres [i.e., Demeter] of her cares and cause her to slumber, for in her weariness she was neglecting the corn. After her refreshing sleep the crops revived, and that is why Ceres is usually represented as wearing a garland of corn mingled with poppies. From this legend the ancients derived a comforting belief that poppies were essential to the health of the corn.

Persephone, herself, is also directly associated with the opium poppy. Evans (1925, 1930) suggested that some Late Bronze Age artifacts show Persephone's ancestral form rising from the earth with "poppy-heads" in her hands. Evans (1925) also pointed out that ". . . on the well-known terracotta relief from Magna Graecia, the resurgent Kore [i.e., Persephone] holds in either hand a triple group of poppy-heads, lilies and ears of barley" (see figure 73).

According to Grose, evidence in the form of artistic designs on at least two ancient Greek coins also associates Persephone with opium poppy capsules. For example, on

FIGURE 73. Persephone rising from the ground. This terracotta relief was recovered from Magna Graecia. It depicts the Goddess of Spring emerging from the underworld with poppy capsules, sheaves of grain, and lilies in both hands; ". . . the chthonic character of the Maiden Goddess is manifested by the snakes on either side coiled round her fore-arm and, in this case, raising their heads toward her" (from Evans 1925:16, figure 18).

the obverse side of a Euboic-Attic standard gold drachma coin from Syracuse (dated to the reign of Hieron II, ca. 274–216 B.C.), the head of Persephone is shown wearing a crown of barley sheaves in her hair with a "poppy-head behind her neck" (Grose 1929:345, #2903). In addition, on the obverse side of a silver "stater" coin from Locri (dated to ca. 322–268 B.C.) the head of Persephone is depicted wearing a single drop earring and a necklace of beads with poppy heads visible behind her neck and long hair (Grose 1929:215, #1815).

Persephone's association with the opium poppy capsule and her abduction into the underworld realm of Hades may be related. According to Cook (1964:vol. 2, p. 164), comparison and analysis of some ancient Greek literary fragments and more recent Greek folk songs have led scholars to believe "that the Greeks of old aspired to an actual marriage-union with the deities of the underworld, a union to be foreshadowed here in mystic rites and consummated hereafter in every truth." In other words, the early Greeks believed that each man would "one day enter the bridal chamber of Persephone, every woman that of Hades" (Lawson 1910:543–606). Cook also argued that the many artistic portrayals of "Hero-feasts" in ancient Greece were "naive representations of the dead man's marriage-banquet," and, in some cases, even indicated one of the ways in which the opium poppy could have helped the deceased overcome fatigue during his everlasting banquet in the bridal chamber of Persephone: "if the new immortal tired of it, he had at least the poppy-capsule to lull him to sleep and to renew his generative powers" (Cook 1964:vol. 2, p. 1164).

In Aristophanes' play *The Birds*, written in 414 B.C., there is a reference to certain fruits of the garden such as "white sesame, myrtle-berries, poppy [heads], and [water] mint" (line 160). Cook (1964:vol. 2, pp. 1164–65) suggested that these were the "appropriate" contents of the garlands worn and cakes eaten by the deceased in the everlasting banquet feast because sesame fruits, myrtle-berries, and poppy capsules all contain a multitude of seeds and are thus symbolic of fecundity and abundance, while mint was revered "on account of its aphrodisiac properties." According to Cook, the ancient Greeks believed that these special plant materials provided the "magic means" by which the dead man "was empowered to impregnate his bride" (Persephone).

Cook (1964:vol. 2, p. 1165) also emphasized that the opium poppy produces a large amount of small seeds: "Here it made for fertility, and became the attribute of various mother goddesses." Coates (1968:196) referred to another myth, according to which "the poppy became the emblem of the earth-goddess Cybele (often identified with Ceres) because it signified fecundity, on account of its abundant seeds." I think it is worth nothing that Linnaeus ("with exemplary patience") reportedly counted 32,000 seeds in a single opium poppy capsule (Perry 1972:221). Although this capsule was obviously a large cultivated specimen, it gives us an idea of the great number of seeds which the opium poppy is capable of producing.

It is also necessary to note here that the pomegranate also produces a multitude of seeds. According to Cook (1964:vol. 2, p. 814), the ancient Greeks "were struck by two characteristics of the [pomegranate] fruit—its red colour and its numerous seeds." He argued that the red hue (of both the fruit's exterior surface and interior pulp) "certainly suggested blood"; and the multitude of seeds "possibly suggested fertility" (see Cook 1964: vol. 2, pp. 814–19, for a detailed discussion of the pomegranate as a symbol in ancient Greece).

The problem of identifying which of the two prolific seed-producing species (either the opium poppy or the pomegranate) is associated with a specific ancient Greek myth is complicated by the fact that, in some cases, both seem to have a connection with the same legend. If it is remembered that certain artifacts have been variously identified with both species, then the additional problem of mutual associations with some myths and cult traits only underscores the difficulties of this research.

For example, according to the *Homeric Hymn to Demeter* (lines 360–404), Persephone was eventually able to return to her mother from the underworld, but only for a portion of each year (the growing season) because she ate a pomegranate. Cook (1964: 815) argued that a number of artifacts show Persephone (or dead women) holding pomegranate fruits; he claimed that such fruits were a "desirable food for the dead," as indicated by their frequent representation "on funeral monuments of the archaic period." In any event, apparently both the pomegranate and the opium poppy were associated with the important agricultural cult of Demeter and her daughter, Persephone, or perhaps more generally with the concepts of fertility and rebirth.

In the case of the opium poppy, some artifacts (e.g., the capsules found in Cueva de los Murciélagos, as well as others described in later sections of this study) can be interpreted as "food for the dead," or at least as having a connection with afterlife. I believe that the association between the goddesses of fertility and the opium poppy can even be found in the core of the Demeter Cult—in the Eleusinian Mysteries.

The Eleusinian Mysteries

> Because they vivified the belief in a future life, while at the same time they met the religious needs of the individual, the Eleusinian mysteries were perhaps the most important form of worship in ancient Greece. (Fairbanks 1910: 137)

The Eleusinian Mysteries celebrated both the winter descent of Persephone into the underworld realm of Hades and her springtime return to her mother, Demeter. This myth involving the Earth Mother Goddess and her daughter symbolized the yearly cycles of spring, summer, fall, and winter, or more abstractly, the concept of life, death, and rebirth. First practiced sometime in the second millennium B.C. (Wasson et al. 1978: 36), these sacred mysteries involved two festivals. The first, the "lesser mysteries," took place at Agrae early in March as ritual observance of Persephone's return from Hades' lower world of the dead, in essence a celebration of the rebirth of spring. However, from a wider perspective, the lesser mysteries were necessary preparation for the "greater mysteries" which took place in September in observance of late summer's harvest and the beginning of the fall season. The greater mysteries were held in honor of Demeter and symbolized the participants' appreciation for subsistence and good health; thus Demeter is often shown holding grains (the symbols of sustenance) and opium poppy capsules (the symbols of fertility and protective medicine).

There is at least one additional way in which these "symbols" may be interpreted. We know that the opium poppy capsule is a source of psychoactive alkaloids. The grains, indirectly, may also be such a source, for they can become infected by a fungus which produces hallucinogenic alkaloids. The suggestion that mind-altering drugs were used in the rites of the ancient Greek mysteries is both speculative and controversial. However, as we shall see in my subsequent discussion, it forms the basis of a very recent, thoroughly researched, and certainly challenging thesis that may, in my opinion, involve both infected grains *and* the capsules of the opium poppy. Before I describe this new perspective I want to outline a few key aspects of the principal mystery rites.

We know that the greater mysteries of Eleusis involved ritual fasting, sacrifices, and a huge procession of potential initiates who marched slowly almost thirteen miles from Athens to Eleusis while chanting hymns to Demeter and paying homage to various shrines along the way (Fairbanks 1910: 132; Herodotus Book 8, line 65). "Each year new candidates for initiation would walk that Sacred Road, people of all classes, emperors and prostitutes, slaves and freemen, an annual celebration that was to last for upwards of a millennium and a half, . . ." (Ruck, in Wasson et al. 1978: 35–36).

A general truce among the antagonistic Greek states allowed the devout participants of the mysteries the right to travel unharassed to sites of worship in Attica (Fairbanks 1910: 132). This was a manifestation of the importance of the mysteries in ancient Greece and their wide appeal among the populace. According to the *Homeric Hymn to Demeter*, Eleusis was a major site of the mysteries because of several interrelated factors. For one thing, Eleusis was the place where the goddess of the harvest (Demeter) came veiled as an old lady and later revealed her divinity. Eleusis was also the place where she consecrated the sacred rites and was eventually reunited with her abducted daughter, Persephone. Furthermore, Eleusis was the place where Demeter allowed the gift of the grain to be restored (see Danny Staples's translation of the *Homeric Hymn to Demeter* in Wasson et al. 1978: 59–73).

In accordance with the cult, worshipping at Eleusis involved acts of purification and the essential initiation into the sacred mysteries themselves. The new initiates were known as the *mystae*, persons who had neither seen the sacred ritual nor would speak about it after solemn firsthand experience. Before the *mystae* could observe the mysterious rites in which certain *sacra* were secretly exhibited, they had to undergo the special purification ceremony. According to Jane Harrison (1922: 154), the sacred ritual was "not the main gist of 'a mystery': the essence of it all primarily was purification in order that you might safely eat and handle certain *sacra*."

After fasting and purging their digestive tracts with strong emetics such as hellebore, the specially selected *mystae* were further purified in a bath of pig's blood and baptismally cleansed in sea water (Malcolm 1971: 33–34). Thereafter, with the preliminaries of purification completed, the *mystae* were led down into underground dens of cult worship where they were initiated into the mysteries and were able to handle the "magical *sacra* and . . . the partaking of first fruits" (J. E. Harrison 1922: 162).

The major part of the ritual, which occurred in the subterranean labyrinths of the sacred temples, was never exactly described:

> The Greeks' reticent attitude toward their own religious beliefs constitutes in itself an obstacle to knowledge. Firsthand information about the details of ritual, such as the precise working of the Delphic oracle, is lacking; and for some important religious events, like the mysteries at Eleusis, little is known, since the initiates were bound to secrecy; and late Christian accounts are naturally biased. (Pollard 1975:407)

Indeed, the secret experiences of the Eleusinian rites were "called a mystery because no one, under pain of death, could reveal what happened within the sanctuary" (Wasson et al. 1978:36). Naturally, much confusion has arisen over conjectures concerning just what constituted the "mysteries."

Pausanias (2.37.3–4, at Lerna; 3.22.2, to Dionysus) clearly tells us that the mysteries of Greece were acted out with uttered sentences. However, according to Fairbanks (1910:131), there surely would have been some traces of their content had there been a substantial "body of secret doctrine"; he also suggested that the Greek mysteries centered around "a ritual-drama, beheld and shared only by the initiated." On the other hand, Fairbanks (1910:134) also argued that if we refer to the "vision" or "scenes" *(Epopteia)* seen in the "crowning experience" of the Eleusinian Mysteries as drama, we are suggesting "something more pretentious than the evidence warrants"; rather, he suggested that after imbibing the special drink, the *kykeon*, the *mystae* were shown the sacred symbols, the *sacra*, and that these "symbols were sufficient to suggest scenes of the familiar story, while images of the gods showed Persephone with Hades in the lower world, the mourning Demeter, and Persephone reunited with her mother."

Ruck (Wasson et al. 1978:36–37) also argued against the idea that the Eleusinian Mysteries involved drama, in the strict sense of the word. According to his interpretation of the archaeological remains, the sacred temple where the "visions" were witnessed was "totally unsuited for theatrical performances"; furthermore, he pointed out that there is no written record for monies that might have been spent to pay for the services of actors or theatrical materials. Indeed, Ruck suggested that the "sophisticated" Greeks could not have been tricked by clever theatrical stunts and that the testimony of well-known individuals such as Pindar and Sophocles speaks of the "overwhelming value" of the mystic visions experienced in the sacred hall at Eleusis: "What was witnessed there was no play by actors, but *phasmata*, ghostly apparitions, in particular, the spirit of Persephone herself, returned from the dead with her newborn son, conceived in the land of death" (Wasson et al. 1978:37).

In light of the fact that the "incommunicable" visions of the Eleusinian Mysteries followed the consumption of the special *kykeon* drink, Ruck (Wasson et al. 1978:37) assumed that the ingredients of this liquid produced the extraordinary experience: ". . . it seems obvious that an hallucinogen must have induced it." Ruck also argued that the architecture of the Eleusinian temple and the ceremony practiced within its sacred halls indicate that it was not a dramatic performance but an altered state of consciousness which produced the very special visions of an immortal life experienced by the initiates.

The Opium Poppy and Ergot Alkaloids

"At the heart of the Mysteries of Eleusis lay a secret" (Wasson 1972:193). In 1956, R. Gordon Wasson, a well-known student of hallucinogenic mushroom-use "intimated" that this ancient, and as yet unrevealed, secret of Eleusis would be found in the hallucinogenic experience produced by a psychoactive fungus or higher plant (Wasson et al. 1978:12; see also Wasson 1963:27–42). More recently, Wasson, the ethnomycologist, Carl A. P. Ruck, a classical scholar specializing in Greek ethnobotany, and Albert Hofmann, who first experienced psychoactive LSD-25 (lysergic acid diethylamide), all collaborated in an effort to answer this challenging question: "whether Early Man in ancient Greece could have hit on a method to isolate an hallucinogen from ergot that would have given him an experience comparable to LSD or psilocybin" (Wasson et al. 1978:25). Their sensational solution suggests that alkaloids extracted from an ergot fungus such as *Claviceps purpurea* (Fr.) Tul. or *C. paspali* Stev. and Hall played a crucial role in the ancient Greek mystery rites of Eleusis; in other words, they argued that these ergot alkaloids were the active ingredients of the sacred barley or grain water drink *kykeon*.

Although *C. purpurea* is best known for its parasitic association with rye and the principal host of *C. paspali* is the common Mediterranean grass *Paspalum distichum* L., both wheat and barley are also parasitized by these two species of *Claviceps*. In addition, darnel (*Lolium temulentum* L.), a weedy grass species that invades the grain

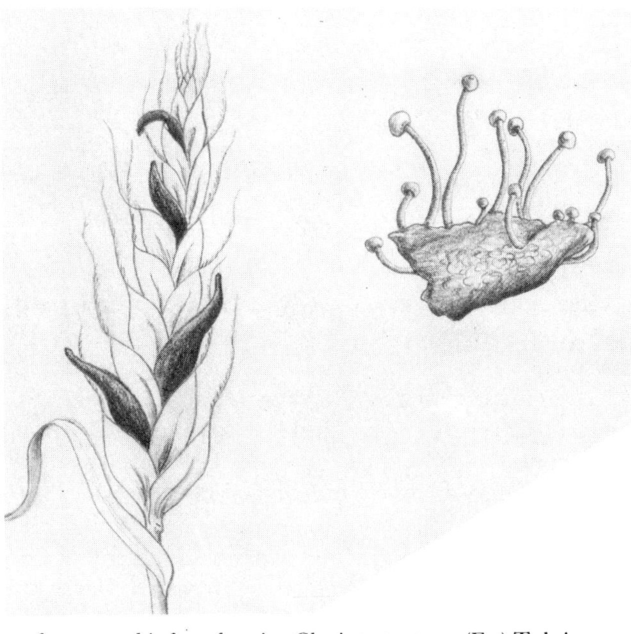

FIGURE 74. Ergot fungus and infected grain. *Claviceps purpurea* (Fr.) Tul. is an ergot fungus that infects a number of grass species: (a) fruiting bodies of the ergot (life size × ca. 2); (b) sclerotium of the ergot on rye *(Courtesy of Denise Peltier, Los Angeles, California).*

fields of Greece and other parts of Europe, is also parasitized by an ergot (Wasson et al. 1978:25–34).

We know that *Claviceps* species contain a variety of alkaloids. Some of these, such as ergonovine (lysergic acid propanolamide) and ergine (lysergic acid amide), are hallucinogenic, while others such as ergotamine and ergotoxine are highly toxic and potentially lethal (Wasson et al. 1978:29). In fact, the unintentional consumption of ergot infected grains has had disastrous effects on some European farming communities, especially during the Middle Ages (Taylor 1965:60; see also Barger 1931). In times of crop shortages, rejection of infected grains was not effective. Consequently, toxic amounts of ergot of rye (*C. purpurea*) were harvested along with the precious food crops. Then, when the contaminated grains were ground into flour, baked into bread, and eaten, the unfortunate consumers often manifested a syndrome of effects which has been referred to as *ignis sacer* ("Holy Fire") or "Saint Anthony's Fire." These symptoms were frequently hideous: "Victims of the disease felt consumed by an inner fire, hallucinations of being chased by wild beasts drove them to madness" (Emboden 1979a:80).

Known today as ergotism, this disease can cut off blood flow and thus indirectly cause gangrene (*Ergotismus gangraenosus*); furthermore, ergotism can have adverse effects on the central nervous system (*Ergotismus convulsivus*), even to the extent of producing death (Taylor 1965:61). In ancient Greece, as in other areas where *Claviceps* species were present, poisonous ergot alkaloids could have been ingested if infected grasses or grains were eaten. However, it is important to point out that the dangerously toxic ergotamine and ergotoxine alkaloids are not soluble, whereas the hallucinogenic alkaloids in the ergot are. This has been demonstrated by Hofmann:

> The separation of the hallucinogenic agents by simple water solution from the non-soluble ergotamine and ergotoxine alkaloids was well within the range of possiblities open to Early Man in Greece. An easier method still would have been to have recourse to some kind of ergot like that growing on the grass *Paspalum distichum*, which contains only alkaloids that are hallucinogenic and which could even have been used directly in powder form. (Hofmann, in Wasson et al. 1978:33)

Hofmann goes on to suggest that the ergot parasitizing *Paspalum distichum* may have been the ultimate psychoactive source utilized by the ancient Greek priests in their mystery rites.

> As I said before, *P. distichum* grows everywhere around the Mediterranean basin. During the many centuries when the Eleusinian Mysteries were thriving and holding the antique Greek world enthralled, may not the hierophants of Eleusis have been broadening their knowledge and improving their skills? For the Greek world as for us, the Mysteries are linked to Demeter and Kore, and they and Triptolemus are famed mythical progenitors of cultivated wheat and barley. But in the course of time the hierophants could easily have discovered *Claviceps paspali* growing on the grass *Paspalum distichum*. Here they would be able to get their hallucinogen direct, straight and pure. (Hofmann, in Wasson et al. 1978:33)

In the latter sections of his recent work dealing with the Eleusinian Mysteries, the

Greek scholar Ruck (Wasson et al. 1978:35–126) presented a series of corroborative evidence gleaned from the written record (especially the *Homeric Hymn to Demeter*). He argued that this testimony and the biochemical research done by Hofmann strongly support the idea, first suggested by Wasson, that an hallucinogenic fungus (probably an ergot) played a crucial role in what was one of the most important cult rites of ancient Greece.

Although the interpretation of Wasson, Ruck, and Hoffmann (1978) is convincing, and does represent a significant addition to the ethnopharmocological literature (Emboden 1978:3), it does not preclude the possibility that the opium poppy had its own role in the Eleusinian Mysteries. As I have shown, there are a number of indications that Demeter was closely connected with both cereal grains and the opium poppy. I believe that grains and poppy capsules, held in the hands or worn in garlands by either Demeter or her daughter, Persephone, not only symbolized the bountiful harvest, but also the cult experience itself.

According to Cook (1964:vol. 1, p. 425), poppy capsules "played their part in the

FIGURE 75. *Atropa belladonna* L., Known commonly as belladonna or deadly nightshade, this member of the solanceous family produces the psychoactive alkaloid atropine, which is sometimes mixed with papaverine (from the opium poppy) to treat ergot poisoning (from Rhind 1868).

Eleusinian initiation." I think both the capsules and the seeds of the opium poppy probably had a role in the mystery rites of Eleusis. In fact, I would like to offer what I believe is an original suggestion concerning the possible connection between the drug plant and the Eleusinian initiation. I have found evidence indicating that opium and/or opium poppy seeds may have been mixed with another drug plant derivative and, for precautionary purposes, either added to a ceremonial drink containing the psychoactive alkaloids of an ergot, or taken after the ergot-based concoction was drunk: "Two of the most useful drugs to offset ergot poisoning are papaverine, one of the alkaloids of opium, and atropine, derived from belladonna or henbane" (Taylor 1965:65; Emboden 1979a:82). In other words, since both the seeds and the capsules of the opium poppy contain papaverine (Duke 1973:392), either could have been part of an antidotal safeguard against the possible effects of the toxic ergot alkaloids described earlier. Although Hofmann (Wasson et al. 1978:25–34) has demonstrated that the toxic ergot alkaloids are not soluble, these poisonous substances might have contaminated the sacred beverage by dropping into the hallucinogenic drink along with broken fragments of the fungus during the processing of the potent *kykeon*. Such a dangerous possibility would have made the use of the suggested antidote a necessity.

This intriguing, and I believe revealing, relationship between some of the alkaloids in the ergot and the opium poppy may tell us something about the development of these species, both in terms of their natural and cultural evolution. As I noted above, the grass species which serve as hosts for the ergot are adapted to the same open, pioneer environment as the opium poppy, henbane, belladonna, and other heliophytic annuals. Therefore, the plants in question could be found in the same habitat and thus associated in a spatial as well as ritual or medical context. In any event, I believe that the concoctions used in cult ritual contained special mixtures of several drugs which, in combination, complemented each other and, in some cases, even may have counteracted the negative qualities of the components.

Several artifacts and ancient literary references may be used to support this hypothesis. One artifact is the so-called Lovatelli Hydria, a 294-centimeter-tall marble funerary urn (now held in the Museo Nazionale Romano). This urn is named after the Countess E. C. Lovatelli, who referred to the connections between the sculptured scenes portrayed on the exterior surface and the mystery rites of Eleusis (Roussel 1930:58–65). In sequence, the scenes show a sacrifice, a purification, and a vision of Demeter and Persephone (see figure 76).

Mylonas (1961:205) suggested that the "initiate," shown on the far right, is sacrificing a pig while a bearded priest facing him pours a libation with one hand and holds a plate offering ("a phiale"), with three poppy capsules on it, in the other: "There can be little doubt that here we have the beginning of the rites held for the purification of Herakles."

According to Ruck (Wasson et al. 1978:105), the initiation of Heracles into the mystery rites of Eleusis is depicted on another ancient specimen of funerary art, "the so-called 'Torre Nova sarcophagus'" (now held in the Palazzo Spagna in Rome), which shows the initiate pouring "wine over a sacrificial pig while holding a platter with mushroom-shaped plants in his other hand." In an attempt to identify the forms, Ruck (Wasson et al. 1978:105–6) pointed out that ". . . the objects in the platter have

FIGURE 76. The scene on the Lovatelli Hydria urn. On the right "the initiate is sacrificing"; in the middle "the purification is represented"; and on the left "he sees the vision of the goddesses, Demeter and Persephone" (Fairbanks 1910:133). Note the objects (poppy capsules?) on the platter held by the second person from the right.

been thought to be poppy capsules; in one depiction, however, the one on the so-called Lovatelli urn, the stems of the plants appear from photographs to be rather too thick to be real stems, while the same scene as it is shown on the Torre Nova sarcophagus depicts the objects actually standing on their stems, more in the manner of stipes." According to Ruck, the objects depicted on both platters "presumably should be the same, but it is always possible that they are not and that in the latter instance they are simple cakes of some kind." Interestingly, Athenaeus (3.113) described a bread which was made out of poppy seeds and shaped in the form of a mushroom (Seeberg 1969:11). Perhaps the form of this "bread" (cake?) did not represent a mushroom, as interpreted by the Classical Greek writer, but rather was a depiction of an opium poppy capsule, the source of the seeds themselves?

I would also like to point out that Fairbanks (1910:107) referred to the ancient Greek "propitiatory sacrifice," in which symbolic offerings such as libations or special cakes were presented: "Instead of cakes such as men ate, a peculiar cake or porridge made of meal, honey, and sometimes poppy-seed." According to Fairbanks, this cake or porridge "was never tasted, but burned on the low altar of the spirits of the deep." If the "cakes" used in the "sacrifice" were actually burned in a subterranean environment, then perhaps the parts of the poppy that were burned were not the seeds, but the alkaloid-rich sap (i.e., opium) contained in the seed capsules. As unlikely as this speculation regarding the early use of opium (the sap) may be, it seems to be supported by some extraordinary artifacts found in Late Bronze Age sites on Crete and Cyprus. These artifacts include a terracotta statue crowned with models of the opium poppy capsule, a few burning and vapor-collecting devices, and a unique "opium pipe." Descriptions and discussions of these artifacts follow later in this chapter.

It should also be noted that artifacts from Eleusis itself indicate that the opium poppy was connected to the famous mystery rites. For example, two mammoth female figures used as columns (Caryatides), which stood on the inner side of the doorway that faces the Eleusinian Sanctuary, depict maidens bearing mystic cylindrical *kistai* on their heads. Each *Kiste* (i.e., "sealed hamper of the mysteries"; Ruck, in Wasson et al. 1978:84) is decorated with an emblematic relief "of Demeter's cult" and thus shows the sacred items including "the poppy" (Mylonas 1961:154). According to Ruck (Wasson et al. 1978:84), the special revelation of the *sacra* (i.e., the sacred first fruits in the *kernos* and the holy objects contained in the *kistai*) took place after the sacred *kykeon* brew was drunk. This implies that the *sacra* were observed under abnormal psychological and/or physiological conditions since the *kykeon* drink was "accompanied by such symptoms of the drug experience as a cold sweating and a sense of vertigo" (Ruck, referring to Plutarch frg. 178 [Stobaeus 4.52.49, p. 1089 H] and Themistius, *Oration* 20.235). According to Ruck (Wasson et al. 1978:84), the hallowed contents of the *kistai* included "sacred cakes" and "poppies."

Poppies are supposedly also depicted on a tablet from Eleusis (now in the Greek National Archaeological Museum) which shows three scenes dealing with the mystery rites. "The torches, the bacchoi, the wreaths of myrtle, the kernos, even the poppies strewn in the field indicate that the scenes are connected with Eleusis and its cult" (Mylonas 1961:216). It should be pointed out that the *kernos* was a well-known

FIGURE 77. Minoan kernos located in the archaeological remains at Malia, Crete.

"cult vessel" used in the "worship of Demeter during Classical times"; according to Alexiou (1973:98), similar containers were familiar even during the Minoan period and were "evidently used for the *panspermia* offerings: a presentation of small quantities of all kinds of grain and agricultural produce." Alexiou (1973:99) referred to an excerpt from Athenaeus which "informs us that during the Classical period small amounts of all kinds of grain; wheat, barley, oats; various legumes, such as lentils, beans and vetch; liquids, such as oil, milk, wine and honey; white poppies (probably opium poppy-seed); and unwashed sheep's wool, were placed in the little cups." At Eleusis, poppies were also represented "on the architrave of the Lesser Propylaea where the emblems of the cult of Demeter were depicted" (Mylonas 1961:217; see Kerényi 1967:56, 74, 130–44 for a discussion of the many symbolic representations of the poppy on the monuments of Eleusis).

According to Ruck, ". . . poppies are a frequent motif in Eleusinian decoration, occurring usually with their four petals doubled to form roses of eight petals, whereas the true rose would have five petals or its double: the poppy capsule and the pomegranate itself were symbolic both of the marital abduction and of the fertile resurrection from death" (Wasson et al. 1978:115). However, Ruck indentified this poppy as the "pomegranate poppy" *(Papaver rhoeas)* and assumed that it was a common weed in the grain fields (cf. Theophrastus, *History of Plants*, 9.12.4). Although this species of poppy can still be found in great abundance in the cultivated and uncultivated fields of Greece during the spring season, I do not think that its capsule resembles a fruit of a pomegranate, as Ruck has suggested. Nevertheless, Ruck has raised an important question concerning the indentification of the poppies associated with the cult of Earth Mother at Eleusis (and elsewhere in the Greek realm). Even if there are actual representations of poppies on the various artifacts referred to, were these supposed to have represented opium poppies *(Papaver somniferum)* or the corn (i.e., red or field) poppy *(P. rhoeas)*?

I believe that both of these species were available in ancient Greece, and that by the time of the Late Bronze Age, they could be found as volunteers in or near agricultural fields. In addition, both could have been used to make medicinal, perhaps even psychoactive, drugs (see Duke 1973:390–400 and El Masry et al. 1981:61–64). However, only the opium poppy could have provided the nutritious seeds used in the "cakes." Furthermore, only the opium poppy capsules or seeds could have provided papaverine alkaloids, which may have been administered to the initiates of the mystery rites so that they might not be affected by the toxic alkaloids in the ergot-based *Kykeon* drink.

Thus I have suggested that by the Mycenaean period, the opium poppy was known in the Greek realm, if not as a common folk medicine among the masses, then within the mystic circle of the shaman seers who interpreted divine messages and were in charge of spiritual guidance, ritual observance, and dreams. Although the case that I have attempted to build involving the opium poppy's mind-altering qualities and its use in cult ritual is admittedly a tenuous one, I suggest that the corroborative evidence be appraised and the mysterious nature of the cult(s) be considered before my hypotheses are dismissed. It should be remembered that the spiritual association of the opium poppy with somniferous effects was often referred to in the ancient literature of the Mediterranean world. For example, according to the Roman poet Virgil, the opium poppy was pervaded with Lethean sleep ("Lethaeo perfusa papavera somno," *Georgica* 1. 78; see also 4. 545). These words of Virgil are reflections of a tradition that reached back into the prehistoric period, perhaps to the very beginnings of the Hellenistic culture. According to Greek mythology, ". . . Lethe is a river that flows through the regions of the dead, the waters of which, if drunk by anyone, cause oblivion in regard to their past existence" (Anderson 1922: 83).

In this chapter I have already presented various kinds of evidence to support the idea that the opium poppy first appeared in the eastern Mediterranean region during the Late Bronze Age and that it was introduced as a result of trade or other forms of cultural contact. In addition, I have discussed some aspects of the plant's role as an exotic weed in or near cultivated fields and its possible use as a ritual source of intoxicating drugs. In the following sections, I will describe additional indirect archaeological evidence for the opium poppy's presence on the mainland of Greece, as well as in Crete, Cyprus, and Egypt during the Mycenaean period. Furthemore, I will discuss the cultural implications that can be drawn from this evidence, implications which I believe support the idea that the opium poppy played a role in both mind-altering ritual and therapeutic medicine.

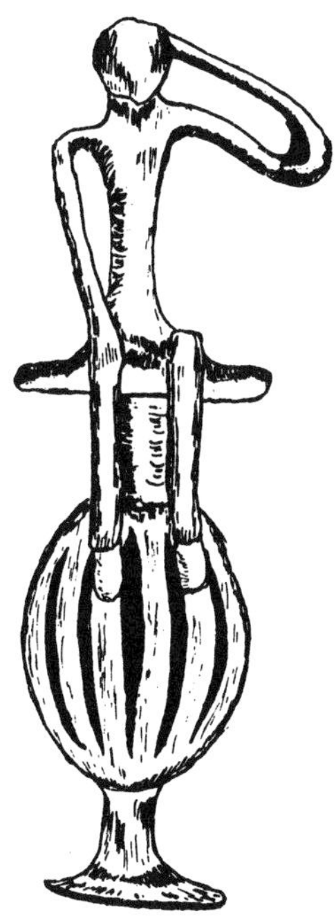

FIGURE 78. Early Greek bronze artifact found at Kozani. It appears to represent an inverted, incised opium poppy supporting a human figure whose gesture emphasizes the psychoactive effects of the opium exudate. The height of the artifact is ca. 8 cm (see Daux 1960: 777, figures 8 and 9; *Courtesy of Professor William Emboden, Department of Biology, California State University, Northridge*).

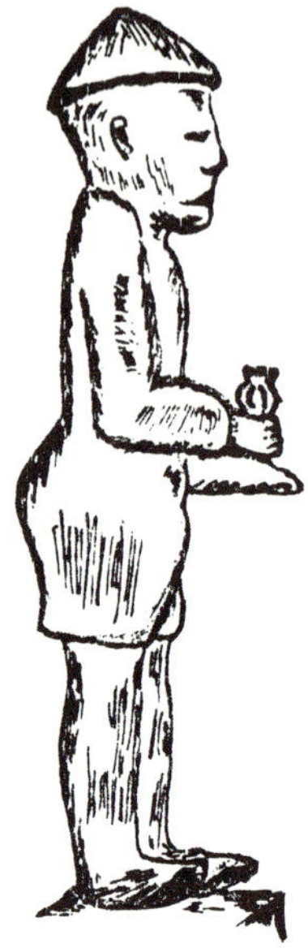

FIGURE 79. Line drawing of an ancient Greek bronze statuette found in southern Greece. According to Kritikos and Papadaki (1967a:32) this artifact depicts "a shepherd or peasant holding the fruit of the poppy in his right hand."

Pinheads and Poppy Capsules

According to Gimbutas (1965:113), the "Late Bronze Age chronology of Europe is chiefly based on intimate contacts between central European early Urnfield culture and Greece during the end of the Late Helladic period." These contacts have been used to explain the presence of the central European styled weapons and pins in various parts of the Balkan Peninsula, including Mycenaean Greece. A number of the Late Bronze Age pins found in Greece have spherical heads and have been identified with poppies, pomegranates, and apples.

If the connections between the opium poppy capsule and some ancient Greek pins are correct, and if the prototype for these pins did come from the north in central Europe during the Late Bronze Age, then we would have another indication of the central European origin of the opium poppy in Greece and the eastern Mediterranean region. According to Alexander (1964:177), in the latter part of the Bronze Age, ". . . pins of Central European type were scattered through Jugoslavia and like the other objects show that the country formed an outlying province of Central Europe"; he also referred to the distributions of two central European pin types which "appear to show the routes taken by Late Bronze Age groups moving south in Macedonia and peninsular Greece." Some relevant pins are noted below.

One possible central European prototype for the Greek "poppy pins" is an unburnt bronze, "late type of vase-headed pin," which was found in a late Bronze Age grave at

Neszmely in western Hungary (Gimbutas 1965:346). The inspiration for its shape may well have been a vase or the concept of one; however, we might consider its abstract similarity with some plant fruit such as the opium poppy capsule.

Among the artifacts recovered from the royal tomb III at Mycenae were pins with spherical heads; according to Schliemann (1878:201), these pins were "Scepters," but Kritikos and Papadaki (1967a:27) referred to them as "broach pins" with "capsule-shaped" heads. Somewhat similar pins of the same age were found in Grave Omicron at Mycenae (Mylonas 1957:145).

One spherical-headed pin from royal tomb III at Mycenae is illustrated in figure 80; it is 27.7 cm long, has a head made out of "cloudy-brown mountain crystal," and is dated to about the sixteenth century B.C. (Kritikos and Papadaki 1967a:27). According to Jacobsthal (1956:185), this pin and similar ones "resemble apples and might be apples, but they are stylized too strongly to be verified with certainty." On the other hand, Kritikos and Papadaki (1967a:27–28) argued that this pin is grooved and therefore represents an opium poppy capsule, not an apple. They have interpreted the grooves as "the notches for the extraction of opium" and suggest that the top of the crystal model of an opium poppy capsule has lost its metal "star-shaped crown" probably because of oxidation.

Kritikos and Papadaki also described other, younger, prehistoric Greek pins made out of silver with notched heads from Ephesus and Vaphio in Laconia. They believe that these pins are abstract representations and comparable to the crystal pinhead from the Mycenaean royal grave.

FIGURE 80. Spherical-headed pin from royal tomb III at Mycenae (see Jacobsthal 1956, figure 140). Size = 27.7 cm.

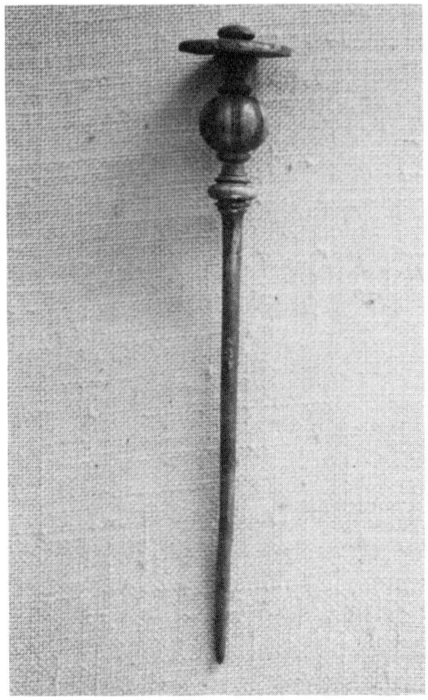

FIGURE 81. Silver pin, inscribed with a dedication to Hera (found near the Argive Heraeum, seventh century B.C.).

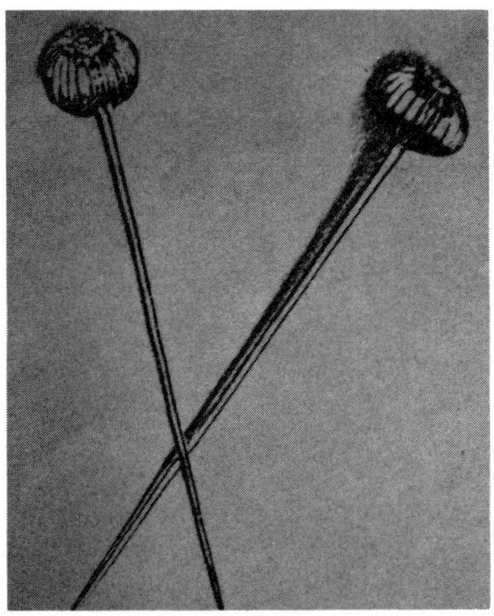

FIGURE 82. Two gold pins from Anatolia with "poppy capsule" heads. According to Merrillees (1979:168 and plate XXXVI A), the hollow, oval heads of these Late Bronze Age pins (ca. 1400–1200 B.C.) are representations of dried opium poppy capsules. The pins belong to the Norbert Schimmel Collection *(Photo by O. E. Nelson; Courtesy of Professor Robert S. Merrillees, Department of Foreign Affairs, Canberra, Australia).*

A recent analysis of two identical gold pins (see figure 82) found in Anatolia is especially noteworthy. A number of morphological features of these interesting objects, apparently fabricated during the time of the Hittite Empire (i.e., ca. 1400–1200 B.C.), indicate that they represent opium poppy capsules:

> Making allowance for a certain artistic licence, the resemblance of these pins to opium poppy capsules on stalks is unmistakable. Not only have the thalamus and stigma been reproduced, albeit with a degree of stylization, but the ribbed exterior of a dried poppy capsule has been faithfully copied, even down to the metallic inclusions representing the seeds loose in the organic original. The inspiration for these pins cannot therefore be doubted (Merrillees 1979: 168).

According to Muscarella (see Merrillees 1979: 168), the loose objects located within the hollow, oval heads give "the pin the quality of a rattle when it is shaken." This is of interest because dried opium poppy capsules and their stalks are still utilized as baby rattles in some parts of the world where this plant is available (Duke 1973: 393).

Jacobsthal (1956: 38) described a "short pin" from Delos which has a gold-plated silver shank and a solid gold head. He referred to the design as "Doric sub-Geometric" and suggested that it could represent either the fruit of a "pomegranate or *Papaver somniferum*."

Kritikos and Papadaki (1967a: 28) also referred to a bone amulet recovered from the ruins of the acropolis at Mycenae (and now held in the archaeological museum at Mycenae) which they believe has the same form as a grooved poppy capsule. The authors argued that these grooved artifacts (and others), in conjunction with corroborative evidence (discussed in the following section of this chapter), indicate that ". . . the poppy and its incision for the extraction of opium were known at Mycenae at least in the sixteenth century B.C. (Late Mycenaean I Age)" (Kritikos and Papadaki 1967a: 28). Furthermore, they referred to the suggestions of "other scholars" that "because of their great length," the pins in question may *not* have been used for hairdressing or holding clothes together in a decorative manner; instead they suggested that the pins may have been used "in the process of smoking opium"!

Although this hypothesis may be wildly inaccurate, especially in light of the general absence of smoking in the Old World until after the discovery of tobacco and the smoking of its leaves in the New World, there is some supportive evidence from Late Bronze Age Crete and Cyprus which should stimulate serious consideration of the possibility of narcotic inhalation in the prehistoric eastern Mediterranean region. The evidence pertinent to this question is presented in the following section.

The Goddess Statue from Crete and Her Special Paraphernalia

In 1936, E. Blegen published a short report in the *American Journal of Archaeology* briefly describing two terracotta idols found on Crete at Gazi, an archaeological site about six kilometers west of the major Minoan palace center of Knossos. The peasants who uncovered the remarkable figurines "unfortunately destroyed the building in which they lay" (Blegen 1936: 372). However, after careful archaeological investigation, Marinatos (1937:279) was able to describe and illustrate the "sacred" room in which the "goddess idols" were found. The two "goddess" idols recovered from the "sacred" room both have uplifted hands and half-length bodies rising from cylindrical bases. This posture of a female idol with uplifted hands is a well-known goddess motif of the Myceanaean Age (i.e., the Late Bronze Age; see Taylour 1969).

The smaller terracotta idol (ht. 530 mm) "has on her head a pair of doves and the so-

FIGURE 83. The large terracotta goddess idol from Gazi, Crete. This intriguing 3,500-year-old artifact bears three moveable, incised pins in the head which some scholars have identified as representations of opium poppy capsules tapped for their alkaloid-rich sap.

called 'horns of consecration'; the larger one (ht. 800 mm.) "wears in her hair three pins with representations of pomegranates and has remarkably Greek features" (Blegen 1936:372; see figure 83). Although Cook (1964:vol. 2, p. 818) also referred to these pins as pomegranates, the three "moveable" pins have since been studied in detail by other scholars who are convinced that they may represent opium poppy capsules (compare Marinatos 1937:278–84; Zervos 1956:47; Kritikos 1960:54–73; Butterworth 1966:175; Kritikos and Papadaki 1967a:23; Alexiou et al. 1968:198; Hawkes 1968:126; Hood 1971:232). Marinatos interpreted the three moveable pins on the tallest figurine as capsules of the opium poppy and referred to the idol as "the Poppy Goddess, Patroness of Healing" (Kritikos and Papadaki 1967a:23):

> We may note that the poppy-heads are cut in the same way that they are today to extract opium, before they ripen. The deduction is that the women of Crete understood the effect of opium which they tended, as a mystery of the goddess, and dispensed in cases of suffering or despair. (Marinatos and Hirmer 1960:153)

Three additional terracotta goddess idols with uplifted hands were found at Gazi; Marinatos (1937:278–91) described all five of them, including the "Poppy Goddess," which was the tallest of the group and is now on display (inventory item 9305) with the others in the Late Minoan section of the Iráklion Museum on Crete (see figure 84). According to Alexiou (1968 et al.: 198), the "Poppy Goddess" and the other five terracotta figurines from Gazi "are among the largest and most interesting examples of female divinities from the post-Palace period" (1400–1150 B.C.), and were found in a "consecrated row of private homes."

In 1959, Kritikos made a study of the goddess idols from Gazi and came to the following conclusions regarding the tallest figurine:

FIGURE 84. Five terracotta statues found in a subterranean sanctuary at Gazi, Crete. These statues are now held in the Iráklion Museum on Crete. The "Poppy Goddess" is the largest of these statutes, which all have the characteristic Mycenaean feature of uplifted hands.

1. The three pins on the head of the goddess do in fact represent heads of the opium poppy.
2. The vertical notches in the capsules, which are more deeply coloured, belong to one of the typical forms of poppy used for producing opium.
3. The carving of the capsules, then executed vertically, is different from the circular carving usual today, or the mixed pattern, circular around the top and then vertical as that of the figurine (Dioscorides, Pliny).
4. Of especial interest is the artist's rendering of the colouring in the notches: it corresponds with the colour of the dried juice of the poppy.
5. The Minoan goddess appears to have her eyes closed as though asleep.
6. Especially impressive are the folds in the cheeks giving a smiling effect, and the lifelessness of the parting of the lips.
7. . . . that a tubular vase discovered at the same site and belonging, according to S. Marinatos, to the equipment of the goddess, was used for preparing inhalations of opium. This receptacle had a base and a hole on the sides, and bore a remarkable resemblance to those used in Java in earlier times for the inhalation of the vapours of opium [see figure 85].
8. It is to be noted that the goddesses were found in a room presumably sacred, enclosed on all sides, doorless and windowless, and obviously approached from above. On the ground were found the remnants of a heap of [char]coal. (Kritikos and Papadaki 1967a:23–24)

a. b. c.

FIGURE 85. Cult artifacts and opium inhalation device compared. (a) and (b) "objects associated with the cult of the poppy goddess and found at the same site"; (c) "a vase from Java for inhaling the fumes of opium" (from Kritikos and Papadaki 1967a:24; *Courtesy of Archaiologike Ephemeris, Hellenike Archaiologike Hetairia, Athens, Greece*).

Kritikos's interpretation of the artifacts found in the "sacred room" near Knossos is certainly provocative. He offered additional conclusions drawn from his observations of the "Poppy Goddess":

(a) The poppy and the extraction of its opium through a vertical notch was already known at least at the time of the making of the statuette, viz. the Late Minoan III period. . . .
(b) This manner of notching survived in the East Indies up to the beginning of our century.
(c) The use of opium for religious purposes at least, was known in Crete in the same period.

(d) The significance of the poppy and of opium was such that to it should be attributed the special posture of the goddess with uplifted hands.
(e) From the other objects found near the goddess (receptacles and coals), it must be inferred that the opium was taken by the inhalation of vapours.
(f) The goddess appears to be in a state of torpor induced by opium; she is in ectasy, pleasure being manifested on her face, doubtless caused by the beautiful visions aroused in her imagination by the action of the drug. For this reason . . . she should be called "the goddess of ecstasy"; and lastly,
(g) The passivity of her lips is also a natural effect of opium intoxication. (Kritikos and Papadaki 1967a: 24)

Kritikos's references to opium smoking are especially interesting; I have already discussed the long pins with "notched poppy-heads" found at Mycenae on the Greek mainland, which according to Kritikos and Papadaki (1967a: 28), were connected with opium use. These scholars also referred to the discovery of a similar long pin (c. 17 cm), in a grave at Mouliana on Crete, which they believe has a head shaped like an opium poppy capsule (Kritikos and Papadaki 1967a: 25). In addition, it should be noted that another of these "long pins" with a poppy capsule shaped head was found in a "Late Minoan" tomb on Rhodes at Ialysos (see Jacobsthal 1956: plate 3, figure 8). Also of significant interest here is a decorated jar found in a grave at Pachyammos on Crete.

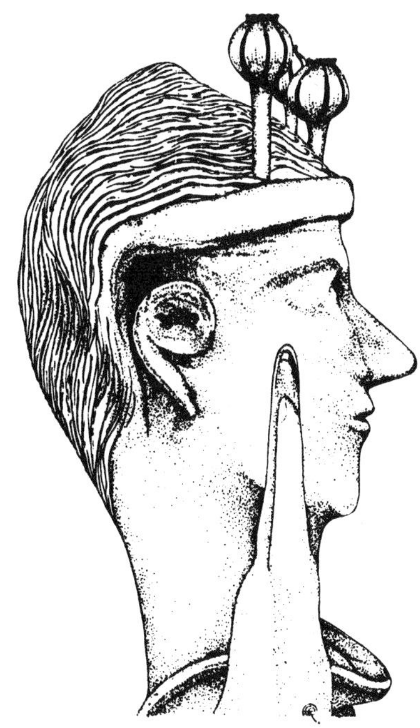

FIGURE 86. Profile of the "Poppy Goddess" from Gazi, Crete (from Emboden 1979: 28, figure 23; *Courtesy of Professor William Emboden, Department of Biology, California State University, Northridge*).

> On the body of the jar the poppy head is portrayed between the sacred horns and is guarded by birds. This reveals the sacred nature attributed to the plant by the Minoans. On the lid of the jar is the picture of a bird tearing apart the capsule of the poppy. (compare Zervos 1956: figures 739 and 740)

The decorated, Late Minoan jar from Pachyammos may have been used as a container of opium and symbolically placed in the grave as a religious offering connected with the cult of rebirth and the importance of dreams discussed in a previous section. Zervos (1956:47) argued that the Minoans considered the opium poppy to be a sacred plant and believed that its capsules symbolized eternal life. He suggested that "The Goddess of the Dead" (represented by the largest terracotta statue from Gazi) puts people to sleep at the time of their death and guides them to the underworld and another existence. Zervos's assumptions regarding the role of the opium poppy on ancient Crete are similar to mine, only I believe that the drug plant did not arrive on the island until the Late Bronze Age (i.e., Late Minoan period) when the influences of the Mycenaeans became significant.

Late Bronze Age Smoking Paraphernalia from Crete and Cyprus

The possibility of opium inhalation in the Gazi sanctuary is supported by the discovery of the "Poppy Goddess" and the interesting implements which were found in this subterranean room. It should also be noted that, in her early report, Blegen (1936:372) referred to the "small plate and one of the well-known tubular cult implements which grow narrower toward the top and which admirably confirm the explanation of Sir Arthur Evans that the objects are derived from clay water-pipes."

Evans (1935:vol. 4, 145ff.) did interpret the tubular vessels (without bottoms) as water pipes and suggested that they were later used as "snake houses" and "libation vessels." However, Kritikos (1960, cited in Karageorghis 1976:128) argued that these cult vessels were strictly for religious use, probably involving vapor inhalation. Referring to archaeological locations east of Crete, Karageorghis (1976:128) pointed out that some ". . . Near Eastern sites and also Enkomi, Cyprus have produced tubular vessels without a bottom, mainly in Late Bronze Age sanctuaries." In addition, Karageorghis reported the recent discovery of a similar tubular vessel (probably having a bottom in its original condition) in the courtyard of "temple 5" at Kition, Cyprus, dated to about 1200 B.C. (see figure 87).

> It certainly formed part of the sacred vessels of the temple; it has two large identical signs of the Cypro-Minoan syllabary inscribed on its outer surface. There is enough evidence to show that it had a bottom, though now broken, but was open at the upper part. Near the bottom there are two opposed holes. The vessel is preserved to a height of 23 cm; it had a diameter at the bottom of 26 cm and at the top of 19 cm. The rim is broken, suggesting that the vessel was taller than its present height. (Karageorghis 1976:127–28)

Karageorghis (1976:128) argued that in addition to its discovery in the temple court-

FIGURE 87. Cylindrical vessel from Kition, Cyprus. This tubular vessel was found in temple 5 of the Late Bronze Age site at Kition. The diameter of the bottom is 26 cm. The cross-section diagram shows the "opposed holes." These openings may have allowed a draught of air to form and thus aid in the cult practice of inhaling the fumes of opium. Note the "large sign of the Cypro-Minoan syllabary"; according to Karageorghis (1976: 128) this "leaves no doubt that it formed part of the sacred vessels of the temple" (from Karageorghis 1976: 127, figure 2 and plate XV, b; *Courtesy of Professor Vassos Karageorghis, director, National Museum of Antiquities, Nicosia, Cyprus*).

yard there is conclusive evidence indicating the vessel's religious nature: "The large sign of the Cypro-Minoan syllabary, engraved on both sides of it, leaves no doubt that it formed part of the sacred vessels of the temple."

Kritikos and Papadaki (1967b: 7) referred to the open and closed tubular vessels as "pipes" and described the ways in which they could have been used in the Late Bronze Age:

> *Open pipes.* (a) These were probably placed, as is popularly done today, over braziers, so that they were kept alight by a current running from bottom to top. In such cases, if a stone or tile were placed on the coals, and when it was well heated the opium were poured upon it, then its vapours would emerge concentrated from the upper mouthpiece without any overflow drip. Thus they would be more satisfactorily inhaled.
>
> (b) Probably on the pipe placed over the fire in the abovementioned manner a tile would be placed on which, when well heated, opium would be poured; the vapours rising from it would be absorbed by nasal inhalation.

Closed pipes. As these have holes in the lower lateral section it may be presumed that:

(a) Lighted coals or a burner were placed within them, upon while a tile was placed. By pouring opium upon it, concentrated vapours would issue and be inhaled through the open end; a current of air would form, coming through the holes and flowing upward.

(b) The tile was not placed on the coals or the burner but on the open mouth-piece where it would become hot, and the opium would be poured on it and inhaled.

As noted previously, Kritikos and Papadaki (1967a:28; 1967b:8) suggested that the long metal pins with "notched poppy capsule" models as heads may have been used, not for holding clothing together or keeping one's hair up, but rather "as an instrument for carrying the opium to the source of heat and producing vapours, as is done today." Sir William Roberts described the use of pins in association with opium vapor inhalation as practiced in parts of Asia during this century:

> Chandu [a Chinese word for "prepared opium"] is smoked in a specially constructed pipe in a peculiar manner. The pipe-stem is composed of a tube of bamboo, about 20 inches long and more than an inch thick. One end is closed, and the other end is fashioned into a thick mouth-piece with a large orifice. About 7 inches from the closed extremity is a round hole, mounted in brasswork, for the reception of the neck of the bowl. The bowl of a chandu pipe is usually made of terracotta and is entirely unlike the bowl of a tobacco-pipe. The diagram will assist the reader to understand the description (see below). The bowl (a) consists of a hollow disc, or chamber, about 2½ inches wide and about 1¼ inches deep. The upper surface is flatly convex, and at its centre (b) is a minute shallow cup, lined with metal, and

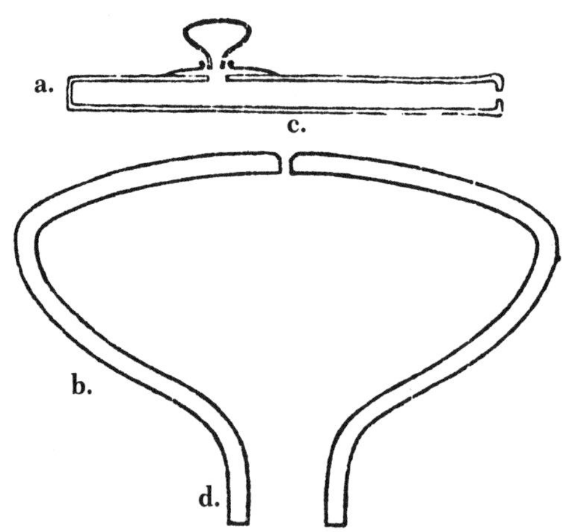

FIGURE 88. Vertical section through an opium pipe and its bowl. (a) vertical section through a pipe used to inhale the vapors of prepared opium; (b) the bowl; (c) the recepticle for the charge of prepared opium; (d) the neck of the bowl (from Addens 1939:104, figures 1 and 2).

> perforated at the bottom with a small orifice. This leads to a short canal running through the terracotta, and is only wide enough to admit a medium-sized pin. The lower side of the disc is continued into a short neck (c) which fits into the opening in the pipe-stem. The chandu-smoker proceeds in the following fashion. Armed with a pointed probe or style he dips the point into his supply of chandu and takes up a suitable quantity which forms a little drop at the point of the style. This he now carefully heats over the flame of a lamp until it dries up into a soft pilule. He next transfers the little pilule to the shallow cup on the upper surface of the bowl. Then, assuming a reclining position, he cautiously applies the pilule to the flame of the lamp and at the right moment, when the chandu begins to bubble and emit smoke, he takes a full inspiration and fills his lungs with smoke, and then slowly discharges it. One or two such inspirations exhausts the charge. A practised smoker will smoke 15 or 20 pipes at a sitting. (Cited in Addens 1939: 103–4)

Kritikos and Papadaki admitted that this hypothesis connecting ancient pins with opium consumption is somewhat weakened by the following evidence:

> 1. The existence of bone pins with notched heads (compare Corinth and elsewhere);
> 2. The discovery in graves of long iron pins with traces of cloth attached. (Kritikos and Papadaki 1967b: 8)

They suggested that more research into this matter is needed and offered an explanation for the lack of information on this subject (an explanation which I have already alluded to in reference to the shaman cult and its association with the opium poppy in the Greek realm during the Late Bronze Age):

> No information has come down to us regarding the method of opium smoking practised in antiquity, possibly because the use of the drug, primarily euphoric, remained hidden behind the walls of sanctuaries, where many capsule-shaped objects have been found.
>
> Perhaps opium vapours were used at prophetic shrines in the process of inducing vaticination. Such a use of opium may provide the reason why physicians of classical antiquity did not describe the method of taking opium "by nasal inhalation." (Kritikos and Papadaki 1967b: 8)

Another more recent discovery in the eastern Mediterranean region further supports the idea that opium vapors were inhaled in some form (if only by an elite, perhaps religious, group) as far back as the Late Bronze Age. In 1974, the "holy-of-holies" ("temple 4") was discovered at Kition on the island of Cyprus. In the following year a "later (Phoenician) wall" was removed from the northeastern corner of the temple and the fill was carefully excavated. According to Karageorghis (1976: 125), a horde of ivory objects was found in the fill which dated to about 1220–1190 B.C. and was laid upon the earliest floor of the "temple": ". . . its construction may coincide with the arrival of the first Achaean [i.e., Mycenaean] colonists in Cyprus."

Among the ivory objects found in "temple 4," two are especially relevant to our discussion of the possible ritualistic inhalation of opium in the Late Bronze Age. First there is the 22-cm-tall vase with a two-sided carving of Bes, the benevolent Egyptian god of the home, who aided pregnant women; and second, there is a very unusual and intriguing cylindrical object (see figure 89).

Karageorghis (1976: 125–27) has described this 13.5-cm-long ivory object in detail, arguing that its use in antiquity is obvious:

> The character of the object and the fact that around one of its two round openings (opening of the upper part) there are distinct traces of burning, leaves no doubt as to its function as a smoking pipe. It is not difficult to imagine the substance which was smoked.

Karageorghis (1976: 127) also pointed out that the "temple 5" courtyard (in which the "cylindrical vase" discussed earlier was found) is adjacent to "temple 4."

Recently Karageorghis (1976: 129) asked for the opinion of the Greek pharmacological specialist, P. G. Kritikos, regarding the "pipe" and "vase" found at Kition. Kritikos, whose analysis of the Cretan artifacts has already been presented, responded to the queries of Karageorghis:

> There is no doubt that the cylindrical ivory object is a pipe for smoking, though it is not possible to prove that it was used for smoking opium. Our first impression from the archaeological discoveries in temples 4 and 5 is that they are comparable to those found in the sanctuary of Gazi, in Crete, which was dedicated to the "Goddess of the Poppies." There, too, a roofed holy-of-holies was found containing a cylindrical vessel with a perforation as well as a pile of charcoal. We suggested that

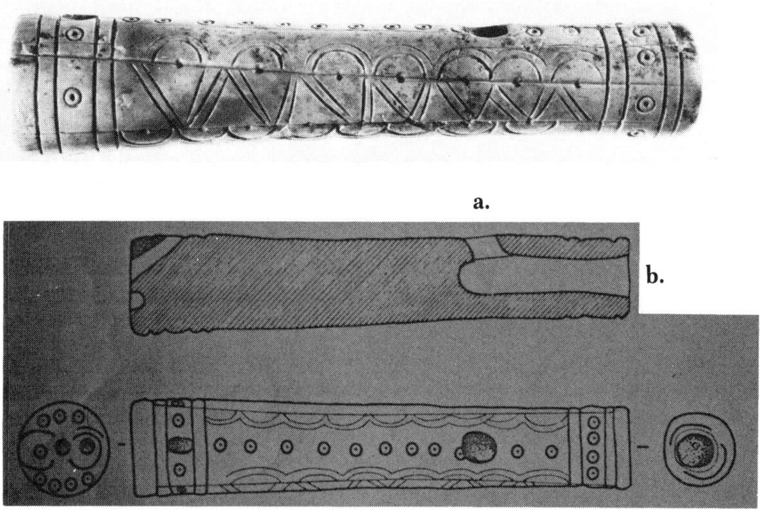

FIGURE 89. Late Bronze Age "opium pipe" from Cyprus. Found in temple 4 of the site at Kirion, Cyprus, this 13.5 cm long ivory object has "distinct traces of burning" at (a). Fumes (of opium?) were apparently inhaled through the open end (b). This object and other implements from sites on Crete and Cyprus have been connected with a Late Bronze Age cult practice of opium intoxication (from Karageorghis 1976: 128, figure 1 and plate XV, a; *Courtesy of Professor Vassos Karageorghis, director, National Museum of Antiquities, Nicosia, Cyprus*).

> the objects discovered at Gazi were used for inhaling smoke of opium and that the image of the goddess discovered in the sanctuary seemed to have an expression of ecstasy, caused under the influence of opium. The sanctuaries of Kition have produced not only the same cylindrical vessel with two perforations at the lower part and the pile of charcoal as at Gazi, but furthermore a new element, suitable for smoking, a cylindrical pipe. We may, therefore, suggest the use of opium at Kition, as at Gazi, and ever further, that the smoke of opium was inhaled at Kition through a pipe which bears obvious signs of dark stains caused by fire round the vertical perforation through which smoke entered. . . . On the surface of the pipe, near the non-perforated terminal, there are also similar signs of dark stains.

Kritikos went on to offer some suggestions concerning the manner in which opium may have been inhaled with the use of the "pipe" and "vase" in this Late Bronze Age context:

> The cylindrical clay vessel with two perforations at its lower part was used as a receptacle for burning charcoal on which opium was placed; we may also suggest that opium was not placed directly on burning charcoal but indirectly, on a clay plaque which, having been heated on this charcoal, could cause smokes from the opium which was thrown on it.The ivory cylindrical pipe was not used for burning opium on it like the smoking of cigarettes; this fashion of smoking is of a much later date. The way of smoking opium to-day is also the same, consisting of inhaling the smoke through a pipe of a more perfected type. It is probable that during the act of smoking the vertical opening of the pipe which has traces of dark stains all round it was brought above the cylindrical clay vessel in which opium was burning; the actual smoking (inhaling of smoke) was done through the perforation at the terminal of the pipe. The function of the two perforations at the lower part of the cylindrical clay vessel was to create a draught of air which would keep the charcoal burning. (P. G. Kritikos; cf. Karageorghis 1976: 129)

Mycenaen Trade and Influence

I have discussed the introduction of the shaman cult into the Greek mainland during the Late Bronze Age, along with its suggested connection with the opium poppy and its effect on the concept of death and afterlife. In addition, I have referred to a group of interesting Late Bronze Age artifacts which indicate that the opium poppy was known and possibly used in religious activity associated with the cult of the dead and involving inhalation of opium vapors. Although the artifacts most strongly associated with smoking were found in Crete and Cyprus, I believe that their cultural inspiration was derived from Mycenaan influences, especially during the Late Minoan II and III periods (i.e., about 1450–1150 B.C.).

> The acceptance by most scholars of Michael Ventris's decipherment of Linear B as an early dialect of Greek has shown one major point of difference between the Mycenaean world and the Cretan. The replacement of the native Linear A script by Linear B at Knossos in Late Minoan II suggests, if not political domination of Knossos at least a strong influence by Mainland elements upon Cretan culture from this period onwards. (Chao 1975: 222)

(A)	Buhen, Egypt: Vasiform beads on necklaces
(B)	Thebes, Egypt: ' Alabaster vase with 'opium', Ebers Papyri
(C)	Biban-el-Molouk, Egypt: Gold earrings and necklace Deir-el-Medineh, Egypt: Poppy capsule in burial tomb
(D)	Tell-el-Amarna: Cypriote and native juglets in opium poppy capsule form, vasiform beads, pendants, etc.
(E)	Saqqara, Egypt: Statuette with vasiform bead earrings
(F)	Kition, Cyprus: 'Opium pipe', vessels, vasiform necklace
(G)	Ras Shamra (Ugarit), Syria: Cypriot juglets
(H)	Mari, Syria: Vasiform beads on necklace
(I)	Beycesultan, Anatolia: Poppy seeds in bowl in burnt palace
(J)	Ialysos, Rhodes: Pin head in shape of opium capsule
(K)	Mouliana, Crete: Pin head in shape of opium capsule
(L)	Gazi, Crete: 'Goddess of Poppies' (clay statue)
(M)	Mycenae, Greece: Gold signet ring, pins, broach, amulets, etc.
(N)	Corinthian Gulf littoral, Greece: Vases in shape of opium poppy capsule

FIGURE 90. Map key (see figure 91). Distribution of evidence for opium poppy use in the eastern Mediterranean region during the Late Bronze Age.

Apparently, sometime after about 1500 B.C., Mycenaean interests established an economic-administrative hegemony over Crete. Eventually, the patriarchal Greek civilization (originally the so-called Achaeans) replaced the matriarchal, non-Greek, maritime-commercial supremacy of the Minoans in the Aegean and beyond, in both east and west directions. It should be pointed out that on the island of Crete, the majority of ritual artifacts which archaeologists and historians have used to interpret Minoan religion have been discovered in contexts dating to the Late Minoan II and III periods:

> The overemphasis by Evans on the exertion of the Minoan culture upon the Mainland obscured for a while any thought regarding the possible contributions

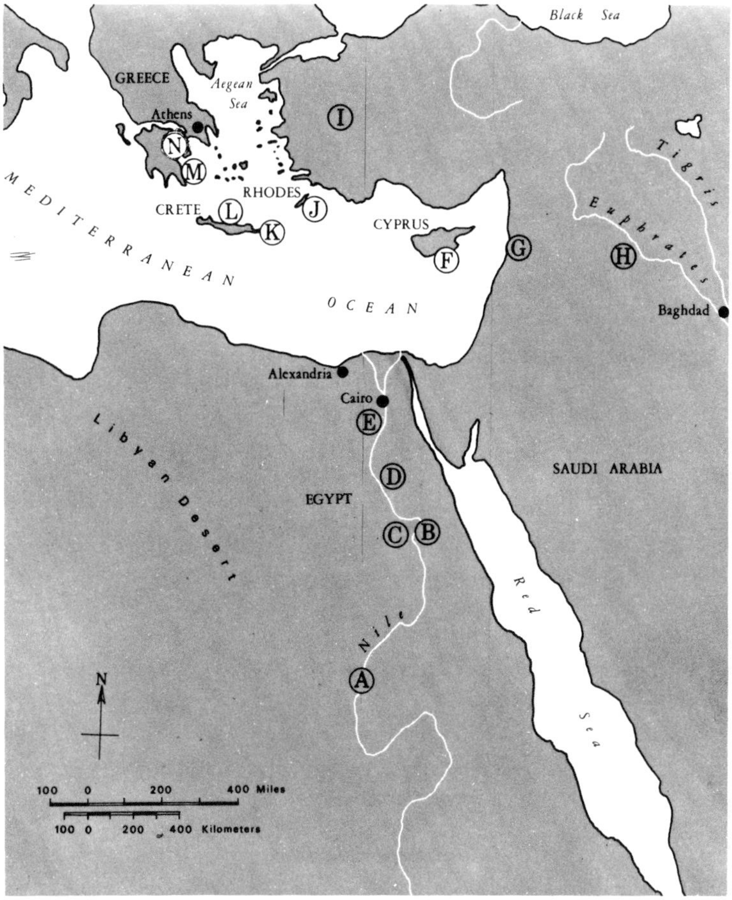

FIGURE 91. Distribution of Evidence for opium poppy use in the eastern Mediterranean region during the Late Bronze Age (see key to map, figure 90).

> that the latter might have made to Crete. In any attempt to investigate the nature of the Minoans' religious beliefs and practices it is necessary to disentangle the present web of confusion surrounding the Minoan-Mycenaean religion. (Chao 1975:222)

Even if the use of the opium poppy in cult ritual was not a direct import from the Greek mainland, its connection with shamanism may be traced back to this region during the Late Bronze Age. Perhaps the opium poppy was introduced into Crete, Cyprus, or Egypt indirectly as a result of Mycenaean influences; and subsequently, one or more of its products (or symbolic representations) were exported.

We know that the Mycenaeans established flourishing colonies on the islands of Kos and Rhodes, as well as at Miletus on the western coast of Asia Minor. In addition, their influence was also important in such far-flung regions as southern Italy, Sicily, Egypt, the Levantine Coast, Troy, and Cyprus (Stubbings 1973:61;

D'Agostino 1975 : 22). Furthermore, as I have pointed out, the Late Bronze Age was a period of relatively intense and extensive trade, linking widespread regions through the exchange of metals, luxury items, and perhaps philosophies and spiritual concepts: "The intangible commerce of the things of the mind is even more elusive, but may have been of great importance. Some themes of Greek mythology that are common to the Near East perhaps first became familiar in this Mycenaean age, and the exchange of religious ideas may have gone deeper than this" (Stubbings 1973 : 67).

In the following chapter, I describe and discuss evidence which indicates that the opium poppy was introduced into Egypt during the Late Bronze Age, possibly, at first, from Cyprus. This evidence includes juglets and bead pendants (amulets?) modeled after the opium poppy capsule, literary references in the papyri records of ancient Egypt, and unconfirmed, but intriguing, reports of an opium residue and an actual opium poppy capsule.

9 • The Opium Poppy in Egypt during the Late Bronze Age

In this chapter I discuss a variety of reasons why I believe the opium poppy and its drug products were brought into Egypt during the era of the New Kingdom (i.e., the Late Bronze Age). I begin this discussion with a description and analysis of R. S. Merrillees's theory that opium drugs were packaged in small ceramic vessels made to look like the poppy capsule and shipped to the Nile Valley from Cyprus or the Levantine Coast.

The Cypriot Juglet Theory: Late Bronze Age Models of Poppy Capsules and Containers of Exported Opium Drugs

> The shape of this juglet is a newcomer to the ceramic repertory of Late Bronze Age Cyprus and has no antecedents in Egypt or elsewhere round the Levant. The origin of this form is therefore of particular interest. . . . (Merrillees 1974:32)

According to Merrillees (1962:287ff., 1968:154ff., 1974:32ff., 1979:169–70), the opium poppy was introduced to the island of Cyprus from the Aegean region during the early part of the Late Cypriot Bronze Age (ca. 1650–1500 B.C.); and, subsequently, the shape of the drug plant's seed capsule served as an organic prototype, inspiring the development of the unique and well-known Cypriot Base-ring juglets (see figure 93).

> It is in fact not inconceivable that the invention of the Base-ring juglet shape and its addition to the traditional ceramic repertory were made in response to the introduc-

FIGURE 92. "Scene from a Feast." This representation of an ancient Egyptian (Late Bronze Age) artwork shows a "maid" filling the "ointment bowl" of a "lady"; note the small juglets (from Erman 1894:405).

a. b. c.

FIGURE 93. Late Bronze Age Cypriot base-ring juglets (average height about 14–15 cm). All three of those shown here are now held in the Ashmolean Museum collection: (a) from Cyprus #1953.293; (b) from Ladish, Israel #1956.605; (c) from Kahun, Egypt #1890.853 *(Courtesy of Michael Vickers, Ashmolean Museum, Oxford, England).*

> tion of the poppy plant from the Aegean at the beginning of the Late Cypriot period. It may be no coincidence that the type originated in the north-west quarter of the island, being best represented round the Morphou Bay area, the first landfall from the Aegean. (Merrillees 1974:7)

The Cypriot Base-ring juglets have been the focus of archaeological attention for some time (see, for example, Wainwright 1920:61; Gjerstad 1926:321ff.; Sjöqvist 1940; Paul Åström 1972; Merrillees 1962, 1968, 1974, and 1979; and Holmes 1975). In fact, among the "vast amounts" of exported Late Cypriot Bronze Age pottery which have been found in Levantine and Egyptian sites, the Base-ring juglets are the most abundant (Holmes 1975:93). These juglets are relatively small, averaging between 13 and 15 centimeters in height, and are rather fragile. In light of their diminutive and somewhat delicate nature, it can be surmised that empty juglets were not in themselves sufficiently desirable to justify their relatively long-distance export, but that their primary commercial value came from the substance(s) that they contained.

> The question which must first be dealt with is why were these juglets imported by Egypt. Was it for their ornamental asset, their intrinsic usefulness, or their contents? There cannot be much doubt that ultimately it was for all three reasons, for not only were they not lacking in proportion and novelty value, but they were also clearly designed to be used as containers, for liquids originally, as their narrow necks would have permitted the passage only of fluid substances when emptied. In this case, however, it must be admitted that from a commercial point of view the principal value of the juglet lay in the last of its three functions, as the other two are obviously of secondary importance. (Merrillees 1962:288)

Merrillees not only suggested that these exported Cypriot juglets contained a precious liquid substance, he also argued that their characteristic shape was a symbolic indication of their contents; in other words, according to him, the specific form of the juglets served as a universal, nonverbal advertisement in the foregin lands to which they were shipped. This would have been especially important since it is highly probable that only a small percentage of the region's inhabitants were literate in their own language; and it is even more unlikely that many could readily read the script of a foreign people.

> If our merchant wished to promote the sale of some substance to people who could not read a label on the outside of its container, the most straightforward way of advertising his product would have been to design a vessel whose shape at a single glance proclaimed the identity of its contents. So that if, for example, the matter was organic in origin, a vessel shaped like the fruit, vegetable or pod, etc., from which it came would have met his requirements ideally. (Merrillees 1962:288)

According to Merrillees, the "organic prototype" for the juglets' shape was the seed capsule of the opium poppy, and the substance held in the tiny pottery vessels was either mixed with water and wine (Merrillees 1962:290) or blended into a syrup of honey (Merrillees 1968:157; 1974:34). He argued that although the solid ceramic base-ring stand differs from the serrated stigmatic disk found on the top of the opium poppy and capsule, its "flaring angle has been noticeably preserved on the juglet"; furthermore, Merrillees suggests that the relatively long, thin neck of the juglet represents the straight and narrow stalk of the poppy capsule:

> To make a handle for the poppy vessel, it would simply have been necessary to pare down one side of the stalk, producing a thin, narrow streamer, which, still hinged to the stem just above the thalamus, could be bent down and attached to the body. But as the exact translation of this feature into pottery would have produced a disproportionately short neck, the thalamus was reproduced further up as two parallel, horizontal plastic belts round the neck, at the junction of the strap handle. The bell or funnel rim is a wholly ceramic addition, and owes nothing to its organic prototype. (Merrillees 1962:289)

It should be noted that Merrillees's theory is not the only one put forward. In fact, in an earlier study, Sjöqvist (1940:79) suggested that the contents of the juglets were probably ointments and perfumes. More recently, Holmes (1975:93) concurred with this earlier interpretation: "Based on the fact that oil and other liquids were known exports from Cyprus to Egypt, the writer is more inclined to support Sjöqvist's theory"; therefore, it is "quite probable that one of the prime Cypriot exports was some type of oil or ointment which was shipped in Cypriot vessels." In a personal communication, A. D. Krikorian, the scholar who recently presented a skeptical review of the opium poppy evidence in the Near East region (see chapter 6), also expressed his doubts regarding the presence of opium in the juglets (Krikorian, personal communication, June 2, 1976).

In support of his theory that the juglets are models of the opium poppy and that they did contain opium, Merrillees argued that additional evidence should be considered. He referred to Kritikos's (1960:71) suggestion that a technique used to release the latex (opium) from the poppy capsules by incising them in their unripe stage is represented on Mycenaean or Late Minoan artifacts such as the moveable pins on the "Poppy Goddess" statue from Crete.

> Kritikos has shown that the method of incising an upripe poppy head to produce opium was known to the Cretans as early as Late Minoan III. So too it would seem that Cypriotes of the Late Bronze Age were acquainted with the practice, if the two parallel, vertical plastic belts, which often appear down the front of the juglet body, represent the curled back edges of a vertical slit made down the face of the poppy head to release its juice [see figure 93–c]. Even the dimensions and variations of the Base-ring I juglet bodies (and this applies to Base-ring II as well) consistently fall within the range of capsule sizes and shapes. Moreover, the buff-brown colour of the desiccated capsule is quite often met with on the surface of these Cypriote vessels. (Merrillees 1962:289)

Furthermore, the later group of juglets (i.e., Base-ring II; see figure 94-c), which appears gradually in the archaeological record towards the end of the fifteenth century B.C., is decorated with sets of three, four, or five parallel lines painted white and set at many angles around the vessels' bodies. According to Merrillees (1962:290), this clearly seems to be a copy of the cultivated opium poppy capsule scored by a special type of knife with many short, closely spaced, metal blades. Such a knife would thus

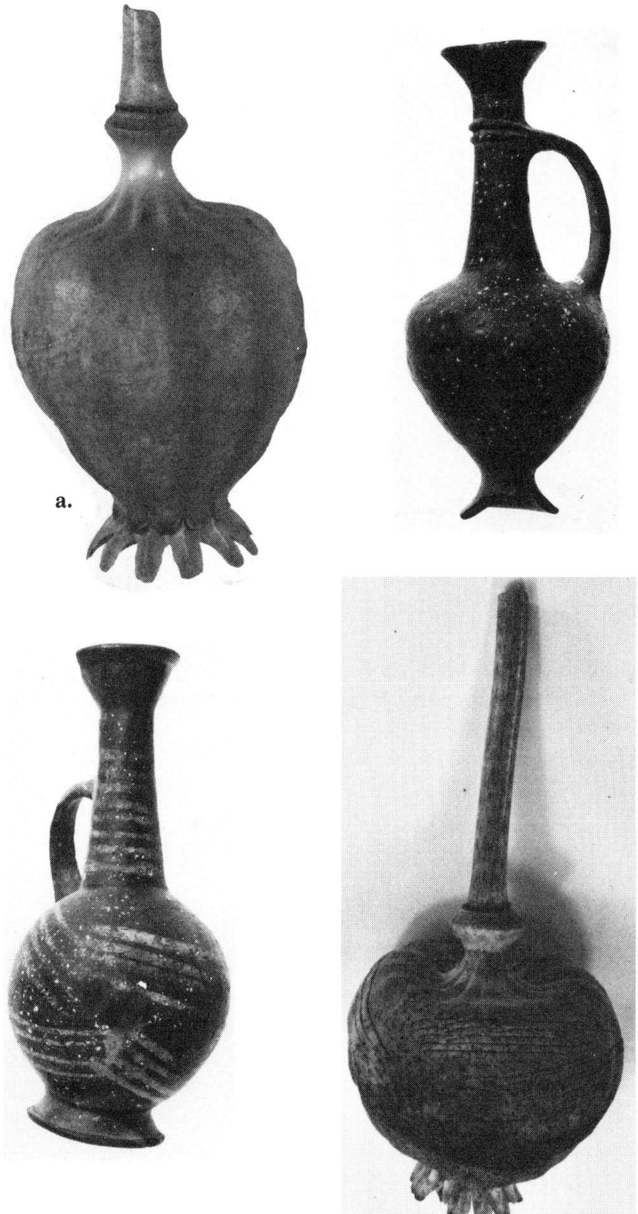

FIGURE 94. Opium poppy capsule and base-ring juglets I and II. (a) unripe opium poppy capsule (*Papaver somniferum* L.); (b) example of Late Bronze Age Cypriot base-ring juglet I; (c) example of later style of the Cypriot juglets, base-ring II (actual size of the juglets = ca. 14 cm tall); (d) unripe opium poppy capsule with parallel incisions cut by a multi-bladed knife *(Courtesy of Professor Robert S. Merrillees, Department of Foreign Affairs, Canberra, Australia).*

be like some still in use today which, when used in an opium harvest, tends to maximize the seepage of the raw, alkaloid-rich latex. This multiple-incision feature is especially useful for commercial production since opium harvesting is a labor-intensive operation and this type of knife increases the efficiency of the individual field worker. Merrillees suggested that the "scratched" parallel lines on the later Cypriot juglets represent the development of this more efficient harvesting technique which came into use shortly before the appearance of the Base-ring II juglets.

> The resemblance is most clearly and unmistakably seen by comparing the scored capsule and Base-ring II juglet illustrated side by side. No one can fail to be struck by their extraordinary similarity. (Merrillees 1962:290)

In his later publications, Merrillees (1968:160; 1974:32ff.) drew upon additional evidence from the archaeological record in Egypt to show that there appears to have been a developing market for the juglets among the "silent majority" of the Egyptian populace during the Late Bronze Age (New Kingdom). Perhaps the multi-incision technique was developed in response to the rising demand for the juglets and their special opium contents. We know that this method had been developed by the time of the famous ancient Greek Physician, Dioscorides (*Materia Medica* 4. 164, 167):

> It was Dioscorides, in the first century A.D., who gave us the first description of how opium was derived from "poppy heads." His account cannot be improved upon, and exactly the same tactics are used today. A few days after the petals fall from the opium poppy plant, there develops a greenish pod about two inches high and nearly as thick. When still quite unripe, this pod is delicately slit with a many-bladed knife. The cuts must not puncture the pod or touch the seeds within it. From these shallow cuts exudes a milky juice the coagulation of which is fairly rapid. This coagulated mass is crude opium. Sometimes the pod may be slit for the second or third time, so long as milky juice continues to flow. (Taylor 1965:208)

In order to confirm the presence of opium in the Late Bronze Age juglets, the necessary direct evidence must be located and objectively verified. In 1940, Sjöqvist indicated that the required evidence had been found; he referred to his oral communication with Dr. A. Westholm regarding the investigation of the "original contents" of two Cypriot juglets recovered from an Egyptian grave of the Late Bronze Age. According to Sjöqvist (1940:200, cited in Merrillees 1968:157), Westholm said that the contents "gave a chemical expert the impression of being crude opium, but the theory could not be definitely verified." However, because the vessels from which the substance was taken have not been identified, Westholm's report has not been corroborated by an additional chemical investigation (these vessels are now held in the Boston Museum of Fine Arts).

Merrillees (1968:157) also presented information he received in a "letter from Mrs. Stewart, dated 10th September 1965" concerning chemical tests done on the contents recovered from two other similar Cypriot juglets found in Egyptian burial contexts dating to the Late Bronze Age (1968:157): "Tests conducted for Stewart at Sydney University on the matter from at least two juglets are also said to have revealed traces of an 'opium derivative, impossible to say in what form.'" Nevertheless, this tentative confirmation is in conflict with other tests carried out by Wallis and Parkinson at London University which "revealed not a single trace of alkaloid in samples submitted for examination" (Merrillees 1974: 34; 1968: 157).

Referring to the advice of a specialist, Merrillees (1968: 157) rejected his own first suggestion "that the solvent used for transporting opium in liquid form was water or wine," since such a solution would have rather quickly become unfit for human consumption due to the development of molds. Merrillees then suggested that a more efficient solvent would have been a syrup of honey in which the opium sap could have been preserved. "That honey was used in antiquity for this purpose, and moreover diluted, presumably with water, to increase its fluidity, can be inferred from Vergil's

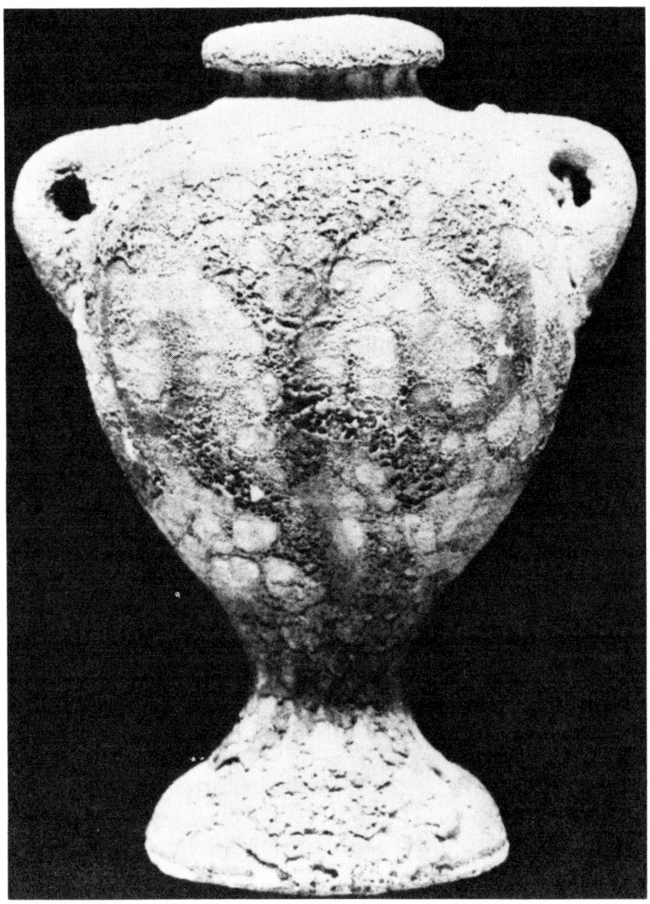

FIGURE 95. Vessel from Knossos. According to Emboden (1979: 28 and 29, figure 24), this "opium vessel" was one of those used to hold the "hypnotic gummy latex dissolved in wine" *(Courtesy of Professor William Emboden, Department of Biology, California State University Northridge).*

famous line: 'spargens umida mella soporiferumque papaver' [treatment of a soporific opium honey liquid] (*Aeneid* 4. 1.486)" (Merrillees 1968: 157). Merrillees also argued that if he is correct in assuming that the opium contents were held in a solvent of honey syrup, ". . . it might be difficult to identify the honey for certain, as chemical decomposition can cause the residue not to respond to the usual tests" (compare with Lucas 1962: 158).

An additional but unverified source of direct evidence comes from the New Kingdom tomb of Cha, an important royal architect during the reign of four successive Pharaohs of the 18th Dynasty. Cha died about 1405 B.C. and was buried inside the Necropolis at Thebes. As was customary for the time, he was entombed with a number of items which could be useful to him in his afterlife (Hall 1928: 203). According to Schiaparelli (1927: 154), one of the funerary objects found in Cha's tomb was an alabaster jar which held a vegetable oil containing iron and opium ("un grasso di natura vegetale continente ferro ed opio"). However, Hall (1928: 205) argued that although Schiaparelli's excavation report is readable and informative, it "is not scientifically precise." For example, the vessel from which the vegetable oil sample was taken was not specified in the excavation report (Merrillees 1962: 290). Although statements regarding the presence of opium in the alabaster jar can be accepted only on a tentative basis, Merrillees went so far as to articulate his own attraction to this enigmatic clue. "How tempted one is to identify the alabaster imitation of a Base-ring I juglet as its original container" (Merrillees 1962: 290–91).

In any case, Merrillees (1968: 158) reminded his reader that "the admittedly unverified detection of opium in some samples does add weight to the main proposition" and went on to refer to the juglets' abundant association with Late Bronze Age Egyptian burials. Here he pointed out that the Cypriot vases recovered from the Late Bronze Age burials of children in Egypt, Syria (at Ras Shamra on the Levantine Coast), and even Rhodes "consist almost exclusively of juglets of the type under discussion":

> Though it is not at all surprising that the juglets had been buried indiscriminately with men and women, since both sexes stood equally to benefit from the medicinal properties of opium, the exclusive association of these juglets with children makes one think of the past and present practice of administrating opium to infants for a variety of everyday purposes. It was used as a sedative for a crying child in Egypt during Pharaonic and more recent times. I learnt that in Cyprus up till the end of the Turkish occupation opium was administered to children to put them to sleep when their parents went to performances of the Dancing Dervishes. In certain societies opium has been given to children to keep them quiet while their mothers are carrying out their household duties. It was also thought to be a protection for children against many illnesses. (Merrillees 1974: 36)

Although no opium alkaloids were found, the chemical investigations in London did reveal the presence of a diversity of residual substances inside the juglets examined; these included wax, fat, and unidentified remains. Perhaps the contents of at least some of the juglets may not have been opium in a solvent but rather poppy seed oil which could have provided a useful medicine for skin treatments, a lubricant, or an edible substance rich in vegetal fats (Duke 1973:391). On the other hand, Merrillees argues that the nonopiate substances found in the juglets could have been put into the vessels after the original contents were emptied: ". . . many of the Cypriote juglets found in Egypt were put to use again in everyday life after their imported contents had been consumed" (Merrillees 1974:34). Addressing himself to the question of "re-use," he pointed out that the archaeological record in Egypt during the Late Bronze Age shows a great sense of thrift among the general populace:

> It could be a commemorative stele which has been re-used at a later period as a door pivot in the threshold of a house or a broken bowl turned into a lid by shaving down the rim. If the ancient Egyptians were as frugal as this, it is not difficult to understand why they would have held on to their juglets and re-used them if they wanted. (Merrillees 1974:36)

However, Merrillees also pointed out that after their original contents had been consumed, the juglets were not only not discarded but were also important enough to be placed in numerous burials. He suggested that this archaeological relationship can be explained in terms of the opium poppy's special therapeutic qualities:

> Eating opium is a household medicine par excellence, and can be used in cases of diarrhoea, dysentery, cold shivers, fits of fever, asthma, chronic coughs, rheumatism, diabetes and pains of all kinds.

Merrillees assumed that the opium poppy was an important medicine in ancient Egypt during the Late Bronze Age and that the organic prototype of the Cypriot juglets was the opium poppy capsule. Moreover, Merrillees believed that the concept of sympathetic magic or the so-called "Doctrine of Similars" (see Petrie 1914:40) played a role in the design and use of the juglets: "According to this theorem, objects which are closely alike have such a connection that power over one, or possession of one, enables the owner to influence the other or command like qualities"; although this speculation may be received with more than a little skepticism, Merrillees (1974:36) suggested that " . . . the ancient Egyptians believed that the juglet itself had in some magical way been invested with the same properties as the substance it was initially designed to hold" (i.e., an opium drug).

In conclusion, Merrillees postulated that it was because of both a belief in sympathetic magic and common frugality that the Egyptian people who purchased the Cypriot juglets did not dispose of them when they had consumed their original contents; in fact, he suggested that they held on to the juglets and refilled them with other valuable substances when a replenishing supply of opium was not readily available. "So they filled the vases with oils, waxes, scented fats, etc. which were just as necessary to the deceased and did not impair the containers' magical influence" (Merrillees 1974:36).

Merrillees (1962:291) suggested that the opium poppy was being cultivated in Egypt at least by the time of Amenhotep III (the father of Akhenaton), and that eventually local production of opium drastically reduced the demand for imports of the drug. This assumption may also be used to explain the declining numbers of Base-ring juglets imported into Egypt following the "Amarna phase" of the 18th dynasty. It may also account for the development and widespread distribution of a "poppy capsule" pendant bead (discussed in the next section of this chapter) which began to show up in foreign sites during the latter part of the 18th Dynasty. Perhaps, by this time, the Egyptians had incorporated the drug products of the poppy into their medical repertoire and were exporting amulets that carried the sympathetic magic of their organic prototype.

Although future finds and chemical tests may provide the direct evidence needed for a positive identification, at present we can only speculate on the possible existence of an opiate drug or poppy capsule amulet trade between Egypt, Cyprus, Crete, Greece, the Levantine Coast, and other areas in the general region bordering the eastern Mediterranean. According to Casson (1977: 178),

> . . . this is the way the matter stands to date: the jugs certainly look as if they had been designed to hold opium, but so far none can be proved to have done so. Whether or not the drug trade goes back to the Bronze Age is still an open question.

The Pendant Poppy Capsule Bead: An Egytian Amulet

In ancient Egypt, amulets were kept, and often worn, with the hope or belief that they possessed supernatural powers that would prevent the "evil spirits" of sickness and disease from entering human bodies (Hayes 1959: 179). "Amulets laid on the body or hung up in the sick man's room were used as a means of healing" (Budge 1926: 198). They were believed to have been endowed with sympathetic magic if their

FIGURE 96. Clay female head with pendant bead amulet. This Greek figurine (dated to the first quarter of the seventh century B.C.) is now held in the museum at Sparta. Note the hair style that reflects Egyptian influence and the bead amulet hanging from her necklace.

shape or substance was similar to that which was known to have possessed supernatural powers.

In the Late Bronze Age, Egyptian artisans frequently made amulets out of precious stones and metals and fabricated them into beads, pendants, earrings, necklaces, and other forms of jewelry. Although we do not know how many of the great quantity of beads and pendants produced during the 18th Dynasty period were actually used as amulets, we can be sure that many served this purpose since vast numbers of them were used to decorate the dead (Rogers and Beard 1940:27). Their placement in tombs, along with food, drinks, furniture, statuary, paintings, jewelry, and other objects, indicated that they were believed to have been useful to the deceased in their afterlife in the "underworld" (Erman 1894:5).

One pendant bead form of this period may have represented a stylized opium poppy capsule. I am here referring to a somewhat common Late Bronze Age (18th Dynasty or New Kingdom) pendant bead which was often made out of carnelian and carved into "the form of a poppy head" (Boardman 1961; see also Wilkinson 1971:152, Pl. LVIII, and Aldred 1971:fig. 130; see figure 97). Identification of this

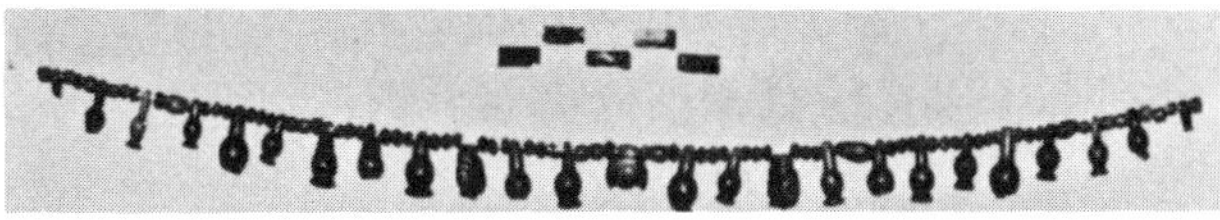

FIGURE 97. Carnelian necklace with pendant beads in the form of "poppy heads." This Late Bronze Age necklace holds several pendant beads, most of which have a common shape referred to by some as resembling the capsule of the opium poppy (from Merrillees 1968, plate XXXVI, figure 6; *Courtesy of Professor Paul Åström, Klassiska Institutionen, Göteborgs Universitet, Götenborg, Sweden*).

form with the capsule of the opium poppy is not necessarily inconsistent with artistic trends of the Late Bronze Age since the shaping of such beads into the "natural forms" of floral prototypes is "characteristic of the jewelry of the late Eighteenth Dynasty and the Ramesside period" (Hayes 1959:179). In fact, the pendant bead form in question here was referred to as a "seed vessel" by Petrie (1914:51) in his book dealing specifically with amulets. He informed us that this pendant bead is often found "on necklaces in carnelian and in glaze."

Merrillees referred to this bead form as "vasiform" and pointed out that it was in fact made out of several materials including carnelian, garnet, jasper, blue glaze, and alabaster; and, although he did note that Petrie (1906:45)described this pendant bead as a "lotus-seed form," Merrillees (1962:291) argued that " . . . its shape conforms too closely with that of the poppy head to deny its relationship" (compare Griffith 1923:137, plate LVIII, 10, 140, plate LXIV, 8). He also referred to the similarities between the pendant bead form and the Base-ring juglets from Cyprus, but pointed out that certain aspects of the bead's shape remain unexplained:

> The Base-ring juglet, shorn of its rim and handle, emphasizes this resemblance, especially as a solid base ring and a neck in proportion are featured on the beads. It does, however, make the original model for the bead somewhat problematical, as

> neither capsule nor pot suits in all respects. Had the poppy head been the prototype, one might have expected the stigma to have been faithfully reproduced on the bead, since naturalism is the key-note of art during this period. (Merrillees 1962:291)

Even though this same criticism can also be used in the case of the juglets, the stylistic inaccuracies in the bead or juglet shape certainly are not in themselves significant enough to negate the suggestion that their prototypic form was that of the opium poppy capsule. Perhaps, as Merrillees has suggested, it was "technical problems" which made it inconvenient for the artisan to make more exact copies of the opium poppy capsule in stone or even clay. In any case, examples of the "poppy head" or "poppy capsule" bead have been found in a number of 18th Dynasty burial tombs in widely separated locations throughout the Nile Valley. A few of these beads are described below.

I refer first to the discovery of the poppy capsule beads in the remote outpost of Buhen, a colony near the southern frontier of Egypt (see map, figure 91). Located near an ancient gold mine and close to the second cataract of the Nile in lower Nubia, Buhen's position is not far from the modern town of Halfa. According to Randall-Maciver and Woolley (1911:2), Buhen was a frontier military settlement established by Egyptian kings of the 12th Dynasty. Because Buhen was geographically isolated from the main centers of Egypt, the ancient cemetery remains of the colony were relatively unexploited before their excavation by archaeologists in the first decade of this century.

The 18th Dynasty tombs of Buhen were especially valuable for archaeological research: " . . . it was the exception and not the rule to open a tomb in which there was not some object of interest and importance" (Randall-Maciver and Woolley 1911:131). Among the many artifacts recovered from these tombs were necklaces carrying many types of beads. Two of these are of importance here because they carried representations of the "poppy capsule bead." One string (no. 10249) was described as comprised of "Carnelian pendants, shape poppy-seed, spaced by carnelian beads, shapes lentoid and spherical"; the other string (no. 10252) is said to be comprised of several kinds of beads, pendants, and "Two small, two medium and one large carnelian amulets" in the shape of "poppy seed" (Randall-Maciver and Woolley 1911:222). The authors of the Buhen excavation report have obviously confused the general shape of the poppy capsule with that of a poppy seed.

I believe that these "amulets" or pendant beads may be associated with poppy capsules, or at least with the characteristic Egyptian "seed vessel" form of the Late Bronze Age. It should also be pointed out that Randall-Maciver and Woolley (1911:131) referred to an imported "late Mycenaean vase" found at Buhen which is an indication of long-distance contact with the Greek realm during this early period.

The ancient Egyptian city now known as Tell el-Amarna, is located approximately halfway between the famous ancient city of Thebes and the Nile Delta (see map, figure 91). Constructed during the reign of Amenhotep IV (i.e., Akhenaton, ca.1365–1350 B.C.) as "an entirely new capital city," it was abandoned early in the reign of the succeeding pharaoh, Tutankhamen (Kemp 1974:157). A few of the many fine pieces of jewelry recovered from the burial tombs of Tell el-Amarna are relevant to this study: "From that site came for certain a blue glaze pendant in the form of a poppy capsule/B.R. juglet" (Merrillees 1968:179; compare with Petrie 1894:470, Pl. XIX). In addition, G. Martin (1974:79) referred to "53 gold pomegranate beads" recovered from the "Royal Tomb" at Tell el-Amarna in 1882. These beads look much like others already described as representations of poppy capsules. Also of interest here are nine "gold hollow husk-shaped pendants recovered from this same tomb, which (according to Martin 1974:79) are very similar to artifacts found in a tomb on Cyprus. I think their form resembles an oblong variety of the opium poppy capsule (see figure 98). In addition, it should be noted that Tell el-Amarna was also one of many Late Bronze Age Egyptian sites where several Base-ring Cypriote juglets were found.

a.

b.

FIGURE 98. Necklace from the "Royal Tomb" at Tell el-Armarna. Discovered in 1882, this necklace holds 53 small gold "poppy capsule" beads and 10 larger hollow gold pendants (from Martin 1974, plate 50, figure 287): (a) nine of these may be representations of an oblong variety of the opium poppy such as that shown here (b) *(Courtesy of the Egypt Exploration Society, London, England).*

An illustration of a floral collar from the late 18th Dynasty was published by Hayes (1959:320, figure 203); it carries five concentric rows of "fruits" modeled in faience. The innermost row is comprised of pieces that Hayes believed were representations of "cornflowers" (*Centaurea cyanus?* or possibly the corn poppy, *Papaver rhoeas?*). These beads may have been made to represent the seed vessel of the opium poppy.

As I noted previously, Merrillees drew attention to the fact that the remains of children and Base-ring juglets have often been found together in Late Bronze Age burials. He argued that these juglets most likely held medicines used in the administration of child illnesses. I have already referred to the use of opium in Egypt to calm crying infants or to protect children from many possible medical disorders in modern times: "In this connection, a delicately carved statuette of unspecified material, which came from Saqqara [and is discussed below], could have particular relevance" (Merrillees 1968:159).

In 1927, Quibell and Hayter published their excavation report for the tomb of the Teti Pyramid, a late 18th Dynasty structure located at Saqqara (see map, figure 91). Among the discoveries that they described was an unimpressive wooden coffin which contained an "aged person" and a "woman." Along with the deceased they found a finely carved statuette of a small naked boy about nine centimeters tall (see figure 99). According to Quibell and Hayter (1927: 37), several features of the statuette are "reminiscent of the Akhenaton period." Of particular interest here are the two "disproportionately large carnelian earrings" which had been attached to the lobes by gold wire. The beads suspended on these wires are similar to the poppy capsule bead form under discussion.

The archaeologists who discovered the statuette commented on the placement of such a finely crafted piece of artwork in such an ordinary casket: "It was strange to find in this poor coffin so charming a work of art, such work can never have been cheap; the statuette may have been a memorial of a son lost by these old people long

FIGURE 99. Three views of Late Bronze Age statuette from Saqqara, Egypt. Found in a wooden coffin of the late 18th Dynasty (i.e., New Kingdom), this 9 cm tall statuette depicting a small nude boy wearing "disproportionately large carnelian earrings"; these have the characteristic form of the "poppy capsule" pendant beads recovered from a number of Late Bronze Age sites in Egypt and other locations in the eastern Mediterranean region (from Quibell and Hayter 1927, plate 20; *Courtesy of the Service des Antiquités de l'Egypte*).

before" (Quibell and Hayter 1927:38). Perhaps the statuette was a sad, but dear, keepsake of a son who had been treated with an opium-based drug during the illness that took his life.

According to Merrillees (1962:291), the concept of sympathetic magic applies to both the Cypriot Base-ring juglets and the poppy capsule pendant beads which have been found in some eastern Mediterranean burials of the Late Bronze Age. In other words, it was believed that both forms could serve the same "prophylactic function" as their alleged organic prototype, the opium poppy capsule. Placement of a juglet or pendant bead in a grave could thus "prolong into the after life the beneficial effects it bestowed on earth through its opiate contents" or its shape which, because of its resemblance to the poppy capsule, gave it special powers (Merrillees 1968:161). In this context, the beads and the juglets may be seen as serving the function of amulets.

The Egyptian Poppy Capsule Bead in Foreign Archaeological Sites

After the foreign rulers of the Hyksos Empire were overthrown and the New Kingdom (18th Dynasty) was established in Egypt at the beginning of the Late Bronze Age (ca. 1570 B.C.), the Egyptian civilization entered into an era of unprecedented contact with other areas.

> Egypt seems to have gained strength like a field that has lain fallow; she now rose to such great prosperity and power as she had neither experienced before nor has since. Her strength showed itself, not in gigantic buildings as in the earlier days of her glory, but in foreign conquest for these Pharaohs carried their arms as far as the Euphrates and into the distant Sudan. (Erman 1894:42)

During this period Egypt also had contact with areas in the eastern Mediterranean region including Crete, Cyprus, and the Levantine Coast. One evidence of this contact is the pendant bead shaped in the form of a poppy capsule. This bead type, like other gems and beads, may have been used as a medium of exchange or as an item for barter. It could have been worn for personal adornment or kept for its supposed amuletic powers. In any case, trade items had to be something valuable but easily transported; and like precious stones, fine jewels, and costly metals, the pendant poppy capsule beads would have fulfilled this requirement, especially if they were exported as products containing the magical influences of the opium poppy.

Although the poppy capsule bead was "a familiar type in Egypt from the Late Bronze Age on," it was "also found in contemporary sites outside of the Nile Valley where it [was] definitely a foreign import" (Boardman 1961). For example, in the Mycenaean realm, three of these carnelian beads were found in the Late Minoan III (i.e., ca. 1400–1200 B.C.) tombs at Episkopi (Hierapetra) on Crete (see figure 100). Another Late Bronze Age carnelian poppy capsule bead was found along with additional "votives" in the "Dictean Cave" above Psychro in central Crete (now held in the collection of Oxford University). In reference to these beads, Boardman (1961) emphasized that their type was not "native to Greek lands."

On Cyprus, two similar polished carnelian pendant beads were found attached to a

FIGURE 100. "Poppy capsule" beads from Episkopi, Crete. These Late Bronze Age pendant beads are made out of carnelian. They were found in a tomb at Episkopi and are now held in the Iráklion Museum on Crete (in the same room as the "Poppy Goddess" found in another tomb at Gazi, west of Knossos).

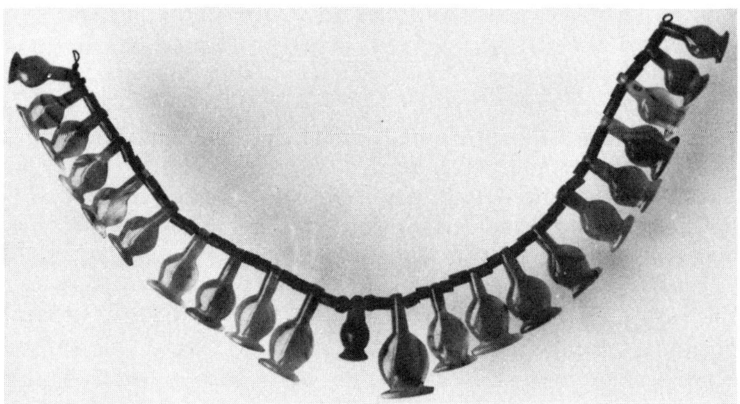

FIGURE 101. Late Bronze Age necklace from Mari, Syria. This necklace holds numerous "poppy capsule" pendant beads made out of polished carnelian *(Courtesy of the Départment des Antiquités Orientales, Musée du Louvre, Paris, France).*

necklace comprising a total of 197 beads (see figure 102). According to Buchholz and Karageorghis (1973 : 166), this necklace was recovered from a Late Cypriote II tomb at Kition (ca. 1400–1230 B.C.). They argued that the carnelian pendant beads attached to the necklace are "certainly imitations of poppy seeds [*sic*] symbolizing fertility." Once again this bead type has been misinterpreted as poppy seeds instead of poppy capsules (or the poppy seed vessel). Perhaps, in this case, the confusion can be attributed to the fact that the original manuscript describing these specific artifacts was written in German, suggesting that a mistake in translation could explain why the organic prototype of a poppy seed vessel was described in English as a poppy seed.

In any case, Buchholz and Karageorghis suggest that the necklace which carried

FIGURE 102. Late Bronze Age necklace from Kition, Cyprus. This necklace holds two "poppy capsule" pendant beads made out of polished carnelian *(Courtesy of Professor Vassos Karageorphis, director, National Museum of Antiquities, Nicosia, Cyprus).*

these pendant beads is especially noteworthy because some of the beads are made out of glass and, in fact, no glass ornaments have been found in any Aegean archaeological site dating earlier than the Mycenaean Age. This artifact and others bearing the poppy capsule pendant bead indicate that the diffusion of innovation through widespread cultural contact occurred in the eastern Mediterranean region during this period. One of the innovations which may have developed and spread during this period is the use of opium drugs or amulets in the generalized form of the opium poppy capsule. According to Buchholz and Karageorghis (1973: 165), use of toilet articles and body care items "reached a degree of sophistication" during the Late Bronze Age; and this included the use of several forms of ornament which "also doubled as amulets." One of these personal forms of adornment was the Late Bronze Age pendant bead "in the shape of poppy seeds [*sic*] or pomegranates" which were fertility symbols "valued as much for their magic properties as for their purely aesthetic appeal" (Buchholz and Karageorghis 1973: 165).

One additional example of the poppy capsule amulet is shown in figure 101, which depicts a necklace found in Mari, Syria (now held in the Museé de Louvre in Paris). Here several carnelian pendant beads are strung along with other beads into a collar or necklace. "The shape of the beads from this collar is mainly of Egyptian origin and can be dated to the New Kingdom"; however, these have been found in the "oriental context" of the "medio-assyrian tombs" of the thirteenth century B.C. (Annie Caubet, Conservateur au Département des Antiquités Orientales, Musée du Louvre, personal communication, April 12, 1977).

The Opium Poppy Capsule as a Motif in Egyptian Artifacts of Gold and Faience

During the Late Bronze Age, Egyptian artisans produced numerous objects made out of gold and faience. A few of these have been associated with the opium poppy capsule. For example, Gabra (1956:40) referred to a necklace and a pair of earrings which were found in the tomb of King Siphtah and Queen Tauosrit at Biban el-Molouk, the narrow desert valley near Thebes where huge galleries were dug out of the valley walls to house the graves of kings (Erman 1894:21). Gabra suggested that the "seven fruits of a plant" depicted on the pair of gold earrings were not models of pomegranates as Mapero had described them earlier (compare with Davis 1908:36), but were actually representations of poppy capsules (see figure 103). Gabra (1956:40–41) listed the "very simple reasons" for his conclusion:

a. *The shape of the serrated top of the fruit is that of the poppy capsule.*
b. *The shape of the fruit with its longitudinal striations is quite identical with the poppy capsule.*
c. *The long stalk of the fruit*
d. *Compared with the ancient Egyptian paintings of pomegranates . . . it is found totally different, simply because these various forms of pomegranate on ancient Egyptian paintings do not show a single one which is similar to the fruits of these earrings.*

Gabra also described a gold necklace of Queen Tauosrit which was found in the same tomb at Biban el-Molouk (see figure 103). He suggests that the pendants (whose mean height is 23 millimeters) were each made into the shape of a poppy capsule.

I have already pointed out that at least one "poppy capsule" pendant bead has been found in the Late Bronze Age site of Tell el-Amarna (Merrillees 1968:179). According to Gabra (1956:43), two additional artifacts from this late 18th Dynasty site represent parts of the opium poppy. He referred to "some ear rings . . . found in the shape of a serrated top of the capsule of papaver somniferum" and "a Vase of blue Faience having a typical shape of a poppy capsule." (See figure 103.)

Although a convincing morphological comparison can be made between these objects, Merrillees (1968:179) has recognized some discrepancies in Gabra's description of this artifact:

> The faience imitation of an opium poppy capsule (Louvre E.14363) . . . was acquired from the Nahman collection in 1933. Though Saber Gabra stated that it came from el-Amarna . . . , the Musée du Louvre Accession Register has no record of its provenance. The head, which was also mistakenly described as a "Vase" . . . , consists of a solid sphere of light blue faience, with dark blue serrated stigmata on top and bottom. The duplication of the latter features can be due only to artistic license.

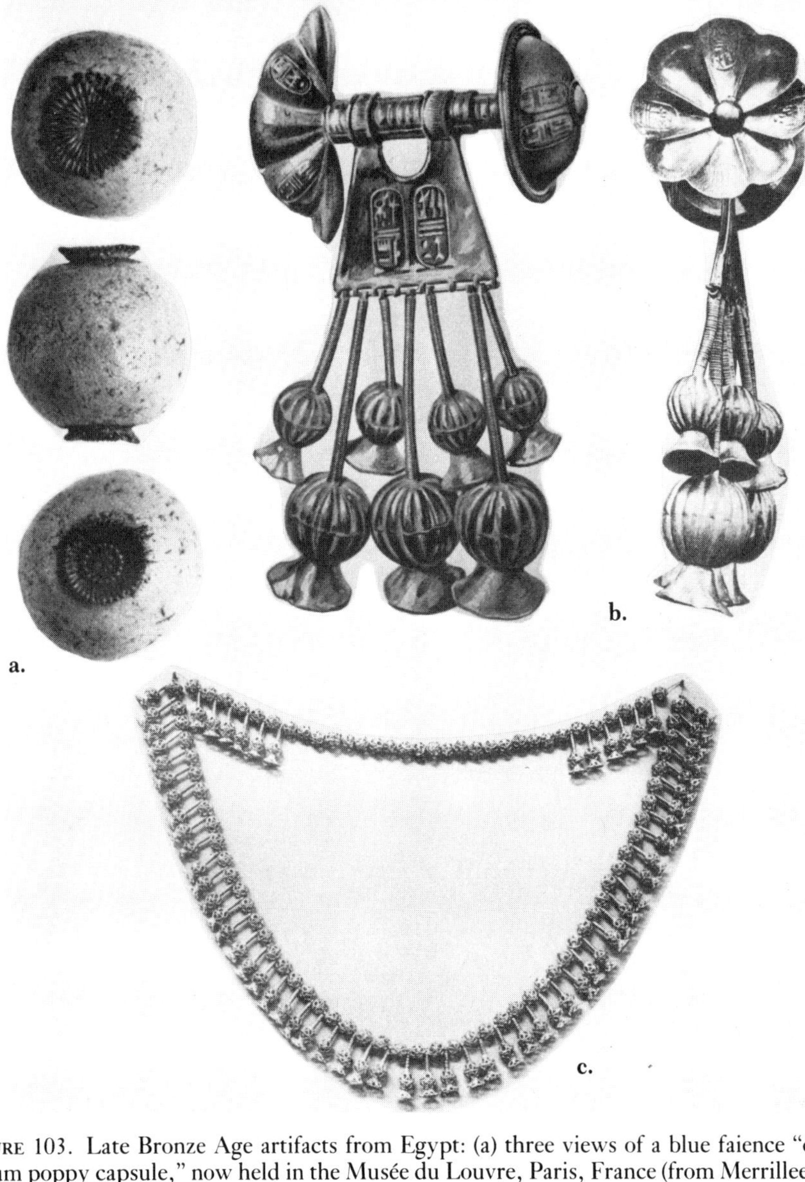

FIGURE 103. Late Bronze Age artifacts from Egypt: (a) three views of a blue faience "copy of opium poppy capsule," now held in the Musée du Louvre, Paris, France (from Merrillees 1968, plate XXXVI, figures 2–5; *Courtesy of Professor Paul Åström, Klassiska Institutionen, Göteborgs Universitet, Göteborg, Sweden*); (b) two views of gold earrings recovered from tomb of King Siphtah and Queen Tauosrit (New Kingdom) near Thebes; the hanging "fruits" may represent "poppy capsules" (from Gabra 1956:41); (c) gold necklace from same tomb as earrings shown above, the numerous pendants (average height 23 mm) may represent "poppy capsules" (from Gabra 1956:42, figure 4).

According to Kritikos and Papadaki (1967:36), a group of native blue porcelain vases found in the 18th Dynasty tombs of Tell el-Amarna also have a form which is an adaptation of the opium poppy capsule. If these, and the other, Late Bronze Age artifacts from Egypt were deliberately carved or fabricated into the general shape of poppy capsules, then obviously the drug plant was known in this country during this early period. In this case, other kinds of supportive evidence should be available.

Poppy Plants in the Paintings of Ancient Egypt

In ancient Egypt, wild and cultivated flowers were sometimes gathered into loosely tied or carefully formed bouquets and presented as gifts. On other occasions a single flower served as an offering. Highly esteemed plants included "the lotus, papyrus, convolvulus, and other favourite productions of the garden. . . ." (Wilkinson 1854:257). In addition, other plants such as fruit trees, sycamores, date palms, pomegranate shrubs, grape vines, acacias, oleanders, jasmine, dwarf chrysanthemums, and red poppies were grown for aesthetic and practical value on palace grounds and within the walls of large Egyptian estates. Many of these plants served as inspiration for the floral designs depicted on numerous ancient Egyptian works of art.

Several New Kingdom (i.e., 18th Dynasty) wall decorations included flowers among the natural designs which were represented. For example, a wide variety of flowers were depicted at Akhenaton's royal city. Many of these were "made in glazed ware and inlaid in the walls or were painted on the walls of floors"; and although the blue centaurea was the most common plant shown, ". . . there were also daisies with white petals and yellow centers, *red poppies*, flowering rushes, and various kinds of foliage" (M. A. Murray 1951:115, emphasis added).

It is imperative at this point in the discussion that I draw attention once again to the important consideration of species identification. Even though I believe that poppies served as a medicinal and ornamental resource in ancient Egypt during the Late Bronze Age, it is not clear whether one or more species was involved. In other words, artistic representation or literary references to poppy plants (or their individual parts) may be related to *Papaver rhoeas* (the red, corn, or field poppy) or *Papaver somniferum* (the opium poppy). It is important to remember that both of these poppy species are annual weeds in grain fields and waste places, popular ornamental garden plants, and sources of medicinal drugs. V. Tackholm (1974:153) has stated that *P. rhoeas* is the most common of all seven species of *Papaver* found growing in Egypt. In 1871, Ibn El-Bitar reported that the capsules of *P. rhoeas* were mixed with a simple syrup and used "to induce a hypnotic effect" (El-Masry et al. 1981:61). According to Popov (1970:486), the dried flowers of *P. rhoeas* "contain rhoeadine and a slight quantity of morphine, often used as a soporific." I should also point out here that although some individuals of *P. somniferum* produce reddish petals, *P. rhoeas* is much more highly correlated with the production of red flowers.

Georgiades (1917:363) referred to representations of the red poppy *(P. rhoeas)* in the paintings of the New Kingdom "Palais de Hawata" (Hawarah). Gabra (1956:48) suggested that a number of 18th Dynasty wall paintings show red poppies *(P. rhoeas)* growing around garden pools. He presented one example recovered from "the tomb No. 1 in the Theban-Necropolis." Gabra (1956:46) also compared a modern specimen of the red poppy with flower designs painted on the paved floors of the palace at Tell el-Amarna (see Petrie 1894: plates II, III, IV) and argued that there is a striking "similarity between the flowers and leaves of the ancient and recent plants."

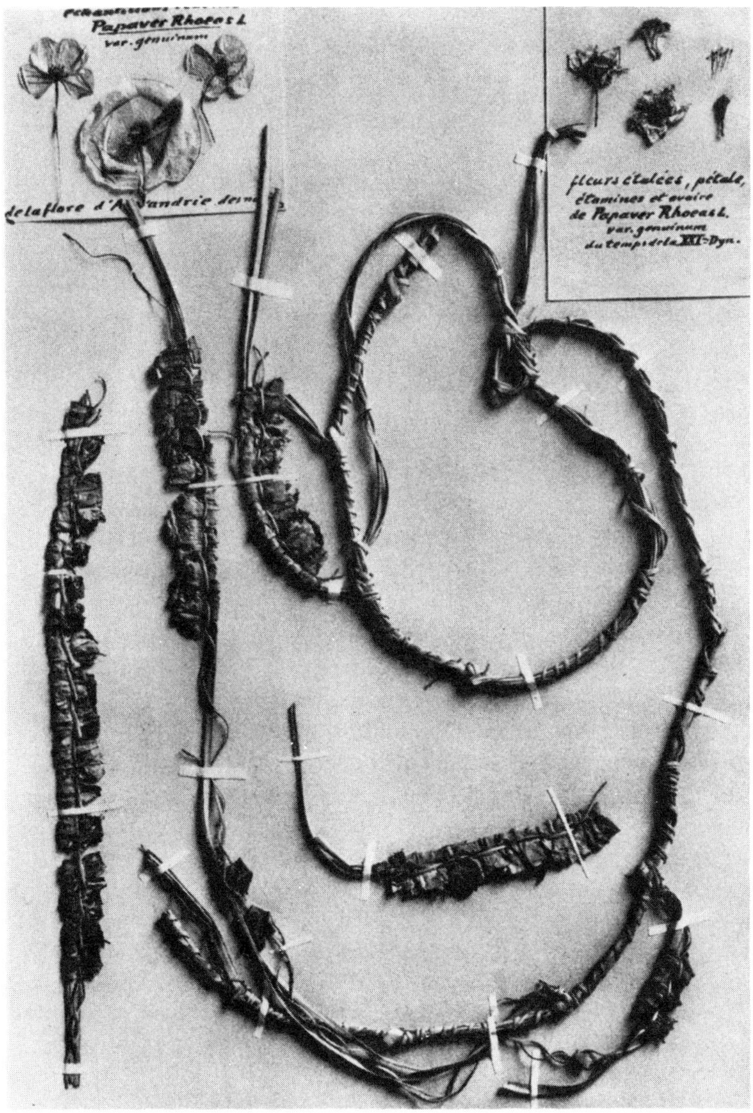

FIGURE 104. Red poppy specimens found in the tomb of an ancient Egyptian princess. These specimens of *Papaver rhoeas* L. were found on the breast of the Egyptian Princess Nzi-Khonsus (21st Dynasty) at Al-Deir Al-Bahari in 1881 (from Laurent-Tackholm 1951:109; *Courtesy of A Bortzells Tryckeri-Aktiebolag Innehavaren av Generalstabens Litografiska Anstalt, Stockholm, Sweden*).

In the latter part of the la.t century, P. E. Newberry was given the task of analyzing the botanical remains excavated from three 12th Dynasty tombs at Kahun, Gurob, and Hawarah in the Fayum (cited in Gabra 1956:45). Among the recovered "weed seeds," Newberry claimed to have identified four seeds of the red poppy *(P. rhoeas)*. He suggested that this species of poppy had been known as a weed in the grain fields of the Fayum as early as the time of the 12th Dynasty in Egypt. According to Georgiades (1917:363), some seeds of *P. rhoeas* have been found in the Fayum tombs erected in the Greek epoch, but he also referred to representations of the red poppy in the paintings of the 18th Dynasty palace of "Hawarah."

Probably the most remarkable evidence of the red poppy's presence in ancient Egypt was found on the breast of the Egyptian Princess Nzi-Khonsus who was discovered along with several other mummies in the 21st Dynasty tombs of Al-Deir Al-Bahari (Bahariya). In 1882, a year after these remains were uncovered, Schweinfurth examined the wide variety of well-preserved botanical specimens, including some flowers which he identified as belonging to *P. rhoeas* (see figure 104). According to Laurent-Täckholm (1951:102), the flowers are approximately 3,000 years old; Popov (1970:486) said they can be dated to "ancient Egypt of 1100 B.C."

Although Gabra (1956:45–46) referred to "numerous representations of *[Papaver rhoeas]* in the tombs and other monuments" of the late 18th Dynasty, he did not believe that the "well known and famous plant among ancient Egyptians" is an indigenous species. Popov (1970:486) described the red poppy as a weed which is "unknown in its natural state," but which supposedly "originated in the Balkan Peninsula, where the related *P. rumelicum* Velen occurs in the wild." Perhaps the opium and red poppies were both introduced into Egypt during the Late Bronze Age.

The Role of the Opium Poppy in Egyptian Medicine during the Late Bronze Age

Ancient Egyptian medicine was "overgrown with superstition and magic" (Erman 1894:364). The physicians attributed ill health and disease to the invasion of the body by wicked spirits who entered and infected the limbs and organs of the sick. In order to overcome these evil supernatural forces, the Egyptian doctors used spells, incantations, and exorcisms in combination with herbal and mineral drug potions. Before administering the material remedies, the physicians would sometimes seek the aid of a helpful god, beseeching him to come and vanquish the demon of disease.

> O Isis, great enchantress, free me, release me from all evil red things, from the fever of the god and the fever of the goddess, from death, and death from pain, and the pain which comes over me; as thou hast freed, as thou hast released thy son Horus, whilst I enter into the fire and go forth from the water, . . . (Erman 1894:356)

The ancient Egyptian physicians sometimes invoked the powers of sacred rods held in their hands over the patients' bodies while reciting antidemonic prayers. In addition, the medical priests utilized amulets or symbolic charms which they placed on the patients' bodies or kept in their quarters. Many of these healing techniques appear to be quite irrational when compared with modern treatments.

However, in certain cases, such sorcery may have relieved psychosomatic symptoms.

Although supernatural beliefs certainly played an important role in ancient Egyptian medicine, it is now generally believed that the empirical aspects of the medical arts were much more important than previously thought (Harris 1871:112–14). The success of the doctor's treatment in many, if not most of the cases, depended in fact upon his knowledge of human anatomy and the efficacy of the drug potions he made out of the substances which comprised the far-reaching and long-lasting medicinal repertoire of ancient Egypt (Hartmann 1923:46–47 and Budge 1926:198):

> A feature of Egyptian medical literature is the richness of the pharmacopoeia, which includes several well-known drugs in their earliest recorded applications. Parallel uses suggestive of direct borrowing have been noted in Hebrew, Syriac, Persian, and Arabic texts, and more important, in the works of many classical writers such as Theophrastus, Celsus, Pliny, Dioscorides, and Galen, and even in the Hippocratic collection—not to mention the compilations of various late-Imperial and Byzantine authorities. It was largely from these sources that physicians of the Middle Ages, like those of the Salerno school, drew their knowledge of *materia medica*, and from them also that much of the plant lore embodied in the popular medical compendiums of the Renaissance and later was ultimately derived. (Harris 1971:115)

One of the more conspicuous aspects of the ancient Egyptian pharmacological tradition was the significant role that herbal medicine played in the treatment of illness. For example, it was widely believed that the remedy for all ills was to be found in some particular plant" (Erman 1894:360); in fact, it seems that most of the ancient Egyptian drugs were derived from vegetable sources, According to Loret (1892):226), over five hundred plants are listed in the medical papyri as having special therapeutic utility. One of the plants which could have been especially beneficial, if it was in fact available during the Late Bronze Age in Egypt, is the opium poppy, particularly if one assumes that the basic, practical goal of ancient Egyptian doctors was the elimination of fever and the many forms of human pain associated with illness (Loret 1892:110, Castiglioni 1941:512, Merrillees 1962:292).

Opium Drugs in the Egyptian Medical Papyri of the Late Bronze Age

The discovery and translation of a few extraordinary medical papyri manuscripts have provided us with valuable information. However, none of these precious documents can really be considered as a book of medicine. Although they do contain descriptions of medical afflictions and remedies, these fragmentary collections of prescriptions are quite often rather rudimentary and unscientific: "The character of the existing material and its relative insufficiency must be borne in mind in any assessment of Egyptian medicine" (Harris 1971:114). Nevertheless, these papyri manuscripts are our only written sources of medicinal information, and thus they offer an important, but as yet limited, insight into the ancient practice of Egyptian medicine.

One of nine principal medical manuscripts available to us today from ancient Egypt is the Ebers Papyrus. This document was recovered in 1872 and is named after its modern editor, the German Egyptologist Georg Moritz Ebers. He allegedly purchased the medical treatise "from an Arab at Luxor, who claimed that it came from between the knees of a mummy at Thebes" (Taylor 1965:153). Authenticated by later scholars, the valuable manuscript lists its date of production in the ninth year of the reign of Amenophis I (i.e., 1550–1553 B.C.; compare with Castiglioni 1941:52).

Like other known papyri, the Ebers Papyrus is not a book in the modern sense, but a collection of medical information extracted from various sources and arranged according to the disposition of its author(s). It is an extraordinary document which many scholars regard as "Egypt's greatest medical document" (Aikman 1977:78). It is, in fact, the most voluminous of the extant papyri, comprising more than eight hundred medical recipes; and, in terms of medical information, it is also the most revealing: "It contains therapeutic recipes, diagnostic notes and, for the first time in history, theoretical considerations on the problems of life, health, and disease devoid of religious or magical considerations " (Ghalioungui 1963:50–51). According to Erman (1894:357), this informative ancient record of medicine probably had less of a magical orientation in its translated form than other Egyptian medical sources because the modern editor, Dr. Ebers "who compiled this book of recipes from older sources had very little liking for magic." At least eighteen drug plant derivatives referred to in the Ebers Papyrus are found today, approximately thirty-five hundred years later, in the *United States Pharmacopoeia* (Taylor 1965:154).

Some students of ancient Egypt have suggested that opium was referred to in the Ebers Papyrus under the name of *shepen* (Erman 1894:362; Hartmann 1923:47; Gabra 1956:48). Kritikos and Papadaki (1967a:37) suggested that the drug was referred to "under the name *seter-seref* or *shepenon* in the Ebers Papyrus." Erman (1894:362) drew attention to a recipe which included *shepen* and was used as a sedative for crying children: "The remedy which worked this miracle was a mixture of the seeds of the plant *Shepen*, and of the everlasting fly-dirt; the second ingredient was of course useless, the first may have been most efficacious, especially if the plant *Shepen* was the same as that now used to quiet children in Upper Egypt, viz. the poppy" (compare with Ebers Papyrus 93.3). Gabra (1956:52) presented this same recipe remedy for weeping youngsters as he found it in Wreszinski's 1913 publication, *Der papyrus Ebers*. Gabra (1956:53) also offered the following translation of this recipe and the effects it produced.

> Remedy to expell [*sic*] cry:
> capsules of poppy, fly's dirt that is on the wall, are mixed, strained and taken for 4 days. It ceases immediately as to 'cry', it is the child that is crying.

According to Gabra (1956:53), people in Egypt and India still give a crying child an "infusion of the capsules of Papaver Somniferum [*sic*] or occasionally they feed him with the seeds of poppy." Collom (1956:614) likewise made note of the remarkable "similarity of the ancients' method of stopping the child's crying over 3,000 years ago to the modern mother's remedy" In another reference to this ancient remedy for weeping youngsters, Aikman (1977:79) also assumed that the mixture contained opium and thus "was a startling precursor of today's modified paregoric that pediatricians frequently prescribe for colicky babies" (see also Scott 1969:109; D. J. Crawford 1973:231; and Casson 1977:175). We should here recall Merrillees's suggestion regarding the association between the Late Bronze Age Cypriote juglets and the burials of children in Egypt during this period. Perhaps the alleged opium contents of the juglets were used in many cases to tranquilize crying children.

In another reference to the Ebers Papyrus, Budge (1926:199) suggested that the prescription on the plate 47, lines 5 through 10, mentions the poppy. C. P. Bryan's English translation (of Joachim's transliteration) of this passage in the Ebers Papyrus is presented below:

> Remedy which the goddess Isis prepared for the god Ra to drive out the pains that are in his head—berry of the coriander, berry [capsule] of the poppy plant, wormwood, berry of sames plant, berry of the juniper plant, honey. Make into one, mix with honey and smear therewith in order to make him well forthwith. When this remedy is used by him against all illnesses in the head and all sufferings and evils of any sort, he will instantly become well. (Compare with Collom 1957:614)

Gabra (1956:48) compared certain remedies he examined in the medical papyri of the 18th Dynasty (especially the Ebers and Edwin Smith Papyri) with the following: (a) similar remedies found in the ancient Coptic medical papyrus translated by E. Chassinat (1921), which (according to Gabra 1956:53) has approximately twenty-two references to opium; (b) the traditional Arab medical use of opium poppy capsules, opium poppy seeds, and flowers of the red poppy *(Papaver rhoeas)* as herbal sedatives (also see Blackman 1968:204); and (c) the references in the *British Pharmacopoeia* (1898:279) and *Médicaments de la pharmacopée française* (1908:177) to the use of fresh petals of the red poppy in the preparation of a soporific syrup. On the basis of striking similarities, it seemed apparent to Gabra that the active ingredient(s) of a number of remedies which can be found in the Ebers and Edwin Smith Papyri contain products of either the opium or red poppy. Gabra (1956:48) referred to three specific drugs mentioned in the medical papyri of the 18th Dynasty. According to Gabra (1956:50), these drugs were used in the treatment of certain "diseases" which he summarized as follows:

a. sedative for wounds and necrotic wounds in the breast.
b. sedative for abscess with prominent head.
c. sedative for exanthema of the scalp.
d. as a dusting powder mixed with some other drugs for the scalp troubles.
e. sedative for the child that is crying (compare with syrupus Rhoedas of B.P. 1898).

Indicating his suggested sequence of introduction into Egypt, Gabra concluded that *shepen*, the ancient Egyptian name for poppy, was originally used for *P. rhoeas* (the red poppy), and then applied to *P. somniferum* (the opium poppy). The word *shepenn* was interpreted by Gabra to mean "the flower, seed, or capsule of Papaver." Furthermore, he believed that *shepenn-dšr* referred to the flowers of *P. rhoeas*, or "red poppy flower." Gabra (1956:51–52) presented a few selected examples "taken from the ancient Egyptian documents to show the various medical actions of poppy plants."

In his summary, Gabra (1956:54) listed seven basic conclusions. First, he believed that the ancient Egyptians were aware of two poppy plants: *Papaver somniferum* and *P. rhoeas.* Second, he argued that the *P. rhoeas* was depicted on ancient Egyptian paintings as a wild plant growing in the marshes of the Nile Delta and ". . . at the same time it was cultivated and acclimatized in their gardens." Third, he suggested that *shepen* was the generic name for poppy ("Papaver") plants. Fourth, he argued that the ancient Egyptians had medical applications for both the capsules of the opium poppy ("shepenn") and the red flowers of the corn poppy ("shepenn-dšr"). Fifth, he suggested that the use of poppy species in ancient Egypt "continued through the ages until we find the name and use of opium during the Greco-Roman period οπιου from which the Copts derived the name οπιου. Sixth, he argued that poppy species and "Opium" were used medicinally in ancient Egypt for the same "sedative and narcotic" purposes that they are used in "the modern reference." Seventh, he claimed that opium "was a well known and famous drug in Egypt and there were certain districts which were famous for its production."

Merrillees (1968:196) suggested that *špn* (i.e., *shepen*) and *špnn* (i.e., *shepenn*) "are generally thought to mean poppy plant and capsule respectively. In a reference to Gabra's conclusion that opium was used as both a sedative and narcotic in ancient Egypt, Merrillees (1962:292) argued that there is little doubt as to the medical importance of this drug in Egypt during the Late Bronze Age:

> . . . it was as indispensable in ancient Egyptian households as aspirins are today in our own homes. It must surely have been treated as nothing more or less than a panacea for all common ills, for which it is admirably suited.

Merrillees (1962:292) also suggested that opium use in ancient Egypt had more than a simple medical importance. He believed that it was used for pleasurable and spiritual purposes as well:

> . . . it would be impossible to believe that advantage was not taken by the ancient Egyptians of the purely sensuous or narcotic effects that opium also produces. Indeed Kritikos has shown that during the Late Minoan III period opium was taken by participants in certain religious ceremonies to induce a state of ecstasy essential for the performance of the sacred rites. Might not opium have been used in the same way in Egypt? How appropriate it would be if the island of Aphrodite could be proved to have introduced Egypt to the drug which served that Goddess so well!

Merrillees thus believed Gabra was correct in assuming that the opium poppy was known and used in Egypt during the Late Bronze Age, and that it has been an important plant in this country ever since. As I noted earlier, Merrillees hypothesized that the opium poppy was introduced into Egypt from Cyprus (or Syria) during the early half of the 18th Dynasty (i.e., about 1575–1367 B.C.).

D. J. Crawford (1973: 231) agreed with Gabra and Merrillees that the opium poppy has a long history in Egypt dating back to the Late Bronze Age period:

> . . . use for opium was recognized by the medical writers and doubtless practised by the people. In this context two main substances were gained from the poppy: *opium*, the air-dried, milky latex which exudes from incisions in the seed pod of the living plant, and *meconium* which is obtained by infusing the whole of the cut plant.

Although Crawford (1973: 231) pointed out that opium poppy drug products were "prescribed in Ptolemaic remedies from the third century B.C.," the main objective of her own research was to describe and explain the abortive effort to establish an ongoing commercial production of opium poppies for seed-oil in Ptolemaic Egypt. Her publication (1973: 223–51) confirms the use and cultivation of the opium poppy in Egypt at least during the third century B.C.

The Absence of Confirmed Evidence from Late Bronze Age Egypt

Other scholars have taken a position that differs from that of Gabra, Merrillees, and Crawford regarding the antiquity of opium poppy use in Egypt. Indeed, there has been a lack of agreement as to the precise age of the drug plant's existence and application among the ancient civilizations of the Nile River Valley. Several scholars have suggested that the opium poppy did not exist in Egypt until many centuries after the Late Bronze Age. For example, in 1857, Unger reportedly "failed to trace any acquaintance with opium in Ancient Egypt" (cited in Macht 1915: 477). In 1886, Woering noted that neither the opium poppy nor its various individual parts could be found in the artistic decorations adorning the early temple monuments and wall inscriptions of ancient Egypt (cited in Kritikos and Papadaki 1967a: 36). In 1895, Buschan argued that there are neither written references nor artistic representations of the cultivated opium poppy in ancient Egypt until the time of Pliny and that this absence indicates that the drug plant was unknown in this region during earlier periods (Buschan 1895: 245).

In 1905, Hoops also claimed that the first evidence of opium poppy cultivation in ancient Egypt comes from the time of Roman occupation (Hoops 1905: 334). As I noted above, Crawford has clearly shown that the opium poppy was not only known in Egypt during the time of the Ptolemies, but was also being cultivated during this same period. In 1917, Georgiades referred to the opium poppy as a plant which was introduced into Egypt from Asia Minor and argued that because of the lack of remedies involving the opium poppy in the medical papyri written prior to the 21st Dynasty, it must have been unknown in the earlier periods (Georgiades 1917: 363). In 1924, Schweinfurth also suggested that the opium poppy was unknown in Egypt before the Roman period (compare Löw 1924: 528–29; Sonnedecker 1962: 276). Ebbell (1937: LXIV, 77, lines 18 & 24; LXV, 77, line 28; and XCII, 108, line 12) left the transliterations of *shepen (špn*' and *shepenn (špnn)* in the Ebers Papyrus untranslated in his own work on this famous manuscript (cited in Sonnedecker 1962: 276). Harris (1971: 117) maintained that even though "Theban opium" (a regional product of ancient Egypt) was one of the better-known drugs during the Ptolemaic period, there is no definite evidence for its previous existence in the Nile River Valley.

More recently, Krikorian has addressed himself to this problem and to the questions raised by Merrillees regarding the introduction of opium into Egypt from Cyprus or Syria during the Late Bronze Age. Krikorian told me his research indicates that the evidence referred to in Merrillees's work is so weak as to be nonexistent. In fact, Krikorian dismissed the idea that the opium poppy was introduced to Egypt from Syria and does not believe that opium was imported from Cyprus during this early period (personal communication, June 2, 1976).

On the other hand, John C. Kramer, a psycho-pharmacologist who has studied some of the ancient Egyptian writings on drugs, proposes that a frequently used ingredient, *d*3r.t (djaret), may be opium. Among many other indications, it is used in remedies for cough, diarrhea, topical anesthesia notably for the anus, inflammation, and eye disorders. (See Thompson 1949:221–22.) He has shown that earlier suggestions that djaret was either colocynth or carob, are unlikely. Kramer has also pointed out that its use in three prescriptions "to open the bowels" can be explained by the fact that each remedy also contained known laxatives as well as djaret, and that as late as the last century, opium was occasionally included in cathartic mixtures in order to provide the patient a night's sleep, and catharsis the next morning after the opium effect had worn off.

I would like to describe one additional piece of direct evidence which, if it can be verified, could lend considerable support to the interpretations of those scholars who believe that the opium poppy was known and used in Late Bronze Age Egypt. I refer here to the unique and intriguing discovery "of an unscored opium poppy head (Dokki) in Deir el-Medina tomb 1389" (Merrillees 1968:159; see figure 105). Bruyere (1937:109, no. 6) described this specimen as a "capsule de pavot *(Papaver somniferum)*." This desiccated, unincised poppy capsule was found in one of five small bags containing, among other things, three small pots of grains and a "scarabée de Menkheperrè" (Thutmose III) mounted on bronze (Bruyere 1937:201, no. 13). Merrillees (1968:155) pointed out that only one poppy capsule was identified among the large quantity of plant remains found in the various undisturbed tombs of the 18th Dynasty cemetery at Deir el-Medina.

> Though at first sight this contrast could be taken to mean that *Papaver somniferum* was a rare exotic, the evidence needs to be cautiously interpreted as remains of all flower seeds were scarce, including those of the lotus (Deir el Médineh (1934–1935) II, p. 109), which can hardly have been uncommon in ancient Egypt. It may also be significant that the capsule shows no signs of having been tapped for its latex. In any case, there is now incontrovertible proof that the opium poppy plant was being grown in Egypt at the time of Thutmose III, even if the quantity and purpose of its cultivation are still imperfectly known. (Merrillees 1968:155)

This "incontrovertible proof" is especially important. However, if this poppy capsule cannot be objectively identified and dated to the Late Bronze Age, then the other evidence alone does not make a strong case for the contentions of scholars such as Gabra (1956) and Merrillees (1962, 1968, 1974, and 1979) who have argued that the

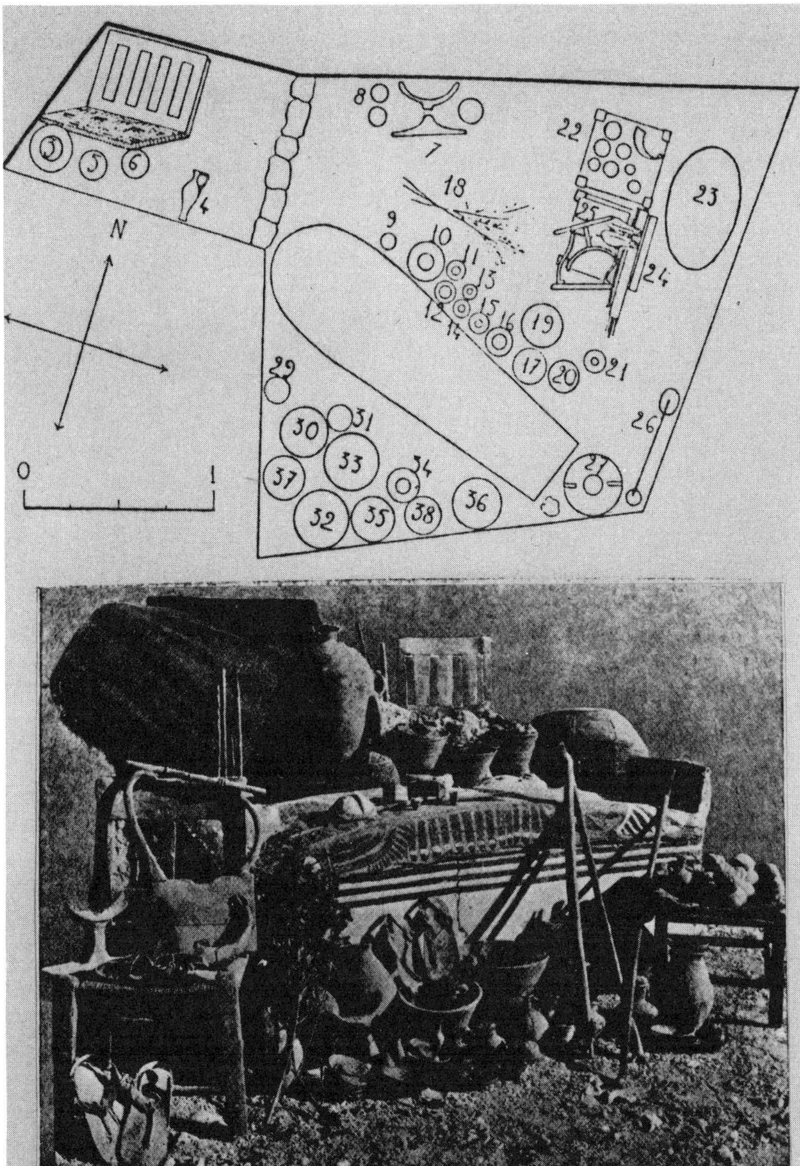

FIGURE 105. Grave gifts and floor plan of Deir el-Medina tomb 1389: (a) contents of the Late Bronze Age Egyptian tomb; (b) floor plan of the tomb, including the in situ location of the container in which an unscored poppy capsule was found (no. 13) (from Bruyere 1937: 200, figures 113, 115, and 198; *Courtesy of Professor J. Vercoutter, directeur d'Imprimerie de l'Institute Francais d'archaologie orientale, Cairo, Egypt*).

opium poppy was known and used in Egypt during this early period. Although many secondary sources refer to the prescriptions allegedly containing opium in the Ebers and other ancient papyri, it must be remembered that these interpretations are certainly not accepted by all students of the subject. Because I am not aware of the details of Krikorian's skeptical examination of Merrillees's theories regarding the Cypriot juglets and the introduction of the opium poppy into Egypt during the Late Bronze Age, I have not been able to comment on them here.

In the final chapter that follows, I summarize the major problems that hinder the kind of research undertaken in this study and refer to some of the major points made in my thesis.

10 • Summary

In much of my research involving the ancient history of the opium poppy, *Papaver somniferum*, I have been left with a feeling of uncertainty. Frequently the limitations of the available information have required the use of considerable speculation and qualifying comments. Obviously, a great deal of research needs to be done before the definitive early cultural geography and history of this very important drug plant can be written. No one person can be expected to accomplish this feat without devoting the greater part of his lifetime to such an effort. In my case, I have spent six years gathering information and synthesizing the data. In retrospect, I would like to believe that my years of research on this subject will serve as a baseline study for future, more intensive efforts. Interested students, such as myself, await the discovery and confirmation of vital new evidence. We hope that new techniques will provide more precise information and thus make it possible to answer some of the questions asked here.

In chapters 2 and 3, I addressed myself to questions regarding the classification and origins of the species *Papaver somniferum* L. An extensive survey of the literature revealed the controversial nature of the taxonomy of the opium poppy at the section level as well as the striking morphological diversity that the species manifests among its numerous varieties. At present, there is no apparent consensus concerning the ancestry of the opium poppy (see table 9). This situation is reflected in the diversity of suggestions regarding the "root form" of the species. During the nineteenth and early twentieth centuries, the opium poppy (*P. somniferum* L.) and the setaceous poppy (*P. setigerum* DC.) were almost always classified as very closely related, but separate species. Although most botanists still choose to split the opium and setaceous poppies into separate species, others prefer to lump them together as subspecies of *Papaver somniferum* L. (i.e., *P. somniferum* L. ssp. *somniferum*, the opium poppy, and *P. somniferum* L. ssp. *setigerum* (DC.) Corb., the setaceous poppy).

The geographical origins of the opium poppy also remain in doubt. Some crop plants can be traced to their native geographical range of distribution and identified with the truly wild progenitors from which they have been derived. Other cultivated species, such as the opium poppy have been artificially selected to such an extent that they have lost the ability to survive under natural conditions and are therefore only known in agricultural fields or environments close to farmlands which have been disturbed by man. This absence of truly wild populations creates special difficulties for students interested in determining the geographical origins of domestication and early cultural histories of cultivated species such as the opium poppy. In fact, many regions including southwest Europe, central Europe, southeast Europe, Asia Minor, the Near East, and central Asia have been suggested as the native home of this famous drug plant (see table 10).

In chapter 4, I suggested that the opium poppy may have provided a number of resources for the early human peoples that knew and used this species. The opium poppy could have furnished human food (i.e., tasty, nutritious seeds and edible seed-oil), animal fodder, lubricating and cooking oil, medicines, and pyschoactive drugs for ritualistic or recreational purposes. I also argued that the creation of disturbed habitats and dump heaps near human settlement areas provided appropriate environments for the opium poppy and indeed probably precipitated the initial relationships between man and this species. Eventually, humans could have learned to utilize the opium poppy plants found growing in their dump heaps or waste areas. Subsequently, they probably encouraged and enlarged the distribution of the species more directly by protecting or cultivating the plant.

In chapter 5, I surveyed the early archaeological record for the opium poppy and described the fossil evidence. Most of this (Neolithic and Early Bronze Age) evidence comes from a relatively small area which I have referred to as west central Europe (i.e., the Swiss Foreland and adjacent areas in southern Germany, eastern France, and northern Italy). However, the geographical concentration of this evidence may be more a reflection of the differential preservation of the macro- and microfossil plant remains and regional enthusiasm for the study of prehistoric plant deposits than the actual distribution of opium poppy use in the prehistoric period. In other words, the relatively abundant waterlogged environments (i.e., lakes and bogs) in west central Europe, as well as the generally more favorable attitude toward the study of plant remains in archaeological sites in central and northern Europe during the nineteenth and early twentieth centuries, probably explains why most of the evidence for opium poppy use has been found in west central Europe. Other regions might yield similar concentrations of fossil evidence if equivalent environments and plant remains were studied as intensively.

Although most of the fossil evidence for the opium poppy comes from west central Europe, we do have a small amount of direct evidence from other Neolithic and Bronze Age locations in southern Spain, the Rhineland, Austria, Czechoslovakia, Poland, and perhaps Turkey. In chapter 6, I evaluated the various possible places of origin for the use of the opium poppy, including those in southeast Europe, the Near

East, northwest Europe, southwest Europe, and Switzerland. Even though more evidence is needed before we can make definitive statements regarding the geographical origins of man–opium poppy relationships, I have suggested that the opium poppy was spread out of central Europe and down into the eastern Mediterranean region during the later part of the prehistoric period (i.e., the Late Bronze Age). In chapters 7, 8, and 9, I presented a good deal of indirect evidence which indicates that the opium poppy was known and used (even if only among a small select class of people) in some areas of southeast Europe and the eastern Mediterranean littoral during the Late Bronze Age.

Admittedly, the idea that both the Cypriote Base-ring juglets and the widely distributed Late Bronze Age pendant bead form (described in chapter 9) were made to represent poppy capsules does remain hypothetical. Nevertheless, I believe that I have built a strong case for my contention that the opium poppy was both present and utilized in Greece, Crete, and Cyprus during the Late Bronze Age. It seems clear to me that the opium poppy was well known in the ancient Greek cultural realm, certainly during the classical period. Indeed, I think it is safe to assume that the associations of the opium poppy with mythology, religion, and medicine in ancient Greece bespeaks a tradition that goes much further back in time. The abundance of literary references and archaeological evidence supports this hypothesis (e.g., Kritikos and Papadaki 1963 and 1967).

The confusion surrounding the identification of the red (i.e., corn) poppy, *Papaver rhoeas* L., and the opium poppy, *P. somniferum* L. (discussed in relationship to the Egyptian evidence) is relevant to all the Late bronze Age material. It is important to remember that the red and opium poppies have some similar ecological characteristics. As weedy annuals adapted to open or disturbed habitats, both species are commonly found growing in association with cultivated grains. In fact, as I noted previously, both species appear to be dependent upon human disturbance for the

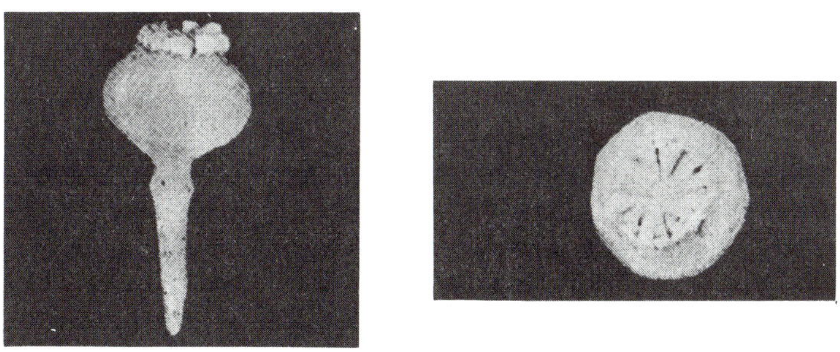

FIGURE 106. Ancient Greek clay model of an opium poppy capsule. This artifact is over 2,500 years old. It was recovered from the archaeological ruins of a sanctuary located near the little village of Galataki (the ancient settlement was named Solygeia). The poppy capsule model "is 6.5 cm in length including the stalk and 3 cm in diameter, and clearly shows the body of the capsule and the point, . . . as well as the stalk with its 'knee'" (Kritikos and Papadaki 1967 : 31, figure 18). According to N. M. Verdelis (1962 : 192), the archaeologist who excavated the artifact, it was a votive object typically associated with the worship of a "chthonic" or underworld diety *(Courtesy of Archaiologike Ephemeris, Hellenike Archaiologike Hetairia, Athens, Greece).*

FIGURE 107. Near East opium harvest in the early 1900s. This rare photograph was reproduced from an old glass slide in the Philadelphia College of Pharmacy and Science *(Courtesy of Professor Ara Der Marderosian. Philadelphia College of Pharmacy and Science).*

creation of appropriate habitats, and are thus known only from cultivated or semi-wild populations.

In the case of the red poppy, the direct evidence referred to earlier in my discussion of the Egyptian material indicates that this poppy species existed in Egypt well before the Ptolemaic period (i.e., at least by the time of the 21st Dynasty, ca. 11 B.C.). If the red poppy is native to southeast Europe as Popov (1970:486) suggested, then perhaps its use as a drug was indigenous to this region. This would have predisposed the inhabitants of this general region toward the use of this species in a very early period (i.e., during or perhaps prior to the Late Bronze Age). However, since the red poppy is not known from truly wild populations, it cannot be said with certainty that it is native to the Balkan region.

In any case, if the opium poppy was introduced into southeast Europe and some littoral areas of the eastern Mediterranean region sometime after the red poppy became a local resource, then I would assume that it either complemented or replaced *Papaver rhoeas* as a source of medicinal and other useful products. The opium poppy produces a greater variety of therapeutic alkaloidal substances than the red poppy (Duke 1973:390–400). In addition, it yields nutritious seeds. Both of these qualities would have made the opium poppy more useful than the red poppy.

I am thus suggesting that the weedy red poppy probably arrived first in the areas included in my discussion of the Late Bronze Age. Subsequently, the cultivated opium poppy was introduced and because of its superior agricultural and medicinal value, it became established as a useful and, at least in some cases, revered plant.

In summation, I must say that I look forward to a significant increase in our body of scientific knowledge and understanding of prehistory. Hopefully these advances will make it feasible to reconstruct precisely the stages of development for some of the more enigmatic crop plants such as the opium poppy. Eventually we may be able to locate the specific habitats of the progenitors of these species and determine the role that humans have played in their domestication, cultural spread, and early histories.

Most importantly, I hope that technological and methodological breakthroughs in the future will allow us to understand how the opium poppy was used or possibly abused in the distant past. The ancient record may someday reveal valuable insights into the beneficial and detrimental relationships between this drug plant and the many human societies which it has long affected.

References

Aaronson, Bernard, and Osmond, Humphrey. *Psychedelics: The Uses and Implications of Hallucinogenic Drugs*. London: Hogarth Press, 1971.

Addens, T. J. *The Distribution of Opium Cultivation and the Trade in Opium*. Haarlem, Holland: J. Enschedé en Zonen, 1939.

Aikman, Lonnelle. *Nature's Healing Arts: From Folk Medicine to Modern Drugs*. Washington, D.C.: National Geographic Society, 1977.

Akehurst, B. C. Tobacco. London: Longmans Green and Co., 1968.

Aldred, Cyril. *Jewels of the Pharaohs*. London: Thames and Hudson, 1971.

Alexander, John. "The Pins of the Jugoslav Early Iron Age." *Proceedings of the Prehistoric Society* 30 (1964): 159–85.

Alexiou, Stylianos. "The Minoan Goddess with Uplifted Hands." *Cretan Chronicles* 12 (1958): 179–299.

———. *Minoan Civilization*. Translated by Cressida Ridley from the Greek. Heraclion, Crete: Spyros Alexiou Sons, 1973.

———; Platon, N.; and Guanella, H. *Ancient Crete*. Translated by D. J. S. Thompson from the German, *Das Antike Kreta*. London: Thames & Hudson, 1968.

Amelineau, E. *Les nouvelles fouilles d'Abydos 1895–1896. Compte rendu in extenso des fouilles, description des monuments et objets découverts*. Paris: E. Leroux, 1899.

Ames, Oakes. *Economic Annuals and Human Cultures*. Cambridge, Mass.: Botanical Museum of Harvard University, 1939.

Anderson, Edgar. *Plants, Man, and Life*. 2nd ed. Berkeley, Calif.: University of California Press, 1967.

Anderson, Roy K. *Drug Smuggling and Taking in India and Burma.* Calcutta: Thacker, Spink, 1922.

Andrews, Alfred C. "The Opium Poppy as a Food and Spice in the Classical Period." *Agricultural History* 26 (1952): 152–55.

Ansley, Clarice, F., ed. *The Columbia Encyclopedia.* New York: Columbia University Press, 1935.

Anslinger, H. J., and Tompkins, William F. *The Traffic in Narcotics.* New York: Funk and Wagnalls, 1953.

Artamanov, M. I. "Frozen Tombs of the Scythians." *Scientific American* 212, no. 5 (1965): 100–9.

Asthana, S. N. "The Cultivation of the Opium Poppy in India." *Bulletin on Narcotics* 6 (September–December 1954): 1–11.

Åström, Lena. *Studies on the Arts and Crafts of the Late Cypriote Bronze Age.* Lund, Sweden: Förf, 1967.

Åström, Paul. *The Swedish Cyprus Expedition.* Vol. 4, Parts IC, ID. Lund, 1972.

Baas, J. "Kultur- und Wildpflanzenreste aus einer romischen Brunnen von Rottweil-Altstadt." In *Verbindung mit prähistorischen Schlehenfunden aus Süddeutschland. Fundberichte aus Baden-Württemberg,* I. 373–416. Berlin, 1974

Bailloud, Gerard. "The First Agriculturists: 400–1800 B.C." In *France Before the Romans,* edited by Stanley Piggot et al., pp. 102–31. Park Ridge, N.J.: Noyes Press, 1973.

Barfield, Lawrence. *Northern Italy before Rome.* New York: Praeger, 1971.

Barger, George, *Ergot and Ergotism.* London: Gurney and Jackson, 1931.

Barnett, R. D. "Early Shipping in the Near East." *Antiquity* 32 (1958): 220–30.

Basilevskaya, N. A. "On the Races of the Opium Poppy Growing in Semiretchie and the Origin of Their Culture." *Bulletin of Applied Botany, of Genetics and Plant Breeding* 19, no. 2. Leningrad, 1928.

———. "Principal Botanic Systematical Groups of the Opium Poppy *(Papaver somniferum).*" *Bulletin of Applied Botany, of Genetics and Plant Breeding* 25, no. 2. Leningrad, 1931.

Baudais-Lundstrom, K. "Plant Remains from a Swiss Neolithic Lakeshore Site: Brise-Lames, Auvernier." *Ber. Deutsch. Bot. Ges.* 91 (1978): 67–83.

Behre, K. E. "Die Entwicklungsgeschichte der natürlichen Vegetation im Gebiet der unteren Ems und ihre Abhängigkeit von den Bewegungen des Meeresspiegels." *Probleme der Kustenforschung im südlichen Nordseegebiet* 9: 13–47. Berlin, 1970.

Bender, Barbara, and Phillips, Patricia. "The Early Farmers of France." *Antiquity* 66 (1972): 97–105.

Bennett, C. L. et al. "Radiocarbon Dating with Electrostatic Accelerators: Dating of Milligram Samples." *Science* 201 (1978): 345–46.

Bentley, Robert, and Trimen, Henry. *Medicinal Plants.* Vol. 1, Nos. 1–69. London: Churchill, 1880.

Bertsch, K. "Paläobotanische Monographie der Federseerieds." *Bibliotheca Botanica* 103: 1–126. Stuttgart: E. Nägele, 1931.

———. "Die Pflanzenreste der Pfahlbauten von Sipplingen und Langenrain im Bodensee." *Badische Fundberichte* 2(1932): 305–20.

———. "Nachträge zur vorgeschichtlichen Botanik des Federseerieds." *Jahreshefte des Vereins für vaterländische Naturkund in Württemberg*, 102–5. 1950.

———, and Bertsch, Franz. *Geschichte unserer Kulturpflanzen*. Stuttgart: Wissenschaftliche Verlagsgesellschaft, 1949.

Bhown, A. S.; Shah, D. K.; and Nath, S. "Studies on Poppy Seed *(Papaver somniferum)*. Part I: Free amino acids." *Naturwissenschaften* 52, no. 18 (1965): 516–17.

Blackman, Winifred S. *The Fellāhīn of Upper Egypt: Their Religious, Social and Industrial Life with Special Reference to Survivals from Ancient Times*. London: Frank Cass, 1968. (1st ed., 1927.)

Blance, B. "Early Bronze Age Colonists in Iberia." *Antiquity* 35 (1961): 192–202.

Blegen, Elizabeth P. "Excavations at Gazi, Crete." *American Journal of Archaeology* 40 (1936): 371–73.

Boardman, John. *The Cretan Collection in Oxford: The Dictaean Cave and Iron Age Crete*. Oxford: Clarendon Press, 1961.

———, and Hayes, John. *Excavations at Tocra 1963–1965: The Archaic Deposits I*. Suppl. vol. 4. London, British School of Archaeology at Athens: Thames and Hudson, 1966.

Bocquet, Aimé. "Lake-Bottom Archaeology." *Scientific American* 240 (1979): 56–64.

Boisacq. Emile. *Dictionnaire étymologique de la langue grecque*. Paris: C. Klincksieck, 1916.

Boissier, Edmond. *Flora Orientalis*, Vol. I. Basel: H. George, 1867.

Bonavia, Emanuel. *The Flora of the Assyrian Monuments and its Outcomes*. Westminster: Archibald Constable, 1894.

Borg, John. *Descriptive Flora of the Maltese Islands*. Malta: Malta Government Press, 1927.

Branigan, Keith. *Copper and Bronze Working in Early Bronze Age Crete* (Studies in Mediterranean Archaeology, 19). Lund, Sweden: P. Åström, 1968.

Braungart, Richard. *Die Urheimat der Landwirtschaft*. Heidelberg: C. Winter, 1912.

Brea, Luigi Bernabo. *Sicily before the Greeks* (Ancient Peoples and Places, 3). London: Praeger, 1966.

Breasted, James Henry. *Ancient Records of Egypt. Historical Documents from the Earliest Times to the Persian Conquest, Collected, Edited and Translated with Commentary: The Eighteenth Dynasty, II*. Chicago: University of Chicago Press, 1923.

Brecher, Edward M., and the editors of *Consumer Reports*. *Licit and Illicit Drugs*. Boston: Little, Brown & Co., 1972.

Brothwell, Don. "Diet, Economy and Biosocial Change in Late Prehistoric Europe." In *Economy and Settlement in Neolithic and Early Bronze Age Britain and Europe*, edited by D. D. A. Simpson, pp. 75–88. Leicester: Leicester University Press, 1969.

———, and Brothwell, Patricia. *Food in Antiquity*. New York: Praeger, 1969.

Brown, Charles. *Flower Lore*. Madison, Wis.: C. E. Brown, 1938.

Bruyere, Bernard. *Rapport sur les fouilles de Deir el Medineh (1934–1935). Deuxième partie; la nécropole de l'est* (Fouilles de l'Institut Français du Caire, XV). Cairo: Imprimerie de l'Institut français d'archéologie orientale, 1937.

Buchholz, Hans-Günter, and Karageorghis, Vassos. *Prehistoric Greece and Cyprus: An Archaeological Handbook.* Translated by Francisca Garvie from the German. London: Phaidon Press, 1973.

Budge, E. A. Wallis. *The Dwellers on the Nile.* London: Religious Tract Society, 1926.

Burleigh, Richard. "Radiocarbon Dating: Some Practical Considerations for the Archaeologist." *Journal of Archaeological Science* 1 (1974)
:69–87.

Buschan, Georg. *Vorgeschichtliche Botanik der Kultur- und Nutzpflanzen.* Breslau: J. U. Kern's Verlag (M. Müller), 1895.

Butterworth, E. A. S. *Some Traces of the Pre-Olympian World in Greek Literature and Myth.* Berlin: Walter de Gruyter, 1966.

Butzer, Karl W. *Environment and Archaeology: An Ecological Approach to Prehistory.* Chicago: Chicago University Press, 1971.

Cain, Stanley. *Foundations of Plant Geography.* New York: Harper, 1944.

Caldwell, Anne E. *Origins of Psychopharmacology.* Springfield, Ill.: Thomas, 1970.

Campbell, Douglas H. *The Evolution of the Land Plants.* Stanford, Calif.: Stanford University Press, 1939.

Camps, G. "The Prehistoric Cultures of North Africa: Radiocarbon Chronology." *Problems in Prehistory: North Africa and the Levant*, edited by F. Wendorf and A. Marks, pp. 181–92. Dallas, Tex.: Southern Methodist University, 1975.

de Candolle, Alphonse. *L'origine des plantes cultivées.* Paris: Baillière, 1883. (American edition, New York: Appleton, 1886.)

de Candolle, Augustin Pyrame. *Prodromus systematis naturalis regni vegetabilis.* Vol. 1 of 17 vols. Paris, 1824.

Casson, Lionel. "Where Did the Ancient Traders Sail?" In *Mysteries of the Past*, edited by Joseph J. Thorndike, Jr., pp. 158–84. New York: American Heritage, 1977.

Castiglioni, Arturo. *A History of Medicine.* Translated from the Italian and edited by E. B. Krumbhaar. New York: Knopf, 1941.

Catkins, Alonzo. *Opium and the Opium Appetite.* New York: Lippincott, 1871.

Ceram, C. W. [pseud.] *Gods, Graves, and Scholars.* Translated by E. B. Garside from the German. New York: Knopf, 1951.

Chadwick, John. *The Mycenaean World.* Cambridge: Cambridge University Press, 1976.

Chard, Chester. *Man in Prehistory.* 2d ed. New York: McGraw-Hill, 1975.

Chassinat, M. Emile. *Un papyrus médical copte.* Cairo: Impr. de l'Institut français d'archéologie orientale, 1921.

Chao, Nancy B. "Perspectives on the Religion of Prehistoric Crete." In *Valcamonica Symposium '72—Actes du Symposium International sur les religions de la préhistoire*, edited by Emanuel Anati et al., pp. 221–25. Brescia, Italy: Centro Comuno di Studi preistorica, 1975.

Chopra, Ram Nath. *Indigenous Drugs of India, Their Medical and Economic Aspects.* Calcutta: U. N. Dhur, 1933.

Cirlot, Juan Eduardo. *Dictionary of Symbols.* Translated by Jack Sage from the Spanish. London: Routledge and Kegan Paul, 1971.

Civil, Michel. "Prescriptions médicales sumeriennes." *Revue d'Assyriologie et d'Archéologie Orientale* 14, no. 2 (1960): 59–64.

Clark, John Grahame Douglas. *Prehistoric Europe: The Economic Basis.* London: Methuen, 1952.

Clarke, Connell Robert. *The Botany and Ecology of Cannabis.* Ben Lomond, Calif.: Pods Press, 1977.

Clay, R. C. C. "An Early Iron Age Site on Fifield Bavant Down." *Wilts Archaeological and Natural History Magazine* 160 (1924): 457.

Coates, Alice. *Flowers and Their Histories.* London: Adam and Charles Black, 1968.

Cohen, Michael M. "The History of Opium and the Opiates." *Texas Medicine* 65 (March 1969): 76–85.

Coldstream, J. N., ed. *Knossos: The Sanctuary of Demeter.* Suppl. vol. 8. London, British School of Archaeology at Athens: Thames and Hudson, 1973.

Collom, Arthur B. "Tears of the Poppy: A Review of the History of Opium." *Kansas Medical Society Journal* 58 (1957): 614–26.

Cook, Arthur Bernard. *Zeus: A Study in Ancient Religion.* 3 vols. New York: Biblo and Tannen, 1964.

Cornford, Francis Macdonald. *Greek Religious Thought.* London: J. M. Dent, 1923.

———. *The Origin of Attic Comedy.* Cambridge: Cambridge University Press, 1934 (ca. 1914).

Crabtree, Keith. "A Review of Methodological Advances in Pollen Analyses." In *Process in Physical and Human Geography, Bristol Essays,* edited by R. Peel, M. Chisholm and P. Haggett, pp. 266–83. London: Heinemann Educational Books, 1975.

Crawford, Dorothy J. *Kerkeosiris: An Egyptian Village in the Ptolemaic Period.* Cambridge: Cambridge University Press, 1971.

———. "The Opium Poppy: A Study in Ptolemaic Agriculture." In *Problèmes de la terre en Grèce ancienne,* edited by M. I. Finley, pp. 223–51. Paris: Mouton, 1973.

Crawford, H. E. W. "Mesopotamia's Invisible Exports in the Third Millennium B.C." *World Archaeology* 5, no. 2 (1973): 232–42.

Culican, William. *The First Merchant Venturers.* London: Thames and Hudson, 1966.

Cullen, J. "Papaveraceae." *Flora Iranica.* Lfg. cont. no. 34/16.10.1966; pp. 1–27. Graz: Akademische Druck und Verlagsanstoet.

D'Agostino, Bruno. *Greece: Monuments of Civilization.* New York: Grosset and Dunlap, 1975.

Dai, Bingham. *Opium Addiction in Chicago.* Montclair, N.J.: Patterson Smith, 1970.

Dales, George F. "Necklaces, Bands, and Belts on Mesopotamian Figures." *Revue d'Assyriologie et d'Archéologie* 57, no. 1: 21–40. Paris: University Press of France, 1963.

Dalev, D.; Llieva, L. H.; and Llieva, R. "Poppy Cult in Bulgaria and the Production of Opium." *Bulletin on Narcotics* 12, no. 1 (January–March 1960): 25–37.

Dammel, L. H. *A Manual of Poisonous Plants.* Cedar Rapids, Iowa: Torch Press, 1911.

Danert, S. "Zur Systematik von Papaver somniferum L." *Die Kulturpflanzen* 6 (1958): 61–88.

Daux, G. "Chronique des fouilles et découvertes archéologiques en Grèce en 1960." *Bulletin de Correspondance Hellénique* 85 (1961).

Davis, Theodore M. *Biban-el-Moluk: The Tomb of Siphtah; The Monkey Tomb and the Gold Tomb*. London: A. Constable, 1908.

Dawkins, Richard M. *The Sanctuary of Artemis Orthia at Sparta*. London: Macmillan, 1929.

Deines, Hildegard von, and Grapow, Hermann. *Wörterbuch der Ägyptischen Drogennamen* (Grundriss der Medizin der alten Ägypter VI). Berlin: Akademie-Verlag, 1959.

Dennell, Robin W. "The Interpretation of Plant Remains: Bulgaria." In *Papers in Economic Prehistory*, edited by E. S. Higgs, pp. 149–59. Cambridge, Cambridge University Press, 1972.

———. "Botanical Evidence for Prehistoric Crop Processing Activities." *Journal of Archaeological Science* 1 (1974): 275–84.

———. "The Economic Importance of Plant Resources Represented on Archaeological Sites." *Journal of Archaeological Science* 3 (1976): 229–47.

———. "Archaeobotany and Early Farming in Europe." *Archaeology* 31, no. 1 (January–February 1978): 8–13.

de Wet, J. M., and Harlan, J. R. "Weeds and Domesticates: Evolution in the Man-Made Habitat." *Economic Botany* 29, no. 2 (Apr.–June 1975): 99–107.

Dimbleby, G. W. *Plants and Archaeology*. New York: Humanities Press, 1967.

———. "Archaeological evidence of environmental change." *Nature* 256 (1975): 265–67.

Diodorus Siculus. *The Historical Library of Diodorus the Sicilian*. Translated by G. Booth from the Greek. 2 vols. London: Davis, 1814.

Dodds, Eric R. *The Greeks and the Irrational*. Berkeley, Calif.: University of California Press, 1951.

Dugas, Charles, "Le sanctuaire d'Alea Athena à Tegée." *Bulletin de Correspondance Hellénique* 45 (1921): 368–85.

Duke, James A. "Utilization of Papaver." *Economic Botany* 27, no. 4 (1973): 390–400.

———. "Notes on Meo and Yao Poppy Cult." *Phytologia* 28, no. 1 (1974a): 5–8.

———. "Papaveraceous Polyclave." *CRC Critical Reviews in Toxicology* (Sept. 1974b): 1–95.

——— et al. *Annotated Bibliography on Opium and Oriental Poppies and Related Species*. U.S. Department of Agriculture, Agricultural Research Service, ARS-NE-28, 1973.

Dunbarin, R. L. "Notes on Lewis and Short." *Classical Review* 49, no. 1 (February 1935): 9–11.

Durrani, F. A. "Stone Vases as Evidence of Connection between Mesopotamia and the Indus Valley." *Ancient Pakistan* 1 (1964): 51–96.

Ebert, Max. *Reallexikon der Vorgeschichte*. Vol. 8 of 15 vols. Berlin: de Gruyter, 1924–32.

Edlin, Herbert. *Atlas of Plant Life*. New York: John Day, 1973.

Elkan, L. *Trentamen Monographiae Generis Papaver*. Grafe et Unzer, Kaliningrad, U.S.S.R.: Regiomontii Borussorum, 1839.

El-Masry, S., El-Ghazooly, M., Omar, A., Khafagy, S. and Phillipson, J. "Alkaloids from Egyptian Papaver rhoeas." *Planta Medica* 41 (1981):61–64.

Emboden, William. "The Sacred Narcotic Lily of the Nile: *Nymphaea caerulea.*" *Economic Botany* 32, no. 4 (October–December 1974):395–407.

———. *Narcotic Plants.* 2d ed., rev. and enl. New York: Macmillan, 1979. (a)

———. "*Nymphaea ampla* and other Narcotics in Maya Ritual and Shamanism." *Mexicon* 1, no. 4 (1979):50–52. (b)

———. Book review of *The Road to Eleusis: Unveiling the Secret Mysteries* by Wasson et al. *The Ethno-Pharmacology Society Newsletter* 2, no. 1 (Summer 1978):2–3.

Erman, Adolf. *Life in Ancient Egypt.* Translated by H. M. Tirard from the German. London and New York: Macmillan, 1894.

———. *Aegypten und Aegyptisches Leben im Altertum.* Tübingen, Germany: Mohr, 1923.

Ernst, Wallace R. *A Comparative Morphology of the Papaveraceae.* Ph.D. Dissertation, Stanford University, 1962.

———. "The Genera of Papaveraceae and Fumariaceae in the Southeast United States." *Journal of the Arnold Arboretum* 43 (1962):315–43.

Evans, Arthur. "Ring of Nestor." *Journal of Hellenic Studies* 45, part 1 (1925):1–75.

———. *The Palace of Minos at Knossos.* 4 vols. London: Macmillan, 1921–36; rpt., New York: Biblo and Tannen, 1964.

Evans, J. A. S. "Redating Prehistory in Europe." *Archaeology* 30, no. 2 (March 1977):76–85.

Fagan, Brian. *Men of the Earth: An Introduction to World Prehistory.* Boston: Little, Brown & Co., 1974.

Fairbanks, Arthur. *A Handbook of Greek Religion.* New York, etc.: American Book Company, 1910.

Farmilo, C. G.; Rhodes, H. L. J.; Hart, H. R. L.; and Taylor, H. "Detection of Morphine in *Papaver setigerum* DC." *Bulletin on Narcotics* (January–March 1953):26–31.

Fedde, F. "Papaveraceae-Hypecoideae et Papaveraceae-Papaveroideae." In *Das Pflanzenreich*, edited by A. Engler. vol. 40. Leipzig: Wilhelm Englemann, 1909.

Feinbrun, N. "Chromosomes of Some East Mediterranean Papaver Species." *Caryologia* 16 (1963):649–52.

Fernandes, A. and Queiros, M. "Sur la caryologie de quelques plantes récoltées pendant la IIIème réunion de botanique péninsulaire." *Memoirs of the Society Broteriana* 21 (1970–71):342–85.

Fournier, P. *Les Quatre Flores de la France.* Paris: University Press of France, 1946.

Fox, A. *South West England* (Ancient Peoples and Places, 41). London: Thames and Hudson, 1964.

Frazer, James. *The Golden Bough.* New York: Macmillan, 1922.

Fredskild, Bent. "Seeds and Fruits from the Neolithic Settlement Weier, Switzerland." *Botanisk Tidsskrift* 72 (April 1978):189–201.

Friedrich, Paul A. *Beiträge zur Kenntnis der Braukohlenflora der Prov. Sachsen*. Berlin: J. H. Newman 1883.

Fritel, P. H. "Sur l'existence de l'oeillette (*Papaver somniferum* var. *nigrum* D.C.), en Provence, a l'époque quaternaire." *Bulletin de la Société Géologique*, 4th ser., 20, no. 46: 207–8. Paris: Société Géologique de France, 1920.

Fritsch, Reinhard. "Zur Samenmorphologie des Kulturmohns (*Papaver Somniferum* L.)" *Kulturpflanze* 27 (1979): 217–27.

Fryxell, Paul. "The Mode of Reproduction in Higher Plants." *Botanical Review* 23 (1957): 135–233.

Fulton, Charles Clarke. *The Opium Poppy and Other Poppies*. Washington, D.C.: U.S. Treasury Department, Bureau of Narcotics, 1944.

Furst, Peter T. *Hallucinogens and Culture*. San Francisco, Calif.: Chandler & Sharp, 1976.

———. ed. *Flesh of the Gods*. New York: Praeger, 1972.

Gabra, Sami. "Papaver Species and Opium Through the Ages." *Bulletin de l'Institut d'Egypte*, 37: 39–56. Cairo, 1956.

Georgiades, N. "Contribution à l'étude de l'opium egyptien." *Bulletin de l'Institut d'Egypte*, 37: 361–400. Cairo, 1917.

Ghalioungui, Paul. *Magic and Medical Science in Ancient Egypt*. London: Hodder and Stoughton, 1963.

Gibbs, R. Darnley. *Chemotaxonomy of Flowering Plants*. Montreal: McGill-Queen's University Press, 1974.

Gillen, Paul. "Overheads and Flower Reading: The Spiritual Demonstration." *Mankind*, September 1976; pp. 269–78.

Gimbutas, Marija. *The Balts*. New York: Praeger, 1963.

———. *Bronze Age Cultures in Central and Eastern Europe*. Paris: Mouton, 1965.

———. *The Gods and Goddesses of Old Europe*. London: Thames and Hudson, 1974.

Gizbert, W. "Nowe stanowiska kipalne maku (*Papaver somniferum* L.) na ziemiach Polskich. Nouvelles stations fossiles de pavot." *Materialy Archeologiczne* 2 (1960): 347–54; addendum 3 (1961): 266.

Gjerstad, Einar. *Studies on Prehistoric Cyprus* (Filosofi, Sprakbetenskap och Historiska Vetenskaper, I). Upsala, Sweden: A.-b. Lundquistska bookhandeln, 1926.

Godolphin, Francis R., ed. *The Greek Historians*. Vol. 1 of 2 vols. *The Historical Works of Thucydides*, translated by Benjamin Jowett. New York: Random House, 1942.

Godwin, Harry. "The Ancient Cultivation of Hemp." *Antiquity* 41, no. 161 (March 1967): 42–48.

Goldblatt, Peter. "Biosystematic Studies in *Papaver* section Oxytona." *Annals of the Missouri Botanical Garden* 61 (1974): 264–96.

Gongora, Manuel. *Antiguedades prehistoricas de Andalucia*. Madrid: C. Moro, 1868.

Good, Ronald. *The Geography of the Flowering Plants*. 4th ed. London: Longman, 1974.

Gow, A. S. F., ed. *Theocritus*. Translated by A. S. F. Gow from the Greek. 2 vols. Cambridge: Cambridge University Press, 1965.

———, and Scholfield, A. F., eds. *Nicander: The Poems and Poetical Fragments.* Translated by A. S. F. Gow and A. F. Scholfield from the Greek. Cambridge: Cambridge University Press, 1953.

Greig, J. R. A., and Turner, J. "Some Pollen Diagrams from Greece and their Archaeological Significance." *Journal of Archaeological Science* 1 (1974): 177–94.

Grieve, M. *A Modern Herbal*, II, edited by Mrs. C. F. Leyel. New York: Hafner, 1959.

Griffith, F. L. "Oxford Excavations in Nubia." *Liverpool Annals of Archaeology and Anthropology* (1923: 137–40.

Grose, S. W. *Fitzwilliam Museum: Catalogue of the McClean Collection of Greek Coins*, I, II, III. Cambridge: Cambridge University Press, 1929.

Grueber, Herbert Appold. *Coins of The Roman Republic in The British Museum.* 3 vols. London: British Museum, 1970.

de Gubernatis, Angelo. *La mythologie des plantes.* 2 vols. Paris: C. Reinwald, 1882.

Gunther, Robert T. *The Greek Herbal of Dioscorides.* Translated by John Goodyear in 1655 into English. Oxford: University Press, 1934. (Later edition, New York: Hafner Pub. Co., 1959.)

Guyan, Walter U., ed. *Das Pfahlbauproblem.* Basel, Switzerland: Birkhäuser, 1955.

Hall, H. R. "Notices of Recent Publications." *Journal of Egyptian Archaeology* 14 (1928): 203–5.

Hammer, Karl, and Fritsch, Reinhard. "Zur Frage nach der Ursprungsart des Kulturmohns." *Kulturpflanze* 25 (1977): 113–24.

Harlan, Jack R. "The Plants and Animals That Nourish Man." *Scientific American*, September 1976, pp. 89–97.

———, and de Wet, J. M. J. "On the Quality of Evidence for Origin and Dispersal of Cultivation Plants." *Current Anthropology* 14, nos. 1–2 (February–April 1973): 51–65.

Harner, Michael J., ed. *Hallucinogens and Shamanism.* London: Oxford University Press, 1973.

Harris, John Richard. "Medicine." In *The Legacy of Egypt*, edited by J. R. Harris, pp. 112–38. 2d ed. Oxford: Clarendon Press, 1971.

Harrison, Jane E. *Prolegomena to the Study of Greek Religion.* Cambridge: Cambridge University Press, 1922.

Harrison, Roland Kenneth. *Healing Herbs of the Bible*, by R. H. [*sic*] Harrison. Leiden, Netherlands: E. J. Brill, 1966.

Hartmann, Fernande. *L'agriculture dans l'ancienne Egypte.* Paris: Libraries-imprimeries réunies, 1923.

Hartwich, C. "Über *Papaver somniferum* und speziell dessen in den Pfahlbauten vorkommende Reste." *Apotheker Zeitung* 14, no. 39 (1899): 278–80; no. 40: 289–91; no. 41: 300–302.

Hawkes, Jacquetta G. *Dawn of the Gods: Minoan and Mycenaean Origins of Greece.* London: Chatto & Windus, 1968.

———. "The Ecological Background of Plant Domestication." In *The Domestication and Exploitation of Plants and Animals*, edited by Peter J. Ucko and G. W. Dimbleby, pp. 17–30. Chicago: Aldine, 1969.

———. *Atlas of Ancient Archaeology*. New York: McGraw-Hill, 1974.

Hayes, William C. *The Scepter of Egypt*, II. New York: Metropolitan Museum of Art, 1959.

Heer, Oswald. "Die Pflanzen der Pfahlbauten." In *Neujahrsblatt der Naturforschunden Gesellschaft in Zürich*, 1866, pp. 1–54.

———. "Plants of the Lake Dwellings." In *The Lake Dwellings of Switzerland and Other Parts of Europe*, edited by F. Keller, pp. 518–36. 2 vols. London: Longmans & Green, 1878.

Heffter, E. R. "Notes on the Culture and Commerce in Opium in Asia Minor." *American Journal of Pharmacy* 40 (1868): 362–68.

Hegi, Gustav. *Illustrierte Flora von Mittel-europa.* Vol. 4 of 7 vols., pp. 33–36. Munich: Lehmann, 1918.

Hehn, Victor. *The Wandering of Plants and Animals from Their First Home.* Edited by James S. Stallybrass. London: S. Sonnenschein, 1885.

Heiser, Charles B., Jr. *Seed to Civilization: The Story of Man's Food.* San Francisco, Calif.: Freeman, 1973.

———. *The Sunflower.* Norman, Okla.: University of Oklahoma Press, 1976.

Helbaek, Hans. "Early crops in Southern England." *Proceedings of the Prehistoric Society* 12 (1952): 194–239.

———. "Notes on the Evolution and History of Linum." In *Kuml*, pp. 103–41. Aarhus: Universitetsforlaget, 1959.

———. "Plant Remains from the Tarxien Cemetery." In *The Prehistoric Antiquities of the Maltese Islands: A Survey*, edited by J. D. Evans, p. 235. London: Athlone Press, 1971.

Hibben, Frank. *Prehistoric Man in Europe.* Norman, Okla.: University of Oklahoma Press, 1958.

Hoffner, Harry A., Jr. *Alimenta Hethaeorum: Food Production in Hittite Asia Minor.* New Haven, Conn.: American Oriental Society, 1974.

Holmes, Y. Lynn. "The Foreign Trade of Cyprus during the Late Bronze Age." In *The Archaeology of Cyprus*, edited by Noel Robertson, pp. 90–110. Park Ridge, N.J.: Noyes Press, 1975.

Hood, Sinclair. *The Minoans.* London: Thames and Hudson, 1971.

Hoops, Johannes. *Waldbäume und Kulturpflanzen im germanischen Altertum.* Strasbourg: K. J. Trübner, 1905.

———. *Reallexikon der germanischen Altertumskunde* 3 (1911–19): 223–34. Strasbourg.

Hort, Arthur, trans. *Theophrastus: Enquiry into Plants*, I & II. London: William Heinemann, 1916.

Hrishi, N. J. "Cytogenetical studies on *Papaver somniferum* L. and *Papaver setigerum* DC. and their hybrid." *Genetica* 31 (1960): 1–30.

Hubbard, R. N. L. B. "On the Strength of the Evidence for Prehistoric Crop Processing." *Journal of Archaeological Science* 3 (1976): 257–65.

Hutchinson, John. *The Families of Flowering Plants.* 3d ed. Oxford: Clarendon Press, 1973.

Isaac, Erich. "Influence of Religion on the Spread of Citrus." *Science* 129 (1959): 179–86.

———. *The Geography of Domestication.* Englewood Cliffs, N.J.: Prentice-Hall, 1970.

Jacobsthal, Paul. *Greek Pins and their Connexions with Europe and Asia.* Oxford: Clarendon Press, 1956.

Jevons, F. R. "Was Plotinus Influenced by Opium?" *Medical History* 9 (Oct. 1965): 374.

Jorgensen, G. "Triticum aestivum s.l. from the Neolithic site of Weier in Switzerland." *Folia Quaternaria* 46 (1976): 7–21.

Joret, Pierre Louis Charles. *Les plantes dans l'antiquité et au moyen âge*, III. Paris: E. Bouillon, 1904.

Karageorghis, Vassos. "A Twelfth-Century B.C. Opium Pipe from Kition." *Antiquity* 50 (1976): 125–29.

Kawatani, T., and Ohno, T. "Chromosome numbers in *Papaver*" (Japanese, summary in English). *Bulletin of the National Institute for Hygienic Science* 83: 127–33. Tokyo, Japan.

Keepax, Carole. "Contamination of Archaeological Deposits by Seeds of Modern Origin with Particular Reference to the Use of Flotation Machines." *Journal of Archaeological Science* 4 (1977): 221–29.

Keller, Ferdinand. *The Lake Dwellings of Switzerland and Other Parts of Europe.* Translated by John E. Lee from the German. 2d ed. London: Longmans, Green, 1878.

Kellman, Martin C. *Plant Geography.* New York: St. Martin's Press, 1975.

Kemp, Barry. "Egypt." In *Atlas of Ancient Archaeology*, edited by J. Hawkes, pp. 146–66. New York: McGraw-Hill, 1974.

Kenneth, J. H., ed. *A Dictionary of Biological Terms.* Princeton, N.J., etc.: Van Nostrand, 1963.

Kerényi, Károly. *The Gods of the Greeks.* London: Thames and Hudson, 1951.

———. *Prometheus: Archetypal Image of Human Existence* (Bollingen Series, 65). Translated by Ralph Manheim from the German. New York: Pantheon Books, 1963.

———. *Eleusis: Archetypal Image of Mother and Daughter.* Translated by Ralph Manheim from the German. New York: Bollingen Foundation (distributed by Pantheon books), 1967.

Kilgour, Frederick G. "Medicine in Art." *What's New Journal of Medicine*, no. 216 (Late Winter 1960). Chicago: Abbott Laboratories.

Kimber, R. G. "Beginnings of Farming? Some Man-Plant-Animal Relations in Central Australia." *Mankind* 10, no. 3 (June 1976): 142–50.

Kirby, G. W. "Biosynthesis of the Morphine Alkaloids." *Science* 155 (1967): 170–73.

Kirchner, Horst. "Ein archäologischer Beitrag zur Urgeschichte des Schamanismus." *Anthropos* 47 (1952): 245–49.

Kitto, Humphrey Davy. *The Greeks.* Middlesex, England: Penguin Books, 1951.

Klein, Ernest. *A Comprehensive Etymological Dictionary of the English Language.* Amsterdam: Elsevier, 1967.

Klichowska, M. "Aus palaoethnobotanischen Studien über Pflanzenfunde aus dem Neolithikum und der Bronzezeit auf polnischem Boden." *Archaeologia Polona* 17 (1976): 27–67.

Knight, Melvin M. *Economic History of Europe to the End of the Middle Ages.* Boston, New York, etc.: Houghton Mifflin, 1926.

Knörzer, Karl-Heinz. "Subfossile Pflanzenreste von bandkeramischen Fundstellen im Rheinland." *Bonner Jahrbucher* 23 (1967): 3–29.

———. "Prähistorische Mohnsamen im Rheinland." *Bonner Jahrbuch* 171 (1971): 34–39. (a)

———. "Eisenzeitliche Pflanzenfune im Rheinland." *Bonner Jahrbuch* 171 (1971): 40–58. (b)

———. "Subfossile Pflanzenreste aus der bandkeramischen Siedlung Langweiler 3 und 6, Kreis Julich, und ein urnenfelderzeitlicher Getreiderund innerhalb dieser Siedlung." *Bonner Jahrbuch* 172 (1972): 395–402.

———. "Pflanzliche Grobreste." In *Der bandkeramische Siedlungsplatz Langweiler 2, Gemeinde Aldenhoven, Kreis Düren*, edited by J.-P. Farruggia, R. Kuper, J. Lüning and P. Stehli, pp. 139–52. 1973. Bonn: Rheinland Verlag.

———. "Bandkeramische Pflanzenfunde von Bedburg-Garsdor, Kreis Bergheim/Erft." *Rheinische Ausgrabungen* 15 (1974): 173–92.

———. "Späthallstattzeitliche Pflanzenfunde bei Bergheim, Erftkreis." *Rheinische Ausgrabungen* 17 (1976): 151–85.

———. "Pflanzliche Grobreste des bandkeramische Siedlungsplatzes Langweiler 9." In *Der bandkeramische Siedlungsplatz Langweiler 9, Gemeinde Aldenhoven, Kreis Duren*, edited by R. Kuper et al., pp. 279–303. Bonn: Rheinland-Verlag, 1977.

Kostrzewski, Jósef. "Biskupin: An Early Iron Age Village in Western Poland." *Antiquity* 12 (1938): 311–17.

———. *Les origines de la civilisation polonaise.* Translated by M. Bernard Hamel from the Polish. Paris: Presses Universitaires de France, 1949.

Koul, A. K., et al. "Genetic Stocks of Kashmir Papaver." *Nucleus* 15, no. 2 (1972): 117–23.

Kramer, John. "Was Opium Mentioned in the Ebers Papyrus?" Unpublished paper presented to the American Institute of the History of Pharmacy, Anaheim, California, April 24, 1979. 15 pages.

Kramer, Samuel N. "First Pharmacopeia in Man's Recorded History." *American Journal of Pharmacy* 126, no. 3 (March 1954): 76–84.

Krikorian, Abraham D. "Were the Opium Poppy and Opium Known in the Ancient Near East?" *Journal of the History of Biology* 8, no. 1 (Spring 1975): 95–114.

———, and Ledbetter, Myron C. "Some Observations on the Cultivation of Opium Poppy (*Papaver somniferum* L.) for its Latex." *Botanical Review* 41, no. 1 (Jan.–Mar. 1975): 30–103.

Kritikos, Pan G. "Der Mohn, das Opium und ihr Gebrauch im Spätminoicum III: Bemerkungen zu dem Gefundenen Idol der Minoischen Gottheit des Mohns." *Praktika es Akademias Athenon* 35, fasc. 1 (1960): 55–73.

———, and Papadaki, Stella P. "The History of the Poppy and Opium and Their Expansion in Antiquity in the Eastern Mediterranean" (in Greek). *Archaiologike Ephemeris*, Hellenike Archaiologike Hetairia, Athens, 1963: 80–150.

———, and Papadaki, Stella P. "The History of the Poppy and of Opium and Their Expansion in Antiquity in the Eastern Mediterranean Area." *Bulletin on Narcotics* (a) 19, no. 3 (July–September 1967): 18–40; (b) 19, no. 4 (October–December 1967): 1–10.

Kühn, L., and S. Pfeifer. "The Genus *Papaver* and Its Alkaloids: A Review." *Pharmazie* 18, no. 12 (1963): 819–43.

Kuzmina, N. E. "Cytology of the Cultivated Poppy in Connection with Its Origin and Evolution." *Trudy po prikeladnoi botanike, genetike i selektsii (Bulletin of Applied Botany, Genetics, Plant Breeding)*, 2d ser. 8: 81–92 (English summary pp. 190–95). Lenin Academy of Sciences, USSR, 1937.

La Barre, Weston. "Old and New World Narcotics: A Statistical Question and an Ethnological Reply." *Economic Botany* 24 (1970): 73–80.

———. "Hallucinogens and the Shamanic Origins of Religion." In *Flesh of the Gods*, edited by Peter Furst, pp. 261–79. New York: Praeger, 1972.

Lamb, Winifred. *Excavations at Thermi in Lesbos*. Cambridge: Cambridge University Press, 1936.

Lamberg-Karlovsky, C. C. "Trade Mechanisms in Indus-Mesopotamia Relations." *Journal of the American Oriental Society* 92, no. 2 (1972): 222–30.

Laurent-Täckholm, Vivi. *Faraos Blomster*. Stockholm: Bokforlaget Natur och Kultur, 1951.

Lawrence, George H. *Taxonomy of Vascular Plants*. New York: Macmillan, 1951.

Lawson, John Cuthbert. *Modern Greek Folklore and Ancient Greek Religion*. Cambridge: Cambridge University Press, 1910.

Layard, Austen Henry. *Nineveh and Its Remains: With an Account of a Visit to the Chaldaean Christians of Kurdistan, and the Yezidis or Devil-Worshippers and an Enquiry into the Manners and Arts of the Ancient Assyrians*. 3rd. ed. 2 vols. London: J. Murray, 1849.

Leach, Maria, ed. *Dictionary of Folklore, Mythology and Legend*. New York: Funk and Wagnalls, 1950.

Leake, Chauncey D. *Letheon: The Cadenced Story of Anesthesia*. Austin, Tex.: The Regents of the University of Texas, 1947.

Leakey, Richard E., and Lewin, Roger. *Origins*. New York: E. P. Dutton, 1977.

LeFebvre, Gustave. *Essai sur la médicine égyptienne de l'époque pharaonique*. Paris: Presses Universitaires de France, 1956.

Legros, J. "Cultivation of the Opium Poppy in Central Europe and the Near East." *International Review of Agriculture* 29 (1938: 29T–41T.

Lenormant, François. *Essai de commentaire des fragments cosmogoniques de Bérose d'après les textes cunéiformes et les monuments de l'art asiatique*. Paris: Maisonneuve et Cie., 1871.

Levey, Martin. "Ancient Chemical Technology in a Sumerian Pharmacological Tablet." *Journal of Chemical Education* 32 (1955): 11–13.

———. "Babylonian Chemistry: A Study of Arabic and Second Millennium B.C. Perfumery." *Osiris* 12 (1956): 368–80.

Lewin, Louis. *Phantastica.* New York: E. P. Dutton, 1964 (rpt.).

Lindley, John, and Moore, Thomas, eds. *The Treasury of Botany.* Rev. ed. 2 vols. London: Longmans, Green, 1889.

Lips, Julius E. *The Origin of Things.* New York: A. A. Wyn, 1947.

Lloyd, Seton and Mellaart, James. *Beycesultan.* Vol. 2. London: British School of Archaeology in Ankara, 1965.

Loret, Victor. *La flore pharaonique.* Paris: E. Leroux, 1892.

Löw, Immanuel. *Die Flora der Juden.* Vol. 2. Vienna and Leipzig: R. Löwit, 1924–34.

Lucas, Alfred. *Ancient Egyptian Materials and Industries.* 4th ed., rev. and enl. by J. R. Harris. London: E. Arnold, 1962.

Lüdi, W. "Beitrag zur Kenntnis der Vegetationsverhältnisse im schweizerischen Alpenvorland während der Bronzezeit." In *Das Pfahlbauproblem*, edited by W. Ú. Guyan, pp. 89–109, 1955.

Lumley-Woodyear, Henry de. *Le paléolithique inférieur et moyen du Midi méditerranéen dans son cadre géologique*, II. Paris: Centre Nationale de la Recherche Scientifique, 1971.

Macht, David I. "The History of Opium and Some of Its Preparations and Alkaloids." *Journal of the American Medical Association* 64, no. 6 (1915): 477–81.

MacDermot, Violet. *The Cult of the Seer in the Ancient Near East.* Berkeley: University of California Press, 1971.

Mackenzie, Donald A. *Myths of Crete and Pre-Hellenic Europe.* London: Gresham Publishing Co., 1917.

Malcolm, Andrew I. *The Pursuit of Intoxication.* New York: Washington Square Press, 1971.

Malinowski, Tadeusz. "An Amber Trading-Post in Early Iron Age Poland." *Archaeology* 27, no. 3 (July 1974): 195–200.

Marinatos, S. "The Minoan Goddess of Gazi." *Archaeologiki Ephemeris* 1 (1937): 278–91.

———. *The Creto-Mycenaean Religion* (in Greek). Athens, 1948.

Marinatos, S. and Hirmer, M. *Crete and Mycenae.* Translated by John Boardman from the Greek. London: Thames and Hudson, 1960.

Marks, Geoffrey, and Beatty, William K. *The Medical Garden.* New York: Scribner, 1971.

Martin, Geoffrey T. *The Royal Tomb at El-Amarna*, I. London: Egypt Exploration Society, 1974.

Martin, Richard T. "The Role of Coca in the History, Religion, and Medicine of South American Indians." *Economic Botany* 24, no. 4 (October–December 1970): 422–34.

Maurizio, Adam. "Brotgewurze." *Naturwissenschaftliche Wochenschrift* 30 (1915): 225–28.

McDonald, W. A., and Rapp, G. R., eds. *The Minnesota Messina Expedition: Reconstructing a Bronze Age Regional Environment.* Minneapolis: Minnesota University Press, 1972.

McIntyre, Loren. "Lost Empire of the Inca." *National Geographic* 144, no. 6 (1973): 729–88.

McKerrell, Hugh. "On the Origins of British Faience Beads and Some Aspects of the Wessex-Mycenae Relationship." *Proceedings of the Prehistoric Society* 38 (Dec. 1972): 286–301.

McMurty, J. "Sicyon in Modern Times." *American Journal of Archaeology*, 5(1889): 268.

Merlin, Mark D. *Man and Marijuana.* Rutherford, N.J.: Fairleigh Dickinson University Press, 1972.

Merrillees, Robert S. "Opium Trade in the Bronze Age Levant." *Antiquity* 36 (1962): 287–92.

———. *The Cypriote Bronze Age Pottery Found in Egypt* (Studies in Mediterranean Archaeology, XVIII). Lund, Sweden: P. Åström, 1968.

———. *Trade and Transcendence in the Bronze Age Levant, Göteborg.* Studies in Mediterranean Archaeology, 39. Göteborg, Sweden: P. Åström, 1974.

———. "Opium Again in Antiquity." *Levant* 11 (1979): 167–71.

Messikommer, H. *Die Pfahlbauten von Robenhausen.* Zurich: Art Institut Orell Füssli, 1913.

Meuli, K. "Scythica." *Hermes* 70 (1935): 121–76.

Milisauskas, Sarunas. "Olszanica: An Early Farming Village in Poland." *Archaeology* 29, no. 1 (January 1976): 30–41.

Moldenke, H. N. and Moldenke, A. L. *Plants of the Bible.* Waltham, Mass.: Chronica Botanica, 1952.

Mook, W. G.; Munaut, A. V.; and Waterbolk, H. T. "Determination of Age and Duration of Stratified Prehistoric Bog Settlements." *Proceedings of the 8th International Conference on Radiocarbon Dating* 2 (1972): 491–503.

Mouterde, Paul. *Nouvelle flore du Liban et de la Syrie*, edited by Dar El-Machreg. Vol. 2. Beirut, Lebanon: Dar El-Machreg et Libr. Orientale, 1970.

Mowat, A. B., and Walters, S. M. "*Papaver* L." In *Lycopodiaceae to Platanaceae:* Vol. 1 of *Flora Europaea*, edited by T. G. Tutin et al., pp. 247–51, Cambridge: Cambridge University Press, 1964.

Much, Matthaus. "Dritter Bericht über die Pfahlbau-Forschungen im Mondsee (1875–1876)." *Mitteilungen der Anthropologischen Gesellschaft in Wien*, 6 (1876): 161–94.

———. "Vorgeschichtliche Nähr- und Nutzpflanzen Europas." *Mitteilungen der Anthropologischen Gesellschaft in Wien* 38 (1908): 217.

Muhly, J. P. "Copper and Tin: The Distribution of Mineral Resources and the Nature of the Metals Trade in the Bronze Age." *Transactions of the Connecticut Academy of Arts and Sciences* 43 (March 1973): 155–535.

Murray, Jacqueline. *The First European Agriculture, a Study of the Osteological and Botanical Evidence until 2000* B.C. Edinburgh: Edinburgh University Press, 1970.

Murray, Margaret Alice. *The Splendor that was Egypt.* London: Readers Union, 1951.

Mylonas, George E. *Ancient Mycenae.* Princeton, N.J.: Princeton University Press, 1957.

———. *Eleusis and the Eleusinian Mysteries.* Princeton, N.J.: Princeton University Press, 1974.

———. *Mycenae and the Mycenaean Age.* Princeton, N.J.: Princeton University Press, 1966.

Nash, Elizabeth T. *One Hundred and One Legends of Flowers.* Boston: Christopher Publishing House, 1927.

de Navarro, J. M. "The British Isles and the Beginning of the Northern Early Bronze Age." In *The Early Cultures of North-West Europe, H. M. Chadwick Memorial Studies,* edited by Sir Cyril F. Fox and Bruce Dickens, pp. 75–105. Cambridge: Cambridge University Press, 1950.

Neal, Marie. *In Gardens of Hawaii.* Honolulu: Bishop Museum Press, 1965.

Neligan, A. R. *The Opium Question: With Special Reference to Persia.* London: John Bale, Sons & Danielsson, 1927.

Netolitzky, Fritz. "Neue Funde prähistorischer Nährungs- und Heilmittel." In *Xenia: Hommage international à l'Université de Grèce à l'occasion du soixante-quinzième anniversaire de sa fondation.* 1912.

———. "Unser Wissen von den alten Kulturpflanzen Mitteleuropas." *Bericht der Romisch-Germanische Kommission* 20 (1931): 17–76.

Neuweiler, E. "Die prähistorischen Pflanzenreste Mitteleuropas mit besonderer Berücksichtigung der schweizerischen Funde." *Vierteljahrsschrift der Naturforschenden Gesellschaft in Zurich* 50 (1905): 72–74.

———. "Die Pflanzenreste aus den Pfahlbauten am Alpenquai in Zürich und von Wollishofen sowie einer interglazialen Torfprobe von Niederweningen." *Vierteljahrsschrift der Naturforschenden Gesellschaft in Zürich,* 64 (1919): 617–48.

———. "Die Pflanzenwelt in der Jungeren Stein- und Bronzezeit der Schweiz." *Mitteilungen der Antiq. Gesellschaft in Zürich* (1924).

———. "Pflanzenreste aus den Pfahlbauten des ehemaligen Wauwilersees (Kanton Luzern)." *Mitteilungen der Naturforschenden Gesellschaft in Luzern* 9 (1924): 301–23.

———. "Pflanzenreste aus den Pfahlbauten vom Hausersee, Greifensee und Zürichsee." *Vierteljahrsschrift der Naturforschenden Gesellschaft in Zürich* 70 (1925): 225–33.

———. "Pflanzenfunde aus dem spätneolithischen Pfahlbau am Utoquai Zurich." *Vierteljahrsschrift der Naturforschenden Gesellschaft in Zürich* 75 (1930): 35–40.

———. "Die Pflanzenreste aus dem spätbronzezeitlichen Pfahlbau 'Simpf' bei Zug." *Vierteljahrsschrift der Naturforschenden Gesellschaft in Zürich* 76 (1931): 116–32.

———. "Nachträge urgeschichtlicher Pflanzen." *Vierteljahrsschrift der Naturforschenden Gesellschaft in Zürich* 80 (1935): 98–122.

———. "Nachträge II urgeschichtlicher Pflanzen." *Vierteljahrsschrift der Naturforschenden Gesellschaft in Aürich* 91 (1946): 122–36.

Nilsson, Martin P. *The Minoan-Mycenaean Religion and its Survival in Greek Religion.* Lund: C. W. K. Gleerup, 1950.

———. *Griechische Feste von Religiöser Bedeutung.* Stuttgart; Teubner, 1957. (1st ed. 1906.)

Noerdlinger, Henry S. *Moses and Egypt: The Documentation of the Motion Picture "The Ten Commandments."* Los Angeles, Calif.: USC Press, 1956.

Oberdorfer, E. *Pflanzensoziologische Exkursionsflora für Süd-deutschland und die angrenzenden Gebiete.* Stuttgart: Ulmer, 1970. [ca. 1962]

Onions, C. T., ed. *The Oxford Dictionary of English Etymology.* Oxford: Clarendon Press, 1966.

O'Rahilly, Thomas Francis. *Early Irish History and Mythology.* Dublin: Dublin Institute for Advanced Studies, 1946.

Pailleux, A., and Bois, D. "Les plantes alimentaires spontanées en Grèce." *Revue des Sciences Naturelles Appliquées* 37 (1890): 1092.

Pellicer, Manuel. *Estratigrafia prehistorica de la Cueva de Nerja.* Excavaciones arqueologicas en España, vol. 16. Madrid: Diana Artes Graficas, 1963.

———. *El Neolitico y el Bronce de la Cueva de la Carigüela de Piñar, Granada.* Trabajos de prehistoria, vol. 15. Madrid, 1964.

Perry, Francis. *Flowers of the World.* London: Crown, 1972.

Persson, Axel W. *The Religion of Greece in Prehistoric Times.* Berkeley: University of California Press, 1942.

Petrie, William Matthew Flinders. *Tell el Amarna.* London: Methuen, 1894.

———. *Hyksos and Israelite Cities.* London: School of Archaeology, etc., 1906.

———. *Amulets.* London: Constable, 1914.

———; Brunton, Guy; and Murray, M. *Objects of Daily Use.* British School in Athens Memoir, XLII. London: British School of Archaeology in Egypt, etc., 1927.

Phillips, Patricia. *Early Farmers of West Mediterranean Europe.* London: Hutchinson, 1975.

Piggott, Stuart. *Ancient Europe, from the Beginnings of Agriculture to Classical Antiquity; A Survey.* Edinburgh: University Press, 1965.

Pinsent, John. *Greek Mythology.* London: Hamlyn, 1969.

Plowman, Timothy. "Botanical Perspectives of Coca." *Journal of Psychedelic Drugs* 11 (1979): 103–17.

Pollard, John R. T. "Greek Religion." In *The New Encyclopaedia Britannica*, pp. 406–10. 15th ed. Chicago: University of Chicago, 1975.

Polunin, Nicholas Vladimir. *Introduction to Plant Geography.* London: Longmans, 1960.

Polunin, Oleg. *Flowers of Europe: A Field Guide.* London: Oxford University Press, 1969.

Polunin, Oleg and Smythies, B. E. *Flowers of South-West Europe: A Field Guide.* London: Oxford University Press, 1973.

Popov, Mikhail Grigor'evich. "Papaveraceae." In *Flora of the USSR*, edited by V. L. Komarov, 7: 437–549. Jerusalem, 1970. (English edition.)

Porter, Cedric Lambert. *Taxonomy of Flowering Plants.* 2d ed. San Francisco, Calif.: W. H. Freeman, 1967.

Porter, Enid. *Cambridgeshire Customs and Folklore.* London: Routledge and Kegan Paul, 1969.

Preaux, Claire. *Les Grecs en Égypte d'après les archives de Zénon.* Brussels: Office de publicité, 1947.

Preston, T. A. *The Flowering Plants of Wilts.* Wiltshire, England: Wiltshire Archaeological and Natural History Society, 1888.

Quibell, J. E. and Hayter, A. G. K. *Excavations at Saqqara: Teti Pyramid, north side* (at head of title: Service des antiquités de l'Egypte). Cair0: Imprimerie de l'Institut français d'archéologie orientale, 1927.

Rackham, Harris, ed. and trans. *Natural History*, by Pliny, XVII-XIX. Vol. 0 of 10 vols. London: W. Heinemann, 1961.

Raglan, Fitz Roy Richard Somerset. *The Origins of Religion.* London: Watts, 1949.

Randall-Maciver, D., and Woolley, C. L. *Buhen* (Eckley B. Coxe, Jr., Expedition to Nubia: vols. 7–8). Philadelphia: University of Pennsylvania Press, 1911.

Reckin, Jurgen. "A Contribution to the Cytology of *Papaver aculeatum* Thunb." *Caryologia* 23 (1970): 401–64.

———. "A Contribution to the Cytology of *Papaver gracile* Auch., Including Proposals for a Revision of the Section *Mecones.*" *Caryologia* 26, no. 2 (1973): 245–51.

Renfrew, Colin. "Colonialism and Megalithismus." *Antiquity* 41 (1967): 276–88.

———. "Trade and Culture Process in European Prehistory." *Current Anthropology* 10 (1969): 151–69.

———. *The Emergence of Civilisation: The Cyclades and the Aegean in the Third Millennium* B.C. London: Methuen, 1972.

Renfrew, Jane M. *Palaeoethnobotany: The Prehistoric Food Plants of the Near East and Europe.* New York: Columbia University Press, 1973.

Renner, H. D. *The Origin of Food Habits.* London: Faber & Faber, 1944.

Rhind, William. *A History of the Vegetable Kingdom.* London: Blackie and Son, 1868.

Robinson, C. E. *Hellas: A Short History of Ancient Greece.* Boston: Beacon Press, 1967.

Robinson, Victor. *Victory Over Pain: A History of Anesthesia.* New York: Schumann, 1946.

Rodden, Robert. "Europe North of the Alps." In *Varieties of Culture in the Old World*, edited by Robert Stigler), pp. 49–77. New York: St. Martin's Press, 1975.

Rogers, Frances, and Beard, Alice, *5000 Years of Gems and Jewelry.* New York: Frederick A. Stokes, 1940.

Ross-Craig, Stella. *Drawings of the British Plants*, Part 2, London: Bell and Sons, 1951.

Rouse, W. H. D. *Greek Votive Offerings: An Essay in the History of Greek Religion.* Cambridge: Cambridge University Press, 1902.

———. *The Story of Odysseus.* New York: Modern Age Books, 1937.

Roussel, P. "L'initiation préalable et les symboles eleusinien." *Bulletin de Correspondance Hellénique* 54 (1930): 58–65.

Rowe, A. *A Catalogue of Egyptian Scarabs, Scaraboids, Seals, and Amulets in the Palestine Archaeological Museum*. Cairo: Government of Palestine Department of Antiquities, 1936.

Rudenko, Sergei I. *Frozen Tombs of Siberia: The Pazyryk Burials of Iron Age Horsemen*. Translated by M. W. Thompson. Berkeley and Los Angeles: University of California Press, 1970.

Russell, W. M. S. *Man, Nature and History: Controlling the Environment*. Garden City, N.Y.: Natural History Press, 1969. [ca. 1967]

Rytz, Walter, "Uber die Früchte and Samen aus dem Pfahlbau Burgäschi-Südwest." In *Das Pfahlbau von Seeberg Burgaschi-Südwest 1945–46*, edited by O. Tschumi. Bern: Historisches Jahrbuch, 1946.

———. "Die Pflanzenwelt." In *Urgeschichte der Schweiz*, edited by O. Tschumi, pp. 15–112. Frauenfeld: Huber, 1949.

Salisbury, Edward James. *Weeds and Aliens*. London: Collins, 1964.

Sameh, Waley-el-dine. *Daily Life in Ancient Egypt*. Translated by Michael Bullock from the German. New York: McGraw-Hill, 1964.

Samuels, Alan. *The Mycenaeans in History*. Englewood Cliffs, N.J.: Prentice-Hall, 1966.

Santavy, F. "Chapter 5," in *The Alkaloids: Chemistry and Physiology*, edited by R. H. F. Manske, vol. 12. New York: Academic Press, 1962.

Santavy, F., M. Maturova, A. Nemeckova, H.-B. Schröter, H. Potesilova, and V. Preiniger. "Isolation of Alkaloids from a Few Poppy Species." *Planta Medica* 8, no. 2 (1960): 167–78.

de Saporta, G. "La flore des tufs quaternaires en Provence." *Congrès Scientifique de France*. Aix-en-Provence, France, 1967.

Sauer, Carl O. "Environment and Culture during the Last Deglaciation." *Proceedings of the American Philosophical Society* 92 (1948): 65–77.

———. *Agricultural Origins and Dispersals*. Bowman Memorial Lectures, ser. 2. New York: American Geographical Society, 1952.

———.*Land and Life: A Selection from the Writings of Carl Ortwin Sauer*. Edited by John Leighly. Berkeley: University of California Press, 1963.

Sauter, Marc R. *Switzerland: From Earlies Times to the Roman Conquest*. Boùlder, Colo.: Westview Press, 1976.

Savory, H. N. *Spain and Portugal: The Prehistory of the Iberian Peninsula*. New York: Praeger, 1968

Sayre, Lucius E. *Manual of Organic Materia Medica and Pharmacognosy*. Philadelphia: P. Blackston's Son & Co., 1917.

Scarborough, John. "The Drug Lore of Asclepiades of Bithynia." *Pharmacy in History* 17, no. 2 (1975).

Schaeffer, Claude F. S. *The Cuneiform Texts of Ras Shamra-Ugarit*. The Schweich Lectures of the British Academy, 1936. London: Oxford University Press, 1939.

———. *Stratigraphie comparée et chronologie de l'Asie occidentale (IIIe et IIe millenaires) Syrie, Palestine, Asie Mineure, Chypre, Perse, et Caucase*. London: Oxford University Press, 1948.

Schiaparelli, Ernesto. *La tomba intatta dell'architetto Cha nella Necropoli di Tebe.* Turin, 1927.

Schliemann, Heinrich. *Mycenae: A Narrative of Researches and Discoveries at Mycenae and Tiryns.* New York: Scribner, Armstrong, 1878.

Schmiedeberg, Oswald. *Über die Pharmaka in der Ilias und Odyssee.* Strasbourg: K. J. Trübner, 1918.

Schrader, Otto. *Reallexikon der indogermanischen Altertumskunde.* 2d ed. 2 vols. Berlin: W. de Gruyter, 1917–29.

Schroeder, K. "Geologisch-palaeobotanische Untersuchung eines romanzeitlichen Brunnen bei Irrel, Kreis Bitburg-Prum (Eifel)." In *Trierer Zeitschrift für Geschichte und Kunst des Trierer Landes und seiner Nachbargebiete* 34 (1971):97–117.

Schultes, Richard E. "Native Narcotics of the New World." *Texas Journal of Pharmacy* 1 (1961):141–67.

———. "The Search for New Natural Hallucinogens." *Lloydia* 29 (1966):293–308.

———. "Hallucinogens of Plant Origin." *Science* 163 (1969):245–54.

———. *Hallucinogenic Plants.* New York: Golden Press, 1976.

———, and Hofman N. Albert. *The Botany and Chemistry of Hallucinogens.* Springfield, Ill.: Charles C. Thomas, 1973 (Revised and Enlarged Second Edition, 1980).

Schultze-Motel, Jürgen. "Literatur über archäologische Kulturpflanzenrest (1974/1975)." *Kulturpflanze* 24 (1976):159–178.

———. "Literatur über archäologische Kulturpflanzenrest (1975/1976)." *Kulturpflanze* 25 (1977):71–88.

———. "Die Urgeschichtlichen Reste des Schlafmohns (*Papaver Somniferum* L.) und die Entstehung der Art." *Kulturpflanze* 27 (1979):207–15.

Schwanitz, Franz. *The Origin of Cultivated Plants.* Cambridge, Mass.: Harvard University Press, 1966.

Schweinfurth, G. "Uber Pflanzenfunde in altägyptischen Gräber." *Sitzber der Botanisches Ver. der Prov. Brandenburg* 23 (1881–82).

———. "Aegyptens Auswärtige Beziehungen hinsichtlich der Culturgewachse." *Verhol. der Berliner Anthrop. Gesellschaft, 1891.*

Scott, John. *The White Poppy: A History of Opium.* New York: McGraw-Hill, 1969.

Seeberg, Axel. "Poppies, Not Pomegranates." *Acta ad Archaeologiam et Artium Historiam Pertinentia* 4 (1969):7–11.

Seyffert, Oskar. *A Dictionary of Classical Antiquities, Mythology, Religion, Literature and Art* (rev. and ed. by H. Nettleship and J. E. Sandys). New York: Meridian Books, 1956.

Sharon, Douglas, "The San Pedro Cactus in Peruvian Folk Healing." In *Flesh of the Gods*, edited by Peter Furst, pp. 114–36. New York: Praeger, 1972.

———, and Donnan, Christopher. "Magic Cactus: Ethnoarchaeological Continuity in Peru." *Archaeology* 30, no. 6 (November, 1977):374–81.

Shuljgin, G. "Cultivation of the Opium Poppy and the Oil Poppy in the Soviet Union." *Bulletin on Narcotics* 21, no. 4 (October–December 1969):1–8.

Sikulová, V. and Opravil, E. "Nález zvírechích kostí a rostlinných zbytků ve Vávrovichíc (okr. Opava). Fund von Tierknochen und Pflanzenüberresten in Vávrovice (Bex. Opava)." *Přěhled Výzkumu* (1974), pp. 124–25.

Simpson, R. Hope. *A Gazetteer and Atlas of Mycenaean Sites* (University of London Institute of Classical Studies. Bulletin supplement no. 16). London: Institute of Classical Studies, 1965.

Singer, Charles; Holmyard, E. J.; and Hall, A. R. *A History of Technology. Vol. 1 of 5 vols. Oxford: Clarendon Press, 1954.*

Siret, Henri, and Siret, Louis. Les premiers âges du métal dans le sud-est de l'Espagne. Brussels: Impr. Polleunis, Ceuterick, et. LeFébure, 1888.

Sjöqvist, Eric. *Problems of the Late Cypriote Bronze Age.* Stockholm: The Swedish Cyprus Expedition, 1940.

Skalet, Charles H. *Ancient Sicyon.* Baltimore, Md.: Johns Hopkins University Press, 1928.

Skeat, W. W. *Etymological Dictionary of the English Language.* Oxford: Clarendon Press, 1879–82.

Slavik, Bohumil. *A Colour Guide to Familiar Wild Flowers, Ferns, and Grasses.* London: Octopus Books, 1974.

Snijder, Geerto Aeilko Sebo. *Kretische Kunst.* Berlin: Gebr. Mann, 1936.

Snyder, Solomon H. "Opiate Receptors and Internal Opiates." *Scientific American* 236, no. 3 (March 1977): 44–56.

Solecki, Ralph S. *Shanidar, the First Flower People.* New York: Knopf, 1971.

———. "Shanidar IV, a Neanderthal Flower Burial in Northern Iraq." *Science* 190 (1975): 880–81.

Sonnedecker, Glenn. *Kremer's and Urdang's History of Pharmacy.* 3rd ed., rev. Philadelphia: J. B. Lippincott, 1957.

———. "Emergence of the Concept of Opiate Addiction." *Journal Mondial de Pharmacie* 3 (1962): 277.

Sordelli, F. "Sulle piante della torbiera e della stazione prestorica della Lagozza." *Atti della Società Italiana di Scienzas Naturales Milano* 23 (1880).

Spleers, Louis. "Le personnage aux pavots." *Bulletin des Musées Royaux d'Art et d'Histoire* (Brussels), 3rd ser. 10 (1938): 122–36.

Sprengel, Kurt Polycarp. *Geschichte der Botanik.* Vol. 1 of 2 vols. Altenburg und Leipzig: F. A. Brockhaus, 1817–18.

Stanislawski, Dan. "Dark Age Contributions to the Mediterranean Way of Life." *Annals of the Association of American Geographers* 63, no. 4 (December 1973): 397–410.

———. "Dionysus Westward: Early Religion and the Economic Geography of Wine." *Geographical Review* 65, no. 4 (October 1975): 427–44.

Staples, Danny. "The Homeric Hymn to Demeter." In *The Road to Eleusis; Unveiling the Secret of the Mysteries,* edited by R. Gordon Wasson, Carl A. P. Ruck, and Albert Hofmann, pp. 59–75. New York: Harcourt Brace Jovanovich, 1978.

Stebbins, G. Ledyard *Flowering Plants:* Evolution above the *Species Level*, Cambridge, Massachusetts: Belknad Harvard, 1974.

Steiner, Rudolf. *Occult Mysteries of Antiquity.* Blauvelt, N.Y.: Rudolf Steiner Publications, 1961.

Stermitz, Frank R. "Alkaloid Chemistry and the Systematics of *Papaver* and *Argemone.*" *Recent Advances in Phytochemistry* 1 (1968): 161–83.

Strong, D. E. *Catalogue of the Carved Amber in the Department of Greek and Roman Antiquities.* London: British Museum, 1966.

Stubbings, Frank H. *Mycenaean Pottery from the Levant.* Cambridge: Cambridge University Press, 1951.

———. *Prehistoric Greece.* New York: John Day, 1973 [c.1972].

Stubbs, S. G. B. and Bligh, E. W. *Sixty Centuries of Health and Physick.* New York: Hoeber, 1931.

Stuiver, Minze. "Carbon-14 Dating: A Comparison of Beta and Ion Counting." *Science* 202 (1978): 881–83.

Sugiura, T. *Botanical Magazine* (in Japanese) 45 (1931): 353.

Täckholm, V. *Students' Flora of Egypt.* Cairo: Cairo University Press, 1974.

Tahara, M. "The Chromosomes of Papaver." *Botanical Magazine* (in Japanese) 29 (1915): 254–57.

Taylor, Norman. *Plant Drugs that Changed the World.* New York: Dodd, Mead, 1965.

Taylour, William. *Mycenaean Pottery in Italy and Adjacent Areas.* Cambridge: Cambridge University Press, 1958.

———. *The Mycenaeans.* Ancient Peoples and Places, 39. London: Thames, 1964.

———. "New Light on Mycenaean Religion." *Antiquity* 43 (1969): 91–97.

Tempir, Z. "Beiträge zur Ältesten Geschichte des Pflanzenbaus in Ungarn." In *Acta Archaeologia* (Academiae Scientiarum Hungaricae, vol. 16), pp. 63–98. 1964.

Terry, Charles E. and Mildred Pellens. *The Opium Problem.* 1928; rpt. ed., Montclair, N.J.: Patterson Smith, 1970.

Thompson, P. A. "Seed Germination in Relation to Ecological and Geographical Distribution." In *Taxonomy and Ecology*, edited by V. H. Heywood, pp. 93–120. New York: Academic Press, 1973.

Thompson, Reginald Campbell. *The Devils and Evil Spirits of Babylonia, etc.* London: Luzac, 1904.

———. "Assyrian Prescriptions for Diseases of the Head." *American Journal of Semitic Languages and Literature* 24 (1907): 327–45

———. *The Assyrian Herbal; A Monograph on the Assyrian Vegetable Drugs.* London: Luzac, 1924.

———. "Assyrian Prescriptions for Diseases of the Urine, etc." *Babyloniaca* 14 (1934): 131.

———. *A Dictionary of Assyrian Botany.* London: The British Academy, 1949.

Thorwald, Jürgen. *Science and Secrets of Early Medicine.* New York: Harcourt, Brace and World, 1963. Tr. by Richard and Clara Winston from the orig. ed. of *Macht und Geheimnis der Frühers Ärzte.* Zurich: Droemer, 1962.

Tristram, Henry Baker. *The Natural History of the Bible.* London: Society for Promoting Christian Knowledge, 1868.

Troels-Smith, J. "Pollenanalytische Untersuchungen zu einigen Schweizerischen Pfahlbauproblemen." In *Das Pfahlbau-Problem*, edited by W. U. Guyan, pp. 1–64. Schaffhausen, Switzerland, 1954.

———. "Neolithic Period in Switzerland and Denmark." *Science* 124 (November 1956): 876–79.

Tschirch, Alexander. *Handbuch der Pharmakognosie*, II. Leipzig: C. H. Tauchnitz, 1912.

Tschumi, O. "Das Pfahlbau von Seeberg Burgäschi-Südwest 1945–46." *Bern-Historisches Museum-Jahrbuch*, 1946.

———, ed. *Urgeschichte der Schweiz*. Frauenfeld: Huber, 1949.

Ucko, Peter J., and Dimbleby, G. W. *The Domestication and Exploitation of Plants and Animals*. Chicago: Aldine, 1969.

Uhlmann, J. "Einiges über Pflanzenreste aus der Pfahlbaustation Möringen am Bielersee (Bronzezeit)." *Mittheilungen der Antiquarischen Gesellschaft in Zürich* 19, no. 3 (1876): 62–65.

Uphof, J. C. *Dictionary of Economic Plants*. 2d ed., rev. and enl. Wurzburg: J. Cramer, 1968.

van Zeist, W. "Palaeobotanical Studies of Settlement Sites in the Coastal Area of the Netherlands." *Palaeohistoria* 16 (1974): 223–383.

———, and Casparie, W. A. "Niederwil, a Palaeobotanical Study of a Swiss Neolithic Lakeshore Settlement." *Geologie en Mijnbouw* 53, no. 6 (1974): 415–28.

———, and Bakker-Heeres, J. A. H. "Evidence for Linseed Cultivation Before 6000 B.C." *Journal of Archaeological Science* 2 (1975): 215–19.

Vavilov, N. I. "The Origin, Variation, Immunity, and Breeding of Cultivated Plants" (selected writings translated from Russian by K. Starr Chester). *Chronica Botanica* 13, nos. 1/6 (1949/1950): 1–364.

Vent, W. "Uber die Stellung von *Papaver glaucum* Boiss. & Hausskn. im System der Gattung *Papaver*." *Wissenschaftlich Zeitschrift Friedrich-Schiller-Universität in Jena* 19, no. 3 (1970): 447–52.

Vercoutter, Jean. *L'Égypte et le monde égéen préhellénique. Étude critique des sources égyptiennes (du début de la XVIII*[e] *à la fin de la XIX*[e] *dynastie)*. Institut français d'archéologie orientale. Bibliothèque d'étude, vol. 22. Cairo, 1956.

Verdelis, Nicholas. "A Sanctuary at Solygeia." Translated by Oscar Broneer. *Archaeology* (1962): 189–92.

Veselovskaya, M. A. *The Poppy: Its Classification and Importance as an Oleiferous Crop*. Translated by A. K. Dhote from the Russian. New Delhi: Amerind Co., 1976 (orig., 1933).

Vidal-Naquet, P. *Le Bordereau d'ensemencement dans l'Égypte ptolémaïque*. Brussels: Fondation égyptologique Reine Elizabeth, 1967.

Viguier, L. G. A. *Histoire naturelle médicale et économique des pavots et des argémones*. Montpellier: J. Martelainé, 1814.

Villaret-von Rochow, Margita. "Die Pflanzenreste der bronzezeitlichen Pfahlbauten von Valeggio am Mincio." In *Bericht über das Geobotanische*. Forschungsinstitut Rübel in Zürich für das Jahr 1957 (edited by E. Rübel and W. Lüdi), (1958): 96–114.

———. "Frucht- und Samenreste aus der neolithischen Station Seeberg, Burgäschisee-Sud." *Acta Bernensia* 2 (1967): 21–64.

———. "Fruit Size and Variability of Swiss Prehistoric *Malus Sylvestris.*" In *The Domestication and Exploitation of Plants and Animals* (Ucko and Dimbleby eds.) Chicago: Aldine Pub. Co., 1969: 201–6.

———. "Samenanalysen aus der spätneolithische Grabung Auvernier (Neuenburger See)." *IIème Congrès International des Musées d'Agriculture Budapest 1971, Résumés des communications présentées, Magyar Mezogazdasagi Muzeum Budapest*, pp. 206–8. 1971.

Wace, A. J. B.; Thompson, M. S.; and Droop, J. B. "The Menelaion (of Sparta)." *The Annals of the British School at Athens* 15 (1908–9): 108–57.

Wainwright, G. A. *Balabish* (Egyptian Exploration Society. Memoir 37.) London: George Allen and Unwin, 1920.

Walker, Winifred. *All the Plants of the Bible.* New York: Harper & Row, 1959 [c.1957].

Wallnöffer, Heinrich, and Rotthausher, Anna von. *Der goldene Schatz der chinesischen Medizin.* Stuttgart: Schuler Verlagsgesellschaft, 1959.

Walton, Alice. *The Cult of Asklepios.* 1894; reprinted., New York: Johnson, 1965.

Ward, W. A. "Egypt and the East Mediterranean from Predynastic Times to the End of the Old Kingdom." *Journal of the Economic and Social History of the Orient* 6 (1963): 1–57.

Warmington, E. H. *The Commerce Between the Roman Empire and India.* Cambridge: Cambridge University Press, 1928.

Wasson, R. Gordon. "The Divine Mushroom of Immortality." In *Flesh of the Gods*, edited by Peter Furst, pp. 185–200. New York: Praeger, 1972.

———; Ruck, Carl A. P.; and Hofmann, Albert. *The Road to Eleusis: Unveiling the Secret of the Mysteries.* New York: Harcourt Brace Jovanovich, 1978.

———. "The Hallucinogenic Mushrooms of Mexico and Psilocybin: a Bibliography" (2nd printing, with corrections and addenda), *Botanical Museum Leaflets*, Harvard University, 20 (1963): 27–73.

Waterbolk, H. T. "Food Production in Prehistoric Europe." *Science* 162 (1968): 1093–1102.

———, and van Zeist, W. "Preliminary Report on the Neolithic Bog Settlement of Niederwil." *Paleohistoria* 12 (1966): 559–80.

Waterbolk, H. T. and van Zeist, W., eds. *Niederwil, eine Siedlung der Pfyner Kultur.* Band I: Die Grabungen, Bern: Verlag Paul Haupt, 1978.

Weidner, Ernst. *Die Reliefs der Assyrischen Könige, erster teil, Die Reliefs in England, in Vatikan-Stadt und in Italien.* Berlin: Im Selbstverlage des Herausgebers, 1939.

Weil, Andrew. "Beyond Drugs—A New Way of Looking at the Higher Consciousness." *Psychology Today*, October 1972, pp. 51–96.

Werneck, H. L. "Ur- und frühgeschichtliche Kultur- und Nutz-pflanzen in den Ostalpen und am Ostrande des Böhmerwaldes." *Schriftenreihe der Landesbaudirektion*, no. 6 (1949): 1–288.

Whitehouse, David, and Whitehouse, Ruth. *Archaeological Atlas of the World.* San Francisco, Calif.: W. H. Freeman, 1975.

Wilbert, Johannes. "Tobacco and Shamanistic Ecstasy Among the Warao Indians." In *Flesh of the Gods*, edited by Peter Furst, pp. 55–83. New York: Praeger, 1972.

Wilkinson, Alex. *Ancient Egyptian Jewellery*. London: 1971.

Wilkinson, J. Gardner. *A Popular Account of the Ancient Egyptians*. New York: Harper and Brothers, 1854.

Willerding, U. "Vor- und Frühgeschichtliche Kulturpflanzenfunde in Mitteleuropa." *Neu Ausgrabungen und Forschungen in Niedersachsen* 5 (1970): 287–375.

Williams, Charles K., and Fischer, Joan. "Corinth 1974: Roman Forum." *Hesperia* 44, no. 1 (March 1975).

Williams, Trevor. *Drugs from Plants*. London: Sigma, 1947.

Wilson, C. Anne. *Food and Drink in Britain: From the Stone Age to Recent Time*. New York: Barnes and Noble, 1974.

Wishart, David. "The Opium Poppy: The Forbidden Crop." *Journal of Geography* (Jan. 1974): 14–25.

Witke, Edward C. "Marcus Aurelius and Mandragora." *Classical Philology* 60, no. 1 (January 1965): 23–24.

Woernig, Franz. *Die Pflanzen des Alten Aegypten*. 2nd ed. Leipzig, 1897.

Wreszinski, Walter. *Der papyrus Ebers*. Leipzig: J. C. Hinrich, 1913.

Wright, Arthur D. "The History of Opium." *Transactions and Studies of the College of Physicians of Philadelphia* 29, no. 1 (1961): 22–27.

———. "The History of Opium." *Medical Biology Illustrated* 18 (1968): 62–70.

Wright, H. E., Jr. "Climatic Change in Mycenaean Greece." *Antiquity* 42 (1968): 123–27.

———. "Vegetation History." In *The Minnesota Messenia Expedition: Reconstructing a Bronze Age Regional Environment*, edited by W. A. McDonald and G. R. Rapp, Jr., pp. 188–99. Minneapolis, Minn.: University of Minnesota Press, 1972.

Wulff, E. V. *An Introduction to Historical Plant Geography*. Translated by Elizabeth Brissenden. Waltham, Mass.: Chronica Botanica Co., 1943.

Yasui, K. "Cytogenetic Studies in Artificially Raised Interspecific Hybrids of *Papaver* IX. On the Bivalents-Association in the Interspecific Hybrids of *Papaver Somniferum*." *Cytologia* 10 (1940): 551–57.

Zehnder, Justin. "Le pavot et son usage chez les Assyriens." *Bull. Soc. Bot. Genève* 20 (1928): 379–82.

Zervos, Christian. *L'art de la Crète néolithique et minoenne*. Paris: Éditions "Cahiers d'art," 1956.

Zeuner, Friedrich E. "Cultivation of Plants." In *A History of Technology*, Vol. 1 of 5 vols., edited by Charles Singer et al., pp. 353–75. London: Oxford University Press, 1954.

Zohary, Daniel. "The Progenitors of Wheat and Barley in Relation to Domestication and Agricultural Dispersal in the Old World." In *The Domestication and Exploitation of Plants and Animals*, edited by P. J. Ucko and G. W. Dimbleby. Chicago: Aldine, 1969.

Index